Knowing Machines

Inside Technology
edited by Wiebe E. Bijker, W. Bernard Carlson, and Trevor Pinch

Wiebe E. Bijker, *Of Bicycles, Bakelites, and Bulbs: Toward a Theory of Sociotechnical Change*

Wiebe E. Bijker and John Law, editors, *Shaping Technology/Building Society: Studies in Sociotechnical Change*

Stuart S. Blume, *Insight and Industry: On the Dynamics of Technological Change in Medicine*

Geoffrey C. Bowker, *Science on the Run: Information Management and Industrial Geophysics at Schlumberger, 1920–1940*

Louis L. Bucciarelli, *Designing Engineers*

H. M. Collins, *Artificial Experts: Social Knowledge and Intelligent Machines*

Pamela E. Mack, *Viewing the Earth: The Social Construction of the Landsat Satellite System*

Donald MacKenzie, *Inventing Accuracy: A Historical Sociology of Nuclear Missile Guidance*

Donald MacKenzie, *Knowing Machines: Essays on Technical Change*

Knowing Machines
Essays on Technical Change

Donald MacKenzie

The MIT Press
Cambridge, Massachusetts
London, England

Set in New Baskerville by The MIT Press.
Printed and bound in the United States of America.

Library of Congress Cataloging-in-Publication Data

MacKenzie, Donald A.
  Knowing machines : essays on technical change
      p.   cm. — (Inside technology)
  Includes bibliographical references and index.
  ISBN 0-262-13315-6 (hc : alk. paper)
  1. Technology—Social aspects. I. Title. II. Series.
T14.5.M32   1996
306.4'6—dc20

95-32334
CIP

in memory of Angus MacKenzie (1915–1995)

Contents

9
Computer-Related Accidental Death   185

10
Tacit Knowledge and the Uninvention of Nuclear Weapons (with Graham Spinardi)   215

Acknowledgments

My most immediate debt is to Boelie Elzen and Graham Spinardi, the co-authors of chapters 6 and 10. Several others have also been involved in the research drawn on here, notably Stuart Anderson, Bernard Carré, Alfonso Molina, Eloína Peláez, Wolfgang Rüdig, and Maggie Tierney. Although the different chapters draw upon work that has had a variety of forms of financial support (acknowledged in those chapters), there has been one central source, the funding provided by the U.K. Economic and Social Research Council's Programme on Information and Communication Technologies (PICT) to the Edinburgh PICT Centre (grants A35250006 and WA35250006). Robin Williams (coordinator of the Edinburgh PICT Centre), Moyra Forrest (information officer), and Barbara Silander (secretary) have all been important in making possible the work described here, as has, in a very different way, Caroline Bamford. Warm thanks to all.

Chapter 2 first appeared in *Technology and Culture* 25 (1984), chapter 3 in *Technological Change and Company Strategies: Economic and Sociological Perspectives,* ed. R. Coombs, P. Saviotti, and V. Walsh (Academic Press, 1992), chapter 4 in *Technology and Culture* 34 (1993), chapter 5 in *Annals of the History of Computing* 13 (1991), chapter 6 in *Jaarboek voor de Geschiedenis van Bedrijf en Techniek* 8 (1991), chapter 7 in *Nature* 352 (August 8, 1991), chapter 8 in *Social Studies of Science* 23 (1993), chapter 9 in *Science and Public Policy* 21 (1994), and chapter 10 in the *American Journal of Sociology* 101 (1995). These chapters are essentially unchanged, bar some editing to minimize overlap, improve clarity, correct errors, and remove the odd infelicity. I am grateful to the copyright holders for permission to reprint them.

Knowing Machines

1

Introduction

4 P.M. on a day in June 1945. A man stands on a bridge over the Santa Fe River. He has been in Santa Fe since lunchtime, doing the things a tourist would do. Now he feels uneasy, standing alone on a bridge in a quiet, un-built-up area. Within a few minutes, though, a battered blue Buick approaches along a gravel road and stops. The driver, who is alone, gets out and joins the man on the bridge, and together they set off walking, talking as they go. When they part, the driver hands the other man a package.

The driver heads back up toward the Jemez Mountains. His destination, 7000 feet high, is a place called simply "the hill" by those who work there, but soon to be better known by the Spanish word for the cottonwood trees in the deep canyon that bisects the mesa on which it stands: Los Alamos.

The other man takes the evening bus to Albuquerque and the next day's train to Chicago. From there he flies to Washington and then takes the train to New York. In a street in Brooklyn, he has another short meeting, passing the package to a man he knows as John.

The driver is a German émigré physicist, Klaus Fuchs. The courier, whom Fuchs knows as Raymond, is a biochemistry technician called Harry Gold. John's real name is Anatolii Yakovlev. Ostensibly the Soviet Union's Vice-Consul in New York, he is actually a senior agent of the Soviet intelligence service. In the package is Fuchs's attempt at a comprehensive description, including a detailed diagram, of the atomic bomb that will shortly be tested in the New Mexico desert and dropped on Nagasaki. The Second World War has still not ended, but the Cold War has already begun.[1]

Late November 1990. Great Malvern, England, a spa town nestling beneath the Malvern Hills. A businessman ponders the future of his small firm. Set up at the start of the free-enterprise economic miracle of the 1980s, the firm is now in deep trouble. Mrs. Thatcher's boom has dissolved as quickly as it materialized, and the recession is biting hard. But the businessman has a more specific concern. He has sunk much of his limited capital into technology, licensed from the British Ministry of Defence, surrounding a new microchip called VIPER: the Verifiable Integrated Processor for Enhanced Reliability.

At the start, the investment seemed an excellent one. VIPER was a response to fears that dangerous "bugs" might lurk unnoticed in computer software or hardware. Other microprocessors on the market were subject to repeated testing, but microprocessor chips are so complex that tests cannot be exhaustive. So one can never be absolutely sure that an undetected bug does not lurk in the chip's design. VIPER was different. Its developers, at the Ministry of Defence's famous Royal Signals and Radar Establishment on the outskirts of Great Malvern, had sought to provide both a formal mathematical specification of how the microprocessor should behave and a formal proof that its detailed design was a correct implementation of that specification.

VIPER had been greeted as a triumph for British computer science in a field dominated by American hardware. The London *Times* wrote that it was "capable of being proved mathematically free of design faults." The *New Scientist* called VIPER "the mathematically perfect chip," with "a design that has been proved mathematically to be correct." It was "failsafe," said *Electronics Weekly*. It had been "mathematically proved to be free of design faults," said *The Engineer*.

Yet, like the 1980s themselves, VIPER has by the end of 1990 turned sour for the businessman. Sales have been far fewer than expected, and computer scientists from Cambridge University and from Austin, Texas, have sharply criticized the claim of mathematical proof. The businessman is looking for a way to recoup his losses, and he instructs his solicitors to sue the Secretary of State for Defence for damages.

So begins a unique legal case. Lawyers have always dealt with matters of proof, but the everyday proofs of the courtroom are examples of worldly reasoning, acknowledged to be less than absolute: "beyond reasonable doubt," not beyond all doubt. What is at stake in the VIPER case is proof of an apparently quite different kind. Mathematical proof, seemingly pristine and absolute, has moved from the abstract realms of logic and pure mathematics into the mundane world of technology, litigation, power, and money.[2]

⤳

Evening, February 25, 1991. A warehouse on the outskirts of Dhahran, Saudi Arabia, whose port and air base are central to the war against Saddam Hussein's Iraq. The warehouse has been turned into temporary accommodations for U.S. Army support staff responsible for stores, transportation, and water purification. Many are reservists from small towns along the Ohio-Pennsylvania border. Some are sleeping, some exercising, some eating dinner, some trying to relax. It is not a comfortable time. Allied bombers relentlessly pound both Iraq and the Iraqi troops in Kuwait, while Iraq is using its Scud missiles to attack Israel and Saudi Arabia. Dhahran is a prime target. So far, the Scuds have carried conventional explosive warheads. These are dangerous enough: hundreds have lost their lives to them in the cities of Iran and Afghanistan. Furthermore, no one can be sure that the next Scud to be fired will not be carrying nerve gas or anthrax spores.

Unlike the citizens of Iran and Afghanistan, however, those of Saudi Arabia and Israel have a defense against the Scuds: the American Patriot air defense system. Although the Patriot's performance will be questioned later, there is no doubt that right now it offers immense psychological reassurance not to feel totally defenseless against the Iraqi missiles and their potentially deadly cargo. Nightly, the world's television screens carry film of Patriot missiles rocketing into the sky to intercept incoming Scuds.

On the evening of February 25, a Scud is fired toward Dhahran. It arches up high into the atmosphere, then plunges down toward its target. American radars detect it as it streaks toward the defensive perimeter of Alpha Battery, protecting the Dhahran air base. But Alpha Battery's Patriots are not launched: the radar system controlling them has been unable to track the incoming missile.

The corrugated metal warehouse offers no protection against the Scud's high-explosive warhead. Blast and fire kill 28 American troops, the most serious single loss suffered by the allies in the Gulf War. Within an hour, the building is a charred skeleton. In the morning, excavators begin searching the ruin, helped by soldiers with picks and shovels. Some survivors still wander around. Many are weeping.

Investigations into why no defensive missile was launched suggest a cause that seems unimaginably tiny: at one point in the software controlling Patriot's radar system, there is an error of 0.0001 percent in the representation of time. By February 25 the error had been found, and corrected software was on its way to Dhahran. It arrived a day too late.[3]

Technology, Society, Knowledge

Three disparate tales; three disparate outcomes. Fuchs's betrayal, or act of idealism, has become part of the history of our times—although only since the Cold War ended have we known for sure what was in the package handed over that June day in Santa Fe and been able to assess its consequences. The legal challenge to VIPER's proof was stillborn. With the Ministry of Defence contesting the suit vigorously, the businessman was not able to keep his company afloat long enough to bring the case to a hearing. While the litigation is, so far, unique, there are strong pressures that may again force mathematical proof into the law courts. The Dharhan deaths are among a relatively modest number that can so far be attributed to computer-system failures, but there is no certainty that in the years to come the number will remain modest.

Three tales; three forms of interweaving. The Cold War accustomed us to the connections among technology, knowledge, and international politics. That interweaving continues, though its form has now changed. Recently, for example, nuclear fears have focused more on the smuggling of fissile materials and on Iraq, Iran, and North Korea than on the East-West confrontation of Cold War days. What kind of knowledge is needed to build a nuclear weapon? How can that knowledge be transferred or controlled? Is it a permanent legacy that humanity must learn to live with, or can it be lost?

The VIPER case points us to an altogether more esoteric interweaving: that of technology with mathematics and even philosophy. Many in computer science feel that the way to keep computer systems under our control is to subject them, like their mechanical and electrical predecessors, to our most powerful form of rigorous thought: mathematics. But what is the status of the knowledge produced by this process of subjection? Can one create a mathematical proof that a machine has been correctly designed? What will happen to "proof" as it moves from lecture theaters and logic texts to the world of commerce and the law?

The Dharhan deaths took place in the most highly computerized war yet fought. But computer systems are increasingly interwoven into our daily peacetime lives as well. Microprocessors proliferate in automobiles and airplanes, in homes and offices, and even in hospitals. Computerization brings undoubted benefits, but certainly there are also risks. What evidence is there about these risks? What is their nature?

T, I, D.

Underpinnings

The essays I have gathered in this book explore a wide range of questions such as these in the relationship between machines and society. The first two are predominantly conceptual, exploring Karl Marx's contribution to the study of technology and the relationship between economic and sociological analyses of technology. The others are more empirical, exploring the interweavings of technology, society, and knowledge in a variety of particular contexts: the "laser gyroscopes" central to modern aircraft navigation; supercomputers (and their use to design nuclear weapons); the application of mathematical proof in the design of computer systems (and arithmetic as performed by computers); computer-related accidental deaths; the knowledge needed to design a nuclear bomb.

These may seem strange topics for a sociologist to explore. One might expect sociology to concentrate on familiar, widely diffused technologies, exploring such subjects as popular beliefs about technology and the societal effects of technology. Instead, this book is concerned mostly with the development of modern, sometimes esoteric, "high" technologies, and the "knowledge" discussed is usually specialized knowledge rather than lay belief. Underlying this choice is a long-standing conviction that the social analysis of technology can make a contribution only if it is willing to tackle the shaping of technologies as well as their adoption, use, and effects, and to grapple with the nature of specialized as well as lay knowledge.

The chapters are diverse in their topics, and they were written at different times for different audiences. There is, however, a shared perspective underpinning them—sometimes explicitly, often implicitly. At a very basic level, this perspective was formed in opposition to the idea that the development of technology is driven by an autonomous, nonsocial, internal dynamic. Although this form of "technological determinism" is no longer prevalent in academic work on the history and the sociology of technology, it still informs the way technology is thought about and discussed in society at large, especially where modern high technologies are concerned. The idea that technological change is just "progress," and that certain technologies triumph simply because they are the best or the most efficient, is still widespread. A weaker but more sophisticated version of technological determinism—the idea that there are "natural trajectories" of technological change—remains popular among economists who study technology.[4]

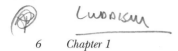

In my experience, the idea of unilinear progress does not survive serious engagement with the detail of the history of technology. For what is perhaps most striking about that history is its wealth, complexity, and variety. Instead of one predetermined path of advance, there is typically a constant turmoil of concepts, plans, and projects. From that turmoil, order (sometimes) emerges, and its emergence is of course what lends credibility to notions of "progress" or "natural trajectory." With hindsight, the technology that succeeds usually does look like the best or the most natural next step.

However—and this is the first argument that underpins these essays—we must always ask "Best for whom?" Different people may see a technology in different ways, attach different meanings to it, want different things from it, assess it differently. Women and men, for example, may view the same artifact quite differently.[5] Workers and their employers may not agree on the desirable features of a production technology.[6]

Such discrepant meanings and interests are often at the heart of what is too readily dismissed as irrational resistance to technological change, such as that of the much-disparaged Luddite machine breakers. We must also ask "Best for whom?" even when we are discussing such apparently "technical" decisions as the best way to automate machine tools or typesetting. These two technologies were the subjects of now-classic studies by Cynthia Cockburn (who focused on the shaping of technology by gender relations) and David Noble (who focused on its shaping by relations of social class); their findings are summarized in chapter 2 below.[7]

Nor is this issue—the different meanings of a technology for different "relevant social groups,"[8] and the consequently different criteria of what it means for one technology to be better than another—restricted to situations of class conflict or other overt social division. The customers for the supercomputers discussed in chapters 5 and 6, for example, were all members of what one might loosely think of as the "establishment": nuclear weapons laboratories, the code breakers of the National Security Agency, large corporations, elite universities, and weather bureaus. Responding to their needs, but far from subservient, were the developers of supercomputers, most famously Seymour Cray. All were agreed that a supercomputer should be fast, but there were subtle differences among them as to what "fast" meant. As a consequence, the technical history of supercomputing can be seen, in one light, as a negotiation—which is still continuing[9]—of the meaning of speed.

Relevant
    Social Groups

at by extension
cultural practices
in relation to technol-

BROADER
EFFECT

We also need to delve deeper even where there is agreement as to what characteristics make a technology the best, and this brings me to the second argument underpinning these essays. Technologies, as Brian Arthur and Paul David point out, typically manifest increasing returns to adoption.[10] The more they are adopted, the more experience is gained in their use, the more research and development effort is devoted to them, and the better they become. This effect is particularly dramatic in the case of "network" technologies such as telephones or the worldwide computer network called the Internet, where the utility of the technology to one user depends strongly on how many other users there are. But the effect can be also be found in "stand-alone" technologies such as the navigation systems discussed in chapter 4.

This means that early adoptions—achieved for whatever reasons—may give a particular technology an overwhelming lead over actual or potential rivals, as that technology enjoys a virtuous circle in which adoptions lead to improvements, which then spawn more adoptions and further improvements, while its rivals stagnate. Technologies, in other words, may be best because they have triumphed, rather than triumphing because they are best.

Hindsight often makes it appear that the successful technology is simply intrinsically superior, but hindsight—here and elsewhere—can be a misleading form of vision. Historians and sociologists of technology would do well to avoid explaining the success of a technology by its assumed intrinsic technical superiority to its rivals.[11] Instead, they should seek, even-handedly, to understand how its actual superiority came into being, while suspending judgment as to whether it is intrinsic. That methodological principle is the third underpinning of this book. It is perhaps most explicit in chapter 4, where I examine the recent "technological revolution" in which the laser gyroscope has triumphed over its mechanical rivals; but other chapters also seek "symmetry" in their explanations of the success and failure of technologies.

As chapters 3 and 4 suggest, expectations about the future are often integral to technological success or failure.[12] Most obviously, a belief in the future success of a technology can be a vital component of that success, because it encourages inventors to focus their efforts on the technology, investors to invest in it, and users to adopt it. These outcomes, if they then bear fruit, can reinforce the original belief by providing evidence for its correctness. Attention to this kind of process—in which beliefs about technology create (or, less commonly, undermine) the conditions to which they refer—is a fourth theme of the book.

Self-validating belief—"self-fulfilling prophecy"—has sometimes been regarded by social scientists as pathological, as permitting false beliefs to become true. The classic example is the way an initially arbitrary belief in the unsoundness of a particular bank can produce a run on that bank and thus cause it to fail.[13] Nevertheless, self-referential, self-reinforcing belief is pervasive in social life, as Barry Barnes has argued eloquently. The most obvious case is money, which can function as a medium of exchange only when enough people believe it will continue to do so; but all social institutions arguably have something of the character of the self-fulfilling prophecy.[14] Some of the most striking phenomena of technological change are of this kind. One example, from chapter 3, is "Moore's Law": the annual doubling of the number of components on state-of-the-art microchips. Moore's Law is not merely an after-the-fact empirical description of processes of change in microelectronics; it is a belief that has become self-fulfilling by guiding the technological and investment decisions of those involved.

Of course, I would not suggest that self-reinforcing belief is all there is to phenomena such as Moore's Law. Expectations, however widespread, can be dashed as technologies encounter the obduracy of both the physical and the social world. As a result, many technological prophecies fail to be self-validating—for example, the prophecy, widespread in the 1960s, that the speed of airliners would continue to increase, as it had in previous decades. In recent years even Moore's Law seems to have lost some of its apparently inexorable certainty, although belief in it is still a factor in the justification of the enormous capital expenditures (of the order of $1 billion for each of the world's twenty state-of-the-art chip fabrication facilities) needed to keep component density growing.[15]

Furthermore, there are some beliefs about technology that have self-negating rather than self-fulfilling aspects. Perhaps the most important example is that of beliefs about the safety or danger of technologies, examined here in the context of computer systems. Belief that a technology is safe may make it dangerous: overconfidence in the correctness of computerized systems seems to have been a major factor in accidents involving such systems. Conversely, a healthy respect for the dangers posed by a technology can be an important factor in keeping it safe. The discussion in chapter 9 suggests that this may be a crucial reason why the number of major computer-related accidents has so far been limited.

Technology and the Sociology of Knowledge

The fifth and perhaps the most general underpinning of these essays is an emphasis on the importance of knowledge (i.e., shared institutionalized belief) in the relations of machines to society.[16] Of course, machines—whether they be spinning mules, gyroscopes, supercomputers, missiles, or radiotherapy systems—are real, physical objects; they are not simply reducible to the ensemble of our beliefs about them. Their obdurate materiality is crucial to their social role (as is discussed in chapter 2), and, as several of the accidents discussed in chapter 9 show, they can behave in ways quite unexpected by anyone involved with them.

Nevertheless, professional and lay knowledge of machines—the *logos* aspects of technology[17]—is utterly crucial. Most obviously, for a machine to be useful to us we need to know how to use it, and the nature and distribution of that knowledge is of considerable importance (see chapter 10). But equally important is our knowledge of the characteristics of the machines we deal with. Are they safe, or dangerous? What effects do they have? Are the characteristics of one machine superior to those of another? What will future machines be like?

The dominant approach taken here to technical knowledge is inherited from the sociology of science.[18] Up to around 1970, the main focus of this field was on issues such as the norms of science, its reward system, and its career structure: it might have been called the sociology of *scientists*. During the 1970s, a new generation of authors (including Barry Barnes, David Bloor, Harry Collins, Bruno Latour, Karin Knorr-Cetina, Michael Mulkay, Steven Shapin, and Steve Woolgar) sought to extend sociological analysis to the *content* of science—to what scientists actually do in their laboratories or other workplaces, and, above all, to the knowledge they produce.[19] These authors differed (and still differ) in how they went about constructing a sociology of scientific knowledge, but there is clear common ground. All have rejected *a priori* divisions between "science" and "ideology," between "good science" and "bad science." All have rejected the restriction of the sociology of knowledge to matter such as political or religious belief and patently "ideological" science, such as Nazi "Aryan physics" or Soviet Lysenkoist biology. David Bloor referred to these restrictions as the "weak program" of the sociology of knowledge. In 1973 Bloor put forward as an alternative what he called the "strong program of the sociology of knowledge,"[20] which would seek symmetrical sociological analysis (indeed sociological *explanation*) of *all* knowledge, irrespective of our current evaluations of its truth or its adequacy. In other words, we should apply the same general explanatory framework to analyze the generation and reception of both "true" and "false" knowledge. We should avoid, for example, explaining "true" knowledge as simply input from the real world and appealing to "social factors" only in the case of knowledge now regarded as false.

Although this "relativism" has been subjected to fierce attack, a significant body of research in the history and sociology of science seems to me to have confirmed both the possibility and the fruitfulness of "strong program" sociological analysis of scientific knowledge. A considerable number of studies have shown the effect upon scientific knowledge of social processes, including both processes internal to the scientific community and those involving the wider society.[21] Several of the chapters that follow reflect the belief that a sociology of technical knowledge, though it has been the subject of much less debate and much less research, should similarly be possible and fruitful.

In particular, such a sociology need not be restricted to lay knowledge of technology; it can encompass professional knowledge, including "correct" professional knowledge, as well as professional knowledge now regarded as erroneous or inadequate. This overall argument finds two particular manifestations in this book.[22] The first concerns the mathematical aspects of computer systems: arithmetic as performed by computers (and by advanced pocket calculators) and efforts (such as VIPER) to apply mathematical proof to computer systems. Aside from their intrinsic importance, these mathematical aspects of computing are of interest because of an imbalance in existing "strong-program" sociology of knowledge. While Bloor's has consistently focused on mathematics and formal logic, nearly all other "strong-program" work has concerned the natural sciences. Yet mathematics and logic arguably constitute the hard case for the sociology of knowledge.

Since the ancient Greeks, our culture has tended to prize the deductive reasoning of mathematics and formal logic more highly than the inductive reasoning of the empirical sciences. The former is taken to be immune from the uncertainty that even the most positivist of philosophers would agree characterizes the latter. Our knowledge that $2 + 2 = 4$ is normally taken to be absolute and therefore different in kind from fallible inductive belief, such as the belief that all swans are white. In his classic of "weak-program" sociology of knowledge, *Ideology and Utopia*, Karl Mannheim tended to place the limits of the sociology of knowledge at the boundaries of mathematics rather than at the boundaries of the natural sciences.[23] Chapter 8 and (implicitly) chapter 7 take up this issue. Drawing on the work of Bloor, these chapters are based upon the assumption that, despite the absolute and pristine aura surrounding mathematics and formal logic, a sociology of these forms of knowledge is possible. These chapters examine the clash between different systems of arithmetic for computers and advanced digital calculators, the controversy over VIPER, and the wider debate over the application of mathematical proof to the design of computer hardware and software.

Of course, only a small part of technological knowledge is of the formal and mathematical kind discussed in chapters 7 and 8. Much more pervasive is tacit knowledge, and that is the second aspect of technological knowledge discussed here. Tacit knowledge is informal "know-how" rather than explicit, systematized belief; it is unverbalized and perhaps unverbalizable. Riding a bicycle and playing a musical instrument are everyday activities that rest in good part upon tacit knowledge: even the most competent cyclist or musician would find it difficult (probably impossible) to give a full verbal description of what these skills consist in. Various authors, from Michael Polanyi on, have argued that tacit knowledge plays a central role not just in the skills of everyday life but also in the practice of science. Harry Collins, above all, has shown how understanding that role is of considerable help in developing a symmetrical sociological analysis of the place of experiment and experimental results in scientific controversies.[24]

Tacit knowledge also plays a central role in technology. Chapter 10 suggests that this is true even for the field of technological endeavor that has probably seen a greater amount of systematic, scientific attention devoted to it than any other: the designing of nuclear weapons. The consequences of the role of tacit knowledge bear on the nature of our knowledge of nuclear weapons, on the mechanisms of their proliferation, and on the possibility of their being "uninvented."

Sociology, Economics, and History

Although I hope it will be of interest to the general public, this book is also meant to contribute to the field of social studies of technology. Like all academic fields, this one has its divides and disagreements. Although these may be of little concern to readers outside the field, they bear on the underpinning themes spelled out above, and therefore they should be introduced at least cursorily.

The only divide directly addressed in this book is one that has not become a full-fledged debate but which certainly should: the divide between sociological and economic explanations of technological change. Too often, sociological analysis of technology—and here, I regret to say, I have to include my own work[25]—is virtually silent on the economic aspects of its subject matter, and yet those economic aspects are both pervasive and crucial.

As chapter 3 notes, part of the reason for this silence is that the dominant "neoclassical" approach within economics rests upon assumptions about the behavior of firms that sociologists typically regard as unrealistic.[26] Yet the neoclassical tradition is by no means all of economics, and chapter 3 explores potential common ground between sociological analysis and forms of economics alternative to neoclassicism. The discussion of Marx in chapter 2 can also been seen as an implicit exploration of this common ground (although, of course, when Marx was writing sociology as we now know it did not exist and the mathematical apparatus of neoclassicism had yet to be created). Marx's work, at its best, simultaneously captures both the centrality of economic phenomena and the historical and social nature of those phenomena. His analysis of the "labor process," for example, avoids counterposing the goals of profit and capitalist control over the work force, as some later Marxist analyses of production technology have done.[27]

Another disciplinary divide—one that has provoked explicit debate recently—is that between the history and the sociology of technology. Angus Buchanan, for example, has contrasted the historian's well-grounded "critical narrative" with the "empty" and "preconceived conceptual boxes" of the social theorist, and has vigorously condemned recent work in the sociology of technology for imposing "an alien conceptual vocabulary on the subject matter of history."[28] Several of the chapters in the present volume seek to straddle precisely this disciplinary divide between history and sociology. Chapters 4–6 are closest to narrative history, although the fact that their subject matter is still subject to commercial confidentiality (and often security classification) means that there are strict limits on the availability of the kind of documentary sources with which historians are typically most comfortable. Even in those chapters, however, theoretical questions are not entirely absent. The other chapters are less narrative than attempts to use historical material to investigate or illustrate theoretical issues.

I leave it to the reader to judge the success of these efforts, but let me say that I see no contradiction between critical narrative and theoretical concerns. I share the sociologist John Law's suspicion that apparently untheoretical narrative may actually rest on implicit (and therefore undebated) theoretical assumptions.[29] By bringing these to the surface, explicit attention to theory can contribute to the goal of *critical* narrative.

Theoretical concerns can also suggest new ways of looking at familiar topics: I hope, for example, that there is at least a degree of novelty in examining the history of nuclear weaponry from the viewpoint of tacit knowledge.[30] Furthermore, theoretical concerns can suggest the interest of hitherto relatively unexplored topics. It was, for example, the strong program of the sociology of knowledge that suggested that it would be interesting to examine computer arithmetic and the (as yet brief) history of program and hardware verification in computer science. Indeed, as will be seen in chapter 8, in this latter area the strong program even led to a broadly successful prediction: that there would eventually be litigation over mathematical proof.[31]   *& hacking won't go away* .

Actors, Networks, and Competing Symmetries

The sociology of technology is, of course, not a homogeneous field. One particular debate that is relevant to this book concerns the validity of a perspective called "actor-network theory," developed especially by the French scholars Michel Callon and Bruno Latour.

The central argument of actor-network theory, in relationship to technology, is that all successful technological innovation involves the construction of durable links tying together humans and nonhuman entities ("actors"). The team that successfully developed the laser gyroscope, for example, had not merely to engineer metal, gas, and ceramics but also to generate commitment to the technology among the managers of their corporations, among the military, and in the world of civil aviation. In the words of another leading contributor to actor-network theory, John Law, they had to be "heterogeneous engineers."[32]

In one sense, of course, this is banal: it is difficult to imagine any serious historical or sociological case study of technological change in which this is not obvious. Nevertheless, the term "heterogeneous engineering," and actor-network theory more generally, usefully remind us *simultaneously* to bear in mind two aspects of technical change that are often treated in isolation from each other. The first is the way that the "physical" aspects of heterogeneous engineering are influenced by the demands of its "social" aspects—for example, the way that production technology can be shaped by the need to create or maintain particular

SST

forms of social relationships between worker and employer or among
workers, the way that domestic technology has to reflect social relationships between and within households, and the way that military technology is shaped by the existing social organization of the armed services.[33]

This "social shaping of technology," however, should not be thought
of simply as unchanging social relationships causing changes to technology, for heterogeneous engineering involves changes to social relations too. This is the second aspect of technical change that
actor-network theory reminds us to keep in mind. "Artifacts have politics," as Langdon Winner puts it.[34] Technologies are not neutral ser-
vants of whatever social or political order chooses to adopt them. Their
adoption and operation often involves changes to that order—changes
that are not automatic consequences of new technology but must themselves be engineered, often in the face of conflict and resistance.

More generally, the actor-network perspective offers a useful critique
of the fact that much social theory conceives of social relations as if they
were simply unmediated relationships between naked human beings,
rather than being made possible and stable by artifacts and technologies. Society can exist without artifacts and technologies, but such societies—whether human or, for example, primate—are typically small.
The actor-network argument is that artifacts and technologies—clothes,
houses, walls, prisons, writing, agriculture—are needed to make larger,
more complex societies possible.[35] Social theory that neglects technology therefore fails to grasp an important part of the answer to its central questions: What is society? What makes social order possible?
"Technology" and "society," the actor-network theorists argue, are not
two independent entities. Each is inextricably part of the other.

These actor-network arguments command, I think, widespread
agreement within the social studies of technology, but there is one particular aspect of the approach that is deeply controversial. It concerns
Callon and Latour's call for an extension to the principle of symmetric
analysis of "true" and "false" belief. This principle, as I suggested above,
is central to the sociology of scientific knowledge. It has also influenced
the sociology of technology, where its analogue is the third underpinning principle noted above: avoiding explaining the success of technologies by their intrinsic superiority.

Callon and Latour's proposed extension is a call for symmetric analytical treatment of human and nonhuman actors. Unlike in conventional sociology, where the term "actor" usually refers solely to human
beings, in actor-network theory "actor" (sometimes, "actant") can refer

SYMMETRY of TREATMENT

both to human beings and to nonhuman entities: electrons, microbes, or whatever. Our analyses, say Callon and Latour, should not privilege human beings by making them, *a priori*, the only active agents. Humans and nonhumans should be treated symmetrically. Callon, introducing a case study of the cultivation of shellfish, puts it this way: "We know that the ingredients of controversies are a mixture of considerations concerning both Society and Nature. For this reason we require the observer to use a single repertoire when they are described." Callon suggests using the same vocabulary "for fishermen, for the scallops and for scientific colleagues": terms such as "problematization, interessement, enrolment, mobilization and dissidence."[36] The meanings of these particular terms are of less importance here than the basic issue of the extension of the principle of symmetry. Harry Collins and fellow sociologist of science Steven Yearley oppose this extension vigorously, arguing that the symmetrical analysis of humans and nonhumans is conducted at the price of asymmetry as regards truth and falsehood. Collins and Yearley point out that the analytical treatment of non-human entities as actors requires us to describe their behavior. To do this, they argue, is to privilege one account of that behavior—normally, the accepted scientific one. "Extended" symmetry, they conclude, can be purchased only by giving up its older sociology-of-knowledge form.[37]

In a way, this recent debate rehearses an old issue: the place of the real, "material" world in sociology-of-knowledge explanations. It seems to me that sociologists of science or of technology have no need to deny that the real world influences our beliefs about it. As David Bloor puts it:

Objects in the world will in general impinge equally on those who have true and those who have false beliefs about them. Consider Priestley and Lavoisier looking at some burning chemicals. They both see the same objects in the world, they both direct their attention and their remarks at the same things. But one says: "In combustion a burning object releases phlogiston into the atmosphere," and the other says: "In combustion a burning object takes oxygen from the atmosphere." There is no question of disqualifying as possible causes the objects before them. Such causes do not however suffice to explain the verbal description that is given of them. This is so both for the versions we ourselves accept as true and for the versions we reject as false.[38]

An example for the case of technology (where there has been analogous debate about the place of the material efficacy of technologies in their sociological analysis)[39] might be the fierce debate that took place in the aftermath of the Gulf War about the efficacy of the Patriot missile system. Actual material events took place in the skies over Saudi

Arabia and Israel, and there is no reason to doubt that these events influenced the beliefs of both the Patriot's defenders and its critics. The two camps, however, drew radically different conclusions from them about the Patriot's efficacy.[40]

The crucial point, it seems to me, is the distinction between "unverbalized reality"[41] and our beliefs—including our verbal descriptions—about that reality. Actor-network theory is right to insist on the independent, causal role of nonhuman entities—"unverbalized reality"—in influencing both scientific knowledge and technological development. The strictures of Collins and Yearley, however, begin to apply when these nonhuman entities become actors and we move from unverbalized reality to particular, verbal accounts of that reality.

The crucial moment of this transition is, typically, when scientific or technological disputes get settled—in Latour words, when "technoscience" (science and technology) moves from being "warm" to being "cold." Latour argues that this is the moment for the analyst to shift from relativism to realism:

When talking about a cold part of technoscience we should shift our method like the scientists themselves who, from hard-core relativists, have turned into dyed-in-the-wool realists. Nature is now taken as the cause of accurate descriptions of herself. We cannot be more relativist than scientists about these parts. . . . Why? Because the cost of dispute is too high for an average citizen, even if he or she is a historian and sociologist of science. If there is no controversy among scientists as to the status of facts, then it is useless to go on talking about interpretation, representation. . . . Nature talks straight, facts are facts. Full stop. There is nothing to add and nothing to subtract. . . . [To go on] being relativists even about the settled parts of science . . . made [analysts of science] look ludicrous.[42]

It is certainly true that virtually all the major empirical, sociological studies of science and technology focus on scientific controversy or on situations where alternative paths of technological development were explicitly available. However, the practical difficulties facing the sociological analysis of established, consensual science or technology should not, I feel, lead us to abandon the effort. There *are* resources available to the analyst.

One such resource is the "insider uncertainty" of those at the heart of knowledge production, even in established fields.[43] Nuclear weapons design is one such field. Although the activity is controversial politically, the technical design of "orthodox" atomic and hydrogen bombs is well-established, almost routine, "technoscience." Yet in the interviews discussed in chapter 10, the designers stressed the dependence of our

"The dependence of deductive knowledge upon trust".

knowledge of the technical characteristics of such weapons upon human judgment, not upon hard empirical fact or sure deduction from established theory. They may well have had specific reasons for doing so, but nevertheless it gives the analyst a way of continuing to be "relativist" even about this settled area of knowledge.

Another resource arises when settled knowledge developed in one institutional setting must be displayed and defended in a different setting—in particular, when scientific knowledge enters the adversarial legal process. For example, the chemical analysis of narcotics and other illicit drugs involves routine, established, empirical procedures about which there is (to my knowledge) no scientific controversy. Yet defense lawyers in drug cases can still undermine the testimony even of expert witnesses who have carefully followed such procedures. In doing so, they lay bare the dependence of the credibility of established, empirical knowledge upon trust.[44] The potential fascination for the sociologist of knowledge of future litigation over mathematical proof applied to computer systems is that this may lay bare the analogous dependence of deductive knowledge upon trust.

Relativism and Indifference; Women and Men

+ when difft groups bring their views of "security" to the party

Another set of debates in the sociology of technology focuses not on extending the principle of symmetry but on the possibility of rejecting it as debilitating. A leading political philosopher of technology, Langdon Winner, argues that symmetrical sociological analysis of "interpretive flexibility" (the variety of interpretations that can be placed on a scientific result, or the different meanings different groups attach to technology) "soon becomes moral and political indifference."[45]

These debates too echo older debates in the sociology of science.[46] My own view is that the satisfactory sociological analysis of scientific or technological knowledge claims does indeed require symmetry, but that this should be seen for what it is: a methodological precept appropriate for a particular, limited, intellectual task.[47] It does not imply moral and political indifference. I hope, for example, that no reader of chapter 10 gets the impression that my co-author and I feel indifferent about nuclear weapons. Nor is relativism necessarily appropriate when the intellectual task is a different one. Chapter 9, for example, attempts, in a wholly nonrelativistic way, to estimate the prevalence of computer-related accidental deaths and to inquire into their causes. It does not attempt a sociology-of-knowledge analysis of controversies over

"interpretive flexibility"

accidents, although that would be possible and indeed enlightening.[48] The chapter's aim is to make knowledge claims about computer-related accidents (with, I hope, due modesty), rather than to seek to understand the generation and reception of such claims. In other chapters, pragmatic considerations (lack of available data, irrelevance to the main narrative, and so on) mean that there are many sets of knowledge claims that I have not sought to subject to sociological analysis, even though such analysis would, in principle, be possible. In chapter 4, for example, I treat the tests of laser gyroscopes as having generated "facts" (as the individuals involved seem to have done); I make no attempt there to probe deeper.[49]

Another aspect of Winner's critique, however, seems to me to have greater force. Winner is right to note that the empirical (perhaps empiricist) methodology of much sociology of technology, focusing on explicit choices and evidently relevant social groups, creates problems for the analysis of processes of structural exclusion. Often, for example, manual workers and women are simply excluded from the arenas within which technological development takes place, never getting the chance to formulate preferences and to struggle to impose these preferences.[50] True, the picture is typically different if one broadens the analysis from technological development to manufacture, distribution, marketing, purchase, and use. But, as Cynthia Cockburn points out, the sociology of technology has tended to focus "upon the design stage and the early development of a technology."[51] My unease about this is greatest in regard to the question of gender.[52] Focusing typically on design rather than on production or use, the essays in this volume deal primarily with the work of white, middle-class men. The women whose different tasks make the work of these men possible generally remain in the background, unexamined.[53] It is difficult to believe that gender is irrelevant to the content of the men's work, but I would not claim to have found an adequate way of analyzing its effects. In this book—and, indeed, in much other writing in the history and sociology of technology—the theme of masculinity is perhaps like the "air tune" described in John McPhee's *The Pine Barrens*: "there, everywhere, just beyond hearing."[54]

The Chapters

Chapter 2, "Marx and the Machine," was written more than ten years ago. The reader may ask why, in the mid 1990s, with Marxism now utterly unfashionable, Marx's writings on technology should be seen as having anything to commend them. I would make three points in response.

First, there is a sophistication to Marx's analysis of technology that can, even now, be helpful. For example, few passages in the writings of the modern actor-network theorists surpass Marx's account of the way the machine made stable and durable the originally strongly resisted condition of wage labor. Furthermore, there is a continuing importance to the effect on production technology of the social relationships within which production takes place. Even if we set aside questions of skill, unemployment, and class conflict, we have here a major and often underestimated determinant of both the shape and the practical success or failure of technological systems.[55]

Second, the collapse of nearly all the regimes claiming allegiance to Marxism has, paradoxically, increased Marxism's relevance. Throughout the twentieth century, the influence of capitalist social relations on technology (and on much else) has been attenuated by the typically different influence of war and preparations for war.[56] In particular, since 1945 much of "high technology" has been nurtured by the entrenched conflict between the Western states and opponents that, though avowedly "socialist," were born in war and molded above all by the exigencies of military mobilization. The end of that entrenched conflict, and capitalism's "triumph,"[57] mean a world in which market forces have unprecedented sway: a world, therefore, in which Marxism may be more, not less, apposite.

A third strength of Marxism is Marx's insistence that in analyzing market forces we should never forget that "capital is not a thing, but a social relation between persons which is mediated through things."[58] The social studies of technology divide too readily into a sociology of technology that emphasizes social relations, but not their mediation through money and the market, and an economics of technology that is too little interested in the social underpinnings of economic phenomena.

Chapter 3 directly addresses this divide between sociology and economics. It does not, I hope, just make the shallow argument that we need to consider "both social and economic factors"; instead, it asks how we could try to transcend the divide. It suggests that one way to do this would be to build on the work of the "alternative" (non-neoclassical) tradition within economics begun by Herbert Simon, a tradition whose view of human behavior is much closer to that of sociology.[59] The chapter calls for "ethnoaccountancy": the empirical study of how people actually reckon financially about technology (as distinct from how economic theory suggests they should reckon). It suggests that we should study how the inherent uncertainty of technical change is (sometimes) reduced to manageable risk: how, out of potential chaos, technologists, workers, managers, and users construct a world in which economics is applicable.

One argument of chapter 3 is that to investigate these phenomena empirically we need to return (with new questions in mind) to an old genre: the "natural history" of innovations, popular in the 1960s and the 1970s. Chapter 4 is a "natural history" of one particular innovation, the laser gyroscope (although, as admitted above, it is only a very partial implementation of the ideas suggested in chapter 3). The chapter begins with the laser gyroscope's conceptual origins in scientific experiments investigating the existence of the ether—a massless substance, pervading the universe, which was held to be the medium of the propagation of light waves (as well as having, in the view of some, a theological significance). The chapter then discusses the fundamental transformations that led to the laser gyroscope's establishment in the 1980s as the dominant technology of inertial (self-contained) aircraft navigation. It describes the heterogeneous engineering needed to achieve that success, discusses how to conceptualize the economic aspects of the device's history, and argues for the crucial role of self-fulfilling prophecies in "technological revolutions" such as this.

Chapters 5 and 6 are also historical in form, but they shift the focus to the technology of high-performance computers. These machines allow the simulation of events too big, too small, too fast, or too slow for experimental investigation to be entirely adequate and too complex to be understood just from theoretical "first principles." They have become fundamental to a range of scientific and technological fields. For example, predictions about coming global warming are based largely on supercomputer simulations of the Earth's atmosphere and oceans. Such simulations raise fascinating issues about how scientific and technological communities, and wider publics, understand the relationship between the model and the reality being modeled.[60] (Some of these issues also arise in chapter 10.)

Chapter 5 and 6 focus on the development of the supercomputers that make the more sophisticated simulations possible. The premier customers for supercomputers have traditionally been nuclear weapons design laboratories. The main question addressed in chapter 5 is the extent to which these powerful organizations have shaped the technology of supercomputing as well as being its primary market. The chapter argues that the weapons laboratories played a key role in defining what we mean by "supercomputing." It also shows, however, that their attempts at more detailed influence on the internal structures or "architectures" of supercomputers were hampered by the diverse and classified nature of the "codes," the computer programs used to simulate nuclear explosions.

Instead, this key modern technology appears at first sight to have been shaped to a striking extent by one man: the American supercomputer designer Seymour Cray. Without in any way belittling Cray's great abilities or his remarkable achievements, chapter 6 (written jointly with Boelie Elzen) attempts a sociological analysis of his charismatic authority, arguing that his apparently extraordinary genius was the expression of a network of social and technical relationships. As with all charismatic authority, this expression was self-undermining: as the network constituting supercomputing developed and grew, it had to find more routine forms of expression.[61]

Chapters 7 and 8 also deal with computers, but their focus is more on the issues from the sociology of knowledge discussed above. Chapter 7 is a brief account of the development of the VIPER microprocessor and of the controversy about whether its design had been proved mathematically to be a correct implementation of its specification. Chapter 8 sets this particular episode in its wider intellectual context, arguing that computer technology offers interesting, counterintuitive case studies in the sociology of mathematical knowledge. It describes the clash between different arithmetics designed for computer implementation. This concern may seem arcane, but it is worth noting that (unknown to me when I was writing the essay) it was an error in this sphere that was the immediate cause of the Patriot failure at Dhahran. Furthermore, in November 1994 there was widespread publicity about an error in the implementation of division in Intel's celebrated Pentium chip.[62] Chapter 8 suggests that, although the litigation over VIPER is so far unique, the controversy around the VIPER proof should not be seen as entirely *sui generis*.[63] The chapter also describes the wider debate among computer scientists and others about whether to class as "proofs" mathematical arguments that rely on computer calculation or manipulation too extensive for humans to check.

The research described in chapter 9 arose as by-product of the interest in mathematical proof as applied to computers. Such work on "formal verification" often makes reference to the risks involved with computer systems upon which lives depend. Colleagues in computer science, however, offered me wildly varying estimates of the prevalence of computer-related accidents,[64] and nowhere could I find a systematic empirical analysis of their frequency or their causes.[65] Chapter 9 is an attempt, not to provide this analysis (that would be an overly grandiose description of the chapter's simplistic contents), but merely to indicate what might be involved in such an enterprise. I have no great confidence in its quantitative findings. Nevertheless, I suspect that a more sophisticated piece of work might find some of that chapter's tentative conclusions to be robust. In particular, my instincts are that it is indeed true that only a small proportion of fatalities are caused solely by "technical" faults in computer systems, and that many computer-related deaths are better attributed to "system accidents" in Charles Perrow's sense,[66] where the "system" involved is human and organizational as well as technical.

Chapter 10 (written with Graham Spinardi) seeks to reverse the focus of many of the preceding chapters and, indeed, of most of the social studies of technology. Its topic is what the processes of the development of technology can teach us about how it might be possible to do away with—to uninvent—particular technologies. The chapter seeks directly to confront the conventional wisdom that the invention of a technology such as nuclear weapons is an irreversible event. Drawing both on historical evidence and on interviews with designers of nuclear weapons, the chapter suggests that the development of nuclear weaponry depends in part upon tacit knowledge embodied in people rather than in words, equations, or diagrams. Therefore, if the designing of nuclear weapons ceases, and there is no new generation of designers to which tacit knowledge can be passed on from person to person, nuclear weapons will have been, in an important sense, uninvented. Their renewed development, though clearly possible, would have some of the characteristics of reinvention rather than mere copying.

There are some important considerations that force us to qualify this conclusion, and chapter 10 does not even mention a variety of other deep problems that would be faced by an attempt to uninvent nuclear weapons. Nevertheless, I hope that the chapter's arguments might help dispel some of the pessimism that too often, even nowadays, surrounds discussion of the future of nuclear weapons. The last few years have seen the sudden, unexpected disappearance of at least two social institutions that seemed permanent features of our world: the Cold War and apartheid in South Africa. Once we start to think about technologies, too, as social institutions—and that, for all the nuances in interpretation and differences in terminology, is the shared underlying theme of the social studies of technology—we can begin to imagine technologies, too, disappearing.

2

Marx and the Machine

As an aside in a discussion of the status of the concepts of economics, Karl Marx wrote: "The handmill gives you society with the feudal lord; the steam-mill, society with the industrial capitalist."[1] The aphorism has stuck; as a succinct précis of technological determinism it has few rivals. Apt and memorable (even if historically inaccurate)[2] as it is, it is nevertheless misleading. There is much in Marx's writings on technology that cannot be captured by any simple technological determinism. Indeed, his major discussion of the subject—occupying a large part of volume 1 of *Capital*—suggests a quite different perspective. Marx argues that in the most significant complex of technical changes of his time, the coming of large-scale mechanized production, social relations molded technology, rather than vice versa. His account is not without its shortcomings, both empirical and theoretical, yet it resonates excitingly with some of the best modern work in the history of technology. Even where these studies force us to revise some of Marx's conclusions, they show the continuing historical relevance of his account of the machine. Its possible political relevance is shown by an interesting connection between the practice of the "alternative technology" movement and an important way of studying the social shaping of technology.

Marx as Technological Determinist

Not so long ago Alvin Hansen's 1921 conclusion that Marxism is a "technological interpretation of history" was still widely accepted. Robert Heilbroner's celebrated 1967 paper "Do Machines Make History?" was headed by the famous "handmill" quotation, and Heilbroner clearly identified "the Marxian paradigm" as technological determinism. In Tom Burns's 1969 reader, *Industrial Man*, the section on Marx had as a head "Technology as the Prime Mover of Industrialization and Social Change."[3]

More recently, things have seemed not quite so clear. Many Marxists—and some non-Marxists—have been profoundly unhappy with the characterization of Marxism as technological determinism.[4] William Shaw complains: "All the friends of old Marx, it seems, have entered into a holy alliance to exorcise this specter [technological determinism]."[5] Yet the book that remains the best discussion of the different varieties of technological determinism, Langdon Winner's *Autonomous Technology,* can still be read as giving (with some crucial reservations) a technological-determinist interpretation of Marx: in changes in the forces of production, Winner writes, Marx believed he had "isolated *the* primary independent variable active in all of history."[6]

To be a technological determinist is obviously to believe that in some sense technical change *causes* social change, indeed that it is the most important cause of social change. But to give full weight to the first term in expressions such as "*prime* mover," a strong version of technological determinism would also involve the belief that technical change is itself uncaused, at least by social factors. The first of these theses we can describe, following Heilbroner,[7] as the thesis that machines make history. The second we might call the thesis of the autonomy of technical change.

The thesis that machines make history is certainly to be found in Marxist writing. Perhaps its most unequivocal statement is in Bukharin's *Historical Materialism,* where we find assertions like the following: "The historic mode of production, i.e. the form of society, is determined by the development of the productive forces, i.e. the development of technology."[8] Bukharin was far from alone in this claim,[9] and there are indeed passages from Marx's own writings that can be read in this way. The best known is the sentence from the *Poverty of Philosophy* quoted above. More weighty, though not so crisp, is the "1859 Preface":

In the social production of their existence, men inevitably enter into definite relations, which are independent of their will, namely relations of production appropriate to a given stage in the development of their material forces of production. The totality of these relations of production constitutes the economic structure of society, the real foundation, on which arises a legal and political superstructure and to which correspond definite forms of social consciousness. The mode of production of material life conditions the general process of social, political and intellectual life. It is not the consciousness of men that determines their existence, but their social existence that determines their consciousness. At a certain stage of development, the material productive forces of society come into conflict with the existing relations of production or—this merely expresses the same thing in legal terms—with the property relations within the framework of which they have operated hitherto. From forms of development of the productive forces these relations turn into their fetters. Then begins an era of social revolution.[10]

And there are several other statements, chiefly from the 1840s and the 1850s, which can be read as claims that machines make history.[11]

Alternative readings of at least some of these are possible. Rosenberg, for example, takes the "handmill" quotation and suggests that in its context it can be seen as not necessarily implying a technological determinism.[12] The "1859 Preface" is, however, where debate has centered. It was explicitly presented by Marx as "the general conclusion at which I arrived and which, once reached, became the guiding principle of my studies."[13] Echoes of it reappear throughout Marx's later works, and it has often been taken as the definitive statement of historical materialism. Anything approaching a careful reading of it quickly reveals two things. First, to make it into a statement that machines make history, the "forces of production" would have to be interpreted as equivalent to technology. Second, to make it into a strong technological determinism in the sense outlined above, the development of the forces of production would have to be taken as autonomous, or at least independent of the relations of production.

Langdon Winner signals his ambivalence about the first point when he writes that "although there is some variation in the manner in which Marx uses these terms, *for our purposes* 'forces of production' can be understood to comprise all of physical technology." Furthermore, Winner also gives a broader definition of forces of production as "the instruments, energy, and labor involved in the active effort of individuals to change material reality to suit their needs."[14] Indeed, even orthodox Marxism has tended to follow the broader meaning. Stalin wrote: "The *instruments of production* wherewith material values are produced, the *people* who operate the instruments of production and carry on the production of material values thanks to a certain *production experience* and *labor skill*—all these elements jointly constitute the *productive forces* of society." The opponents of orthodox Marxism sharply criticized the reduction of the forces of production to technology. Lukács, attacking Bukharin's *Historical Materialism,* wrote: "Technique is a *part,* a moment, naturally of great importance, of the social productive forces, but it is neither simply identical with them, nor . . . the final or absolute moment of the changes in these forces."[15]

Interpretations of Marxism as technological determinism thus rest, in effect, on the equation "forces of production = technology." Yet even

[handwritten note: But technique is a social activity pre-shaped by technical surroundings (Ellul's point)]

defenders of the proposition that Marx was a technological determinist, such as William Shaw, find it difficult to impute this equation to Marx: "For Marx the productive forces include more than machines or technology in a narrow sense. In fact, labor-power, the skills, knowledge, experience, and so on which enable labor to produce, would seem to be the most important of the productive forces." So Shaw concedes that "technological determinism is a slight misnomer since Marx speaks, in effect, of productive-force determinism."[16] But much more is at stake than terminology. For if the forces of production include human labor power, then a productive-force determinism will look very different from a technological determinism as ordinarily understood. From his earliest writings on, Marx emphasized that what was specific about human work was that it was *conscious:*

... free conscious activity is man's species character. . . . In *his work upon* inorganic nature, man proves himself a conscious species being. . . .

A spider conducts operations which resemble those of the weaver, and a bee would put many a human architect to shame by the construction of its honeycomb cells. But what distinguishes the worst architect from the best of bees is that the architect builds the cell in his mind before he constructs it in wax. . . . Man not only effects a change of form in the materials of nature; he also realizes his own purpose in those materials.[17]

The inclusion of labor power as a force of production thus admits conscious human agency as a determinant of history: it is people, as much as or more than the machine, that make history.

The autonomy of technical change is likewise a proposition attributable to Marx only questionably, even if one accepts the equation between productive forces and technology. The "orthodox" position is that the productive forces have a tendency to advance but can be encouraged or held back by the relations of production. Stalin, for example, admitted that the relations of production "influence" the development of the forces of production, but he restricted that influence to "accelerating or retarding" that development. Not all Marxist writers have seen it like this, however. There is a change of terrain in the way the modern French Marxist Etienne Balibar shifts the metaphor away from "accelerate/decelerate": "The most interesting aspect of the 'productive forces' is . . . the *rhythm* and *pattern* of their development, for this rhythm is directly linked to the nature of the relations of production, and the structure of the mode of production." Lukács disagreed with the orthodox interpretation even more sharply: "It is altogether

incorrect and unmarxist to separate technique from the other ideological forms and to propose for it a self-sufficiency from the economic
structure of society. . . . The remarkable changes in the course of [tech-
nique's] development are [then] completely unexplained."[18]

The Difficulties of Determinism

In addition to the unclear meaning and questionable autonomy of the
"forces of production," a further difficulty arises in reading the "1859
Preface" as technological determinism. That is the nature of the middle
terms in the propositions it implies. Just what is the "determination" (or
conditioning, or being the foundation of) exercised by the "totality of
[the] relations of production"? What concept of determination is
implied when it is said that the relations of production themselves are
"appropriate" to "a given stage in the development of [the] material
forces of production"?

On few topics has more ink been spilled. As Raymond Williams has
pointed out, the verb "to determine" (or the German *bestimmen*, which
is what the English translations of Marx are generally rendering when
they write "determine") is linguistically complex. The sense that has
developed into our notion of "determinism"—powerlessness in the face
of compelling external agency—derives, Williams suggests, from the
idea of determination by an authority (as in "the court sat to determine
the matter"). However, there is a related but different sense of "to determine": to set bounds or limits (as in "the determination of a lease").[19]

If the determinative effect of the forces of production on the relations of production or of the relations of production on the "superstructure" can be read in this latter way, then our image of
determination changes radically. It suggests not compelling causes but
a set of limits within which human agency can act and against which it
can push. It is an image fully compatible with another of Marx's apho-
risms, that people "make their own history, but they do not make it just
as they please; they do not make it under circumstances chosen by
themselves, but under circumstances directly encountered, given and
transmitted from the past."[20]

This is not an issue, however, that semantic debate alone can settle.
Dealing with such topics, after all, we approach the conceptual core of
a social science (any social science, not just Marxism). Variant readings
of "determination" *are* possible, from simple cause-and-effect notions to
G. A. Cohen's sophisticated defense of the thesis that the explanations

suggested by the "1859 Preface" are functional explanations ("to say that an economic structure *corresponds* to the achieved level of the productive forces means: the structure provides maximum scope for the fruitful use and development of the forces, and obtains *because* it provides such scope"). Erik Olin Wright argues, indeed, for making a positive virtue of diversity and incorporating different "modes of determination" into Marxist theory. Furthermore, debate on this issue can seldom be innocent. Profound political and philosophical differences entangle rapidly with matters of theory and methodology, as E. P. Thompson's essay "The Poverty of Theory" quickly reveals.[21]

Here we have reached the limits of the usefulness for our purposes of the exegesis of Marx's programmatic statements. The "1859 Preface" and similar passages will no doubt remain a mine, perhaps even a productive mine, for students of Marx's general theory and method. Students of technology, however, can turn their attention to a deposit that is both larger and closer to the surface: Marx's one extended and concrete discussion of technology.[22] Apart from its intrinsic interest (the main focus of what follows), this discussion throws interesting retrospective light on the more summary passages. In particular, it makes the thesis that Marx was a technological determinist in any strong sense extremely difficult to sustain, at least without invoking a peculiar and marked inconsistency between his general beliefs and his particular analyses.

The Labor Process and the Valorization Process

The chapter entitled "The Labor Process and the Valorization Process"[23] is the pivot of *Capital*. Marx, who up to that point had been analyzing chiefly the phenomena of the commodity, exchange and money, employed the full power of his skill as a writer to set the scene for the chapter: "Let us therefore . . . leave this noisy sphere, where everything takes place on the surface and in full view of everyone, and [enter] into the hidden abode of production, on whose threshold there hangs the notice 'No admittance except on business.' Here we shall see, not only how capital produces, but how capital is itself produced."[24] After the chapter, his argument built architectonically to the crescendo of "The General Law of Capitalist Accumulation" some 500 pages further on. While we will not follow him that far, this little chapter is central to an understanding of his discussion of machinery.

First, says Marx, we "have to consider the labor process independently of any specific social formation." He lists the "simple elements" of the labor process: "(1) purposeful activity, that is work itself, (2) the objects on which that work is performed, and (3) the instruments of that work." The labor process is a cultural universal, "an appropriation of what exists in nature for the requirements of man"; it is "common to all forms of society in which human beings live."[25] But it develops and changes through history.

Marx does not, as the technological-determinist reading would lead us to expect, turn now to the development of "the instruments of work." (It is interesting, indeed, that he subsumes technology, in the narrower meaning of "instruments," under the broader head of "the labor process.") Instead, he moves from the labor process in general to the labor process under capitalism, and from labor as a material process of production to labor as a social process. The process of production under capitalism is not just a labor process; it is also a valorization process, a process of adding value. The capitalist "wants to produce a commodity greater in value than the sum of the values of the commodities used to produce it, namely the means of production and the labor power he purchased with his good money on the open market."[26] He wants to produce a commodity embodying surplus value.

The distinction between the labor process and the valorization process is not a distinction between two different types of process, but between two different aspects of the same process of production. Take a simple example, the production of cotton yarn. Looking at that as a labor process means looking at the particular, concrete ways in which people work, using particular technical instruments, to transform a given raw material into a product with given properties. In any society that produces yarn it would be meaningful to examine in this way how it is done. But that is not all there is to the production of yarn under capitalism. The production of yarn as a valorization process is a process whereby inputs of certain value give rise to a product of greater value. The concrete particularities of the inputs and product, and the particular technologies and forms of work used to turn the inputs into the product, are relevant here only to the extent that they affect the quantitative outcome of the process.[27] Capitalist production processes, but not all production processes in all types of society, are valorization processes. The valorization process is the "social form" of the production process specific to capitalism.

Were Marx's theory technological determinism, one would now expect an argument that the labor process—the technology-including "material substratum"—in some sense dominated the "social form."

Quite the opposite. In his general statements on the matter (most of which are to be found in the unpublished chapter of *Capital*, "Results of the Immediate Process of Production"), Marx repeatedly argues that "the labor process itself is no more than the instrument of the valorization process."[28] And in *Capital* itself he presents an extended historical and theoretical account of the development of the capitalist production process—an account in which the social form (valorization) explains changes in the material content (the labor process). From this account let us select one central thread: Marx's history of the machine.

The Prehistory of the Machine

The history begins strangely, in that its central character is absent. The origins of capitalism, for Marx, lay not in a change in technology, but in a change in social relations: the emergence of a class of propertyless wage laborers.[29] "At first capital subordinates labor on the basis of the technical conditions within which labor has been carried on up to that point in history."[30] Archetypally, this took place when independent artisans (say textile workers), who previously produced goods on their own account, were forced through impoverishment to become employees. So instead of owning their spinning wheels or looms and buying their own raw materials, they worked (often in their own homes, under the "putting out" system) on wheels or looms belonging to a merchant, spinning or weaving raw materials belonging to him into a product that would be his property and which would embody surplus value. The social relations within which they worked had thus changed drastically; the technical content of their work was unaltered. This Marx describes as the "formal subordination" of labor to capital.[31] It was formal in that it involved a change in social form (the imposition of the valorization process) without a valorization-inspired qualitative alteration in the content of the labor process—without "real subordination."

Inherited labor processes were, however, severely deficient vehicles for the valorization process. Within their bounds, capitalists could increase surplus value primarily by the route Marx calls "absolute surplus value"—lengthening the working day. But that was not easily achieved. As Marx points out, the earliest statutes in Britain regulating the working day extend it, rather than limit it. But they were largely ineffective. It was often difficult to get workers to turn up for work at all at the beginning of the week (the tradition of "Saint Monday"). The intense, regular work required for valorization was a habit hard to impose. And outworkers without direct supervision had an effective form of disvalorization available in the form of embezzlement of raw materials, as historians more recent than Marx have emphasized.[32]

The ways capitalists sought to overcome these deficiencies in the labor process from the point of view of valorization are the subject of part 4 of volume 1 of *Capital*. The first that Marx discusses is "simple cooperation." This occurs when capital brings individual workers together "in accordance with a plan."[33] There is nothing specific to capitalism about simple cooperation: in all societies it will, for example, offer advantages in the performance of simple physical tasks, two people working together being able to lift a weight each individually could not. Nevertheless, simple cooperation offers definite advantages from the point of view of valorization.

The nature of these advantages highlights an important feature of valorization: it is not simply an economic process; it involves the creation and maintenance of a social relation. Certainly productivity is increased ("the combined working day produces a greater quantity of use-values than an equal sum of isolated working days"[34]), and the centralization of work can lead to savings in fixed capital. But, equally important, the authority of the capitalist is strengthened. For cooperation necessitates coordination. If you are lifting a weight, someone has to say "one, two, three . . . hup." Because the individual workers who are brought together by capital are subordinate to capital, that role of coordination becomes, in principle, filled by capitalist command—by capitalist *management,* to use an anachronism. The consequence Marx describes as follows: "Hence the interconnection between their [the workers'] various labors confronts them; in the realm of ideas, as a plan drawn up by the capitalist, and, in practice, as his authority, as the powerful will of a being outside them, who subjects their activity to his purpose."[35] A form of alienation is involved here—not psychological alienation, nor alienation from a human essence, but the literal alienation of the collective nature of work. That collective nature is here seen as becoming the power of another—of the capitalist. In addition, the physical concentration of workers under the one roof greatly facilitates the down-to-earth tasks of supervision: enforcing timekeeping and preventing embezzlement.[36]

Marx intended "simple cooperation" as an analytic category rather than as a description of a historical period in the development of the labor process (although more recent writers have specified a historical phase in which it was crucial).[37] The form of cooperation typical of the period immediately prior to mechanization Marx describes as "manu-facture."[38] (Marx, of course, uses the term in its literal sense of making by hand.) Crucially, manufacture, unlike the most elementary forms of cooperation, involves the differentiation of tasks, the division of labor. It arises in two ways. One is the bringing together of separate trades, as in the manufacture of carriages, where wheelwrights, harness makers, etc., are brought together under the same roof, and their work special-ized and routinized. The other, and perhaps the more significant, is where the production of an item formerly produced in its entirety by a single handicraft worker is broken down into separate operations, as in the manufacture of paper, type, or (classically) pins and needles.

The division of labor involved in manufacture was often extreme. Marx spends nearly a page listing a selection of the trades involved in the manufacture of watches, and points out that a wire on its way to becoming a needle passes "through the hands of seventy-two, and some-times even ninety-two, different specialized workers." The advantages from the viewpoint of valorization of this division of labor are clear. Labor is cheapened, according to the principle enunciated by Babbage in 1832: "The master manufacturer, by dividing the work to be execut-ed into different processes, each requiring different degrees of skill or of force, can purchase exactly that precise quantity of both which is nec-essary for each process; whereas, if the whole work were executed by one workman, that person must possess sufficient skill to perform the most difficult and sufficient strength to execute the most laborious, of the operations into which the art is divided." Productivity is increased through specialization and the increased continuity and intensity of work, although at the cost of "job satisfaction": ". . . constant labor of one uniform kind disturbs the intensity and flow of a man's vital forces, which find recreation and delight in the change of activity itself."[39]

In addition, the division of labor in manufacture reinforces the sub-ordination of the worker to the capitalist. Craft workers able to produce an entire watch might hope to set up independently; the *finisseurs de charnière*, "who put the brass hinges in the cover," could hardly hope to do so. Even more strikingly than in simple cooperation, under manu-facture the collective nature of work, the interdependence of the dif-ferent labor processes involved, confronts workers as the capitalist's power. The manufacturing worker, unable to perform or even under-stand the process of production as a whole, loses the intellectual com-mand over production that the handicraft worker possessed. "What is lost by the specialized workers is concentrated in the capital which confronts them. It is a result of the division of labor in manufacture that the worker is brought face to face with the intellectual potentialities of the material process of production as the property of another and as a power which rules over him." The alienation of the collective nature of work has advanced one stage further, and the division of head and hand that typifies modern capitalism has begun to open up decisively. Marx quotes from a book written in 1824 a lament that the radical science movement of the 1960s and the 1970s would easily recognize: "The man of knowledge and the productive laborer come to be widely divided from each other, and knowledge, instead of remaining the handmaid of labor in the hand of the laborer to increase his productive powers . . . has almost everywhere arrayed itself against labor. . . . Knowledge [becomes] an instrument, capable of being detached from labor and opposed to it."[40]

And yet manufacture was not a fully adequate vehicle for valorization. The basis of the manufacturing labor process remained handicraft skill, however fragmented and specialized, and that skill was a resource that could be, and was, used in the struggle against capital. So "capital is constantly compelled to wrestle with the insubordination of the workers," and "the complaint that the workers lack discipline runs through the whole of the period of manufacture."[41] But, by one of the ironies of the dialectic, the most advanced manufacturing workshops were already beginning to produce . . . the machine.

Enter the Machine

Up to this point in his discussion, Marx makes effectively no mention of technical change, instead focusing exclusively on the social organization of work. It was not that he was ignorant of the technical changes of the period of manufacture. Rather, his discussion is laid out in the way it is to argue a theoretical point: that preceding organizational changes created the "social space," as it were, for the machine; and that the limitations of those changes created the *necessity* for it.

But what is a machine? Marx's chapter "Machinery and Large-Scale Industry" opens with what appears to be a rather pedantic discussion of the definition of "machine." Yet this little passage is highly significant because of the nature of the definition that Marx chose.

Marx rejected definitions that saw a continuity between the "tool" and the "machine"—definitions typical of "mathematicians and experts on mechanics." While it is true that any machine is analyzable as a complex of more basic parts, "such as the lever, the inclined plane, the screw, the wedge, etc.," this "explanation is worth nothing, *because the historical element is missing from it.*" Nor does it suffice to differentiate the tool from the machine on the basis of the power source (human in the case of the former, nonhuman in the case of the latter): "According to this, a plough drawn by oxen, which is common to the most diverse modes of production, would be a machine, while Claussen's circular loom, which weaves 96,000 picks a minute, though it is set in motion by the hand of one single worker, would be a mere tool."[42]

Instead, Marx offers the following definition: "The machine . . . is a mechanism that, after being set in motion, performs with its tools the same operations as the worker formerly did with similar tools." This *is* a historical definition in two senses. First, Marx argues that of the three different parts of "fully developed machinery"—"the motor mechanism, the transmitting mechanism and finally the tool or working machine"— it was with innovations in the third that "the industrial revolution of the eighteenth century began." Changes in the source of motive power were historically secondary and derivative. Second, and more important, it is a historical definition in that it points up the place of the machine in the process that Marx was analyzing. The machine undermined the basis on which manufacturing workers had resisted the encroachments of capital: "In manufacture the organization of the social labor process is purely subjective: it is a combination of specialized workers. Large-scale industry, on the other hand, possesses in the machine system an entirely objective organization of production, which confronts the worker as a pre-existing material condition of production."[43]

Essentially, in machinery capital attempts to achieve by technological means what in manufacture it attempted to achieve by social organization alone. Labor power is cheapened, for example, by the employment of women and children. This is not merely a technical matter of the simplification of labor or of "machinery dispens[ing] with muscular power." Under manufacture, the division of labor had already created a wealth of jobs requiring neither particular skill nor particular strength; in any case, it is clear that these attributes are not naturally the exclusive preserve of adult males. Rather, the tendency to the employment of women and children had been "largely defeated by the habits and the resistance of the male workers."[44]

In the long run, the machine contributes to valorization crucially through the medium of "relative surplus value": the reduction in the labor time required to produce the equivalent of the worker's wage, with consequent increase in the surplus value accruing to the capitalist. In the short run, however, the machine also sets capital free to accrue absolute surplus value. By undermining the position of key groups of skilled workers, by making possible the drawing of new sectors into the labor market, and by threatening and generating unemployment, the machine "is able to break all resistance" to a lengthening of the working day.[45] And because work can now be paced by the machine, its intensity can be increased.

Most important, the alienation of the collective and intellectual aspects of work, already diagnosed by Marx in simple cooperation and manufacture, achieves technical embodiment in the machine. For "along with the tool, the skill of the worker in handling it passes over to the machine." The machine, increasingly a mere part of an automated factory, embodies the power of the capitalist: "The special skill of each individual machine operator, who has now been deprived of all significance, vanishes as an infinitesimal quantity in the face of the science, the gigantic natural forces, and the mass of social labor embodied in the system of machinery, which, together with these three forces, constitutes the power of the 'master.'"[46]

In the labor process of machino-facture, capitalist social relations thus achieve technical embodiment. It is characteristic of capitalism in all its stages that "the conditions of work," the means of production in their social form as capital, employ the worker, instead of the worker employing the means of production. "However, it is only with the coming of machinery that this inversion first acquires a technical and palpable reality." Before the machine, the worker still commanded the tool—and used this command as a source of countervailing power. From the viewpoint of the worker, the machine is thus a direct threat. It is "capital's material mode of existence."[47]

So class struggle within capitalism can take the form of "a struggle between worker and machine." Workers, of course, directly attacked machines (and still do, even if organized machine breaking has given way to less overt forms of "sabotage").[48] But the struggle, Marx emphasized, is two-sided. Capital uses machinery not only strategically, as outlined above, but also for precise tactical purposes. Where workers' (especially skilled workers') militancy poses a threat to valorization, capital can counter by promoting the invention and employment of machinery to undermine workers' power.

The theorist of this waging of class struggle by technical means was Andrew Ure, who concluded in his 1835 *Philosophy of Manufactures* that "when capital enlists science into her service, the refractory hand of labor will always be taught docility." Marx cited inventions discussed by Ure—coloring machines in calico printing, a device for dressing warps, the self-acting spinning mule—as means of doing this, and he suggested that the work of inventors such as James Nasmyth and Peter Fairbairn had apparently been motivated by the exigencies of defeating strikers. "It would be possible," Marx judged, "to write a whole history of the inventions made since 1830 for the sole purpose of providing capital with weapons against working-class revolt."[49]

Marx's Account and the Historical Record

Capital was published in 1867. How well does Marx's account stand up in the light of over a century of historical scholarship? There is considerable agreement with his characterization of the overall process of the mechanization of production, even from those who would not regard themselves as standing in any Marxist tradition. David Landes writes: "For many [workers]—though by no means for all—the introduction of machinery implied for the first time a complete separation from the means of production; the worker became a 'hand.' On almost all, however, the machine imposed a new discipline. No longer could the spinner turn her wheel and the weaver throw his shuttle at home, free of supervision, both in their own good time. Now the work had to be done in a factory, at a pace set by tireless, inanimate equipment."[50]

The close connection between class conflict and technical innovation in nineteenth-century Britain has been noted moderately often in more recent historical writing. Landes writes that "textile manufacturers introduced automatic spinning equipment and the power loom spasmodically, responding in large part to strikes, threats of strikes, and other threats to managerial authority."[51] Nathan Rosenberg argues that "the apparent recalcitrance of nineteenth-century English labor, especially skilled labor, in accepting the discipline and the terms of factory employment provided an inducement to technical change," and lists particular innovations in which this process can be identified. Rosenberg's list largely follows Marx's, but he adds such items as the Fourdrinier paper-making machine.[52] While denying that the spread of the self-acting mule to America can be accounted for in this way, Anthony F. C. Wallace echoes Ure and Marx on its technical development: "The goal of inventors, from Crompton's time on, was to make the mule completely automatic so as to reduce to a minimum the manufacturer's dependence on the highly skilled, highly paid, and often independent-minded adult male spinners."[53] Tine Bruland argues that, in the case of the mule (and also in those of calico-printing machinery and devices for wool combing), it was indeed true that "industrial conflict can generate or focus technical change in production processes which are prone to such conflict."[54]

For a different historical context (Chicago in the 1880s), Langdon Winner draws on the work of Robert Ozanne to provide another example. Newly developed pneumatic molding machines were introduced by Cyrus McCormick II into his agricultural machinery plant to break the power of the National Union of Iron Molders. "The new machines, manned by unskilled labor, actually produced inferior castings at a higher cost than the earlier process. After three years of use the machines were, in fact, abandoned, but by that time they had served their purpose—the destruction of the union."[55]

The obverse of the capitalists' use of machinery in class struggle, workers' resistance to the machine, is too well known in the case of Britain to require special documentation. Interestingly, though, historians have begun to interpret that resistance differently. Luddism, it has been argued, was neither mindless, nor completely irrational, nor even completely unsuccessful.[56] The working-class critique of machinery, of which machine breaking was the most dramatic concrete expression, left a major mark on British thought. Maxine Berg has shown the extent to which the science of political economy was formed in Britain by the debate between the bourgeois proponents of machinery and its working-class opponents—and also its landed Tory opponents.[57]

Historians are also beginning to find resistance to the machine where it was once assumed that there had been none. Merritt Roe Smith's justly celebrated *Harpers Ferry Armory and the New Technology* shows that the "American system of manufactures"—the distinctive contribution of nineteenth-century America to the development of mechanized mass production—was resisted. The highly skilled armorers, and many of the institutions of the still essentially rural society in which they lived, opposed, often bitterly and on occasion violently, changes which meant that "men who formerly wielded hammers, cold chisels, and files now stood by animated mechanical devices monotonously putting in and taking out work, measuring dimensions with precision gauges, and occasionally making necessary adjustments."[58] The struggle documented by Smith between "the world of the craftsman" and "the world of the machine" at Harpers Ferry significantly modifies the assumption that "American workmen welcomed the American system."[59]

Marx's views on one particular key technology—the steam engine—
have also found confirmation in G. N. von Tunzelmann's recent work.
Marx's analysis, writes Tunzelmann, "is spare and succinct, encapsulating
what emerge in my study as the truly significant links between steam-
power and cotton." Von Tunzelmann finds himself in extensive agreement with Marx's argument that technical changes in the steam engine
resulted from changing capital-labor relations in mid-nineteenth-centu-
ry Britain. It may not have simply been the Ten Hours Act, restricting the
length of the working day, that induced employers and designers to
increase boiler pressures and running speed, but the need "for squeez-
ing out more labor in a given time" was certainly important.[60]

This way of proceeding—comparing Marx's theory with more recent
historical accounts—can, however, too easily become an exercise in
legitimation, or an argument that, to quote Paul Mantoux, Marx's
"great dogmatic treatise contains pages of historical value."[61] It also
ignores real problems of evidence concerning the origins of certain
innovations. It is indeed a fact, as Rosenberg notes, that in early nineteenth-century Britain it was widely agreed that "strikes were a major
reason for innovations."[62] But the extent of that agreement is a different matter from whether it described the actual state of affairs. Neither
the "discovery accounts"[63] of inventors such as Nasmyth nor the anecdotes and inferences of contemporaries such as Andrew Ure or Samuel
Smiles, are necessarily to be taken at face value. Yet, in the still-common
absence of historical research addressing such questions for particular
innovations, more recent writers are often no better placed than Marx
in terms of the sources open to them. Studies such as *Harpers Ferry
Armory*, alive equally to the detail development of particular technologies and to the social relations of production, are still too rare to allow
confident generalization.

Further, it would be quite mistaken to see Marx's account of the
machine as completed. His account contains difficulties and ambiguities, and these need to be clarified in parallel with, and in relation to,
its testing against "actual history." It is actually a theory, not a putative
description of events. It is not a history of the Industrial Revolution, or
even of the Industrial Revolution in Britain, but an attempt to develop
a theory of the social causes of organizational and technical changes in
the labor process. Uniform, unilinear developmental paths cannot
properly be deduced from its premises. Actual history will inevitably be
more complicated. Thus Marx himself had to turn, immediately after
his discussion of machine production, to the very considerable continuing areas of domestic outwork and manufacture. Raphael Samuel's

major survey of the balance between "steam power" and "hand technology" in Marx's time shows the slowness of the process of mechanization. Indeed, Marx was arguably wrong to assume that outwork and small-scale manufacture were necessarily forms "transitional" to "the factory system proper."[64] A century after his death outwork still flourishes, even in some technologically advanced industries.[65] On occasion, valorization may be better served by decentralized rather than centralized labor processes.[66]

This example illustrates a general issue that became important as interest in Marx's theory revived during the 1970s. In the rush of theoretical reflection and empirical research about the labor process, writers sometimes conflated particular strategies that capital employs to further valorization with the goal of valorization itself. Capitalists were seen as *always* pursuing the deskilling of labor, or as *always* seeking maximum direct control over the labor process. But neither assertion is even roughly correct empirically, nor is either goal properly deducible from the imperative of valorization alone. "Skill" is not always a barrier to valorization; only under certain (common but not universal) circumstances does it become one. Direct control over the labor process is not always the best means of valorization.

Marx himself seems on occasion to postulate something close to a thesis of continual deskilling and of the creation of a homogeneous work force: "In place of the hierarchy of specialized workers that characterizes manufacture, there appears, in the automatic factory, a tendency to equalize and reduce to an identical level every kind of work that has to be done by the minders of the machines."[67] The outcome of the extensive research and debate occasioned by Harry Braverman's influential elaboration of the "deskilling" thesis can in part be summarized by saying that deskilling and homogenization are precisely "a tendency"—no more.[68] The imperative of valorization does bring about changes in the labor process that do away with capital's dependence on many human competences that once were necessary, these changes do undermine the position of groups of workers who owe their relatively high wages or ability to resist capital to their possession of these competences, and technology is crucial to this process. But these changes in the labor process also create the need for new competences, create new groups of "skilled" workers, and create types of work that are far from exemplifying the real subordination of labor to capital.[69] The very creation of these is often the obverse of the process of deskilling other occupations: computer programming is a contemporary example.[70]

Similarly with control. From a twentieth-century perspective, too much weight is placed in *Capital* on what Andrew Friedman calls a "direct control" strategy on capital's behalf. This strategy, of which Taylorism is the obvious example for the period after Marx's death, "tries to limit the scope for labor power to vary by coercive threats, close supervision and minimizing individual worker responsibility" and "treats workers as though they were machines." But "direct control" hardly captures the range of strategies for the management of labor power. Management can also involve a "responsible autonomy" strategy, trying "to harness the adaptability of labor power by giving workers leeway and encouraging them to adapt to changing situations in a manner beneficial to the firm . . . [giving] workers status, authority and responsibility . . . [trying] to win their loyalty, and co-opt their organizations to the firm's ideals."[71]

Again, there is nothing in Marx's theory to suggest that capital will seek maximum control over the labor process as a goal in itself, or that capitalists will necessarily prefer direct over indirect forms of control. A degree of control over the labor process is clearly a prerequisite for valorization, but the theory does not lay down how that control can best be achieved, nor does it imply that control should be pursued regardless of its costs. Supervisors, after all, cost money, and techniques of production that maximize direct control over labor power may be fatally flawed in other respects.

To present Marx's theory as hinging around valorization rather than deskilling or control points to the relevance to it of the traditional concerns of those economic historians who have made technology a central focus of their work.[72] The level of wages, the rate of interest, the level of rent, the extent of markets—all these would be expected to influence the choice of technique, and there are passages in Marx that show his awareness of this.[73]

Where the Marxist and the "neoclassical" economic historian would diverge, however, is in the Marxist's insistence that "factor costs" ought not to be treated in abstraction from the social relations within which production takes place. This is a persistent theme throughout *Capital*. Capital, Marx wrote, "is not a thing"; it is not a sum of money or commodities; it is "a social relation between persons which is mediated through things."[74] The relation between capitalist and worker is not simply a matter of wages and hours of work; it is also a matter of law and the state (in, for example, the worker's legal status as "free citizen" or otherwise), of supervision, discipline, culture, and custom, of collective forms of organization, power, and conflict.[75]

William Lazonick, in his study of the choice of technique in British and U.S. cotton spinning, argues that, although factor prices mattered, their effect was conditioned by the very different nature of production relations in such spinning centers as Oldham in Lancashire and Fall River in Massachusetts. Such facts as the preference of Lancashire mill owners for spinning mules and that of their New England counterparts for ring spinning have to be understood in the context of the different historical evolution of relations within the work forces and between workers and capitalists.[76]

Lazonick's work, though, is far from an uncritical confirmation of Marx. Indeed, it points up a major inadequacy in Marx's account—one that ties in closely with the problem of evidence mentioned above. Marx's reliance on sources such as the writings of Ure meant that he had quite plausible evidence for what class-conscious capitalists hoped to achieve from the introduction of the machine. But what they hoped for was not necessarily what happened. Marx quoted Ure's judgment on the self-acting mule: "A creation destined to restore order among the industrious classes." Lazonick's work shows that the mule had no such dramatic effect. In Lancashire, "adult male spinners (now also known as 'minders') retained their positions as the chief spinning operatives on the self-actors," developed a strong union, achieved standardized wage lists that protected their wage levels, and kept a fair degree of control over their conditions of work. Such was the failure of the self-acting mule in increasing capital's control that when ring spinning was introduced in New England it was talked about in precisely the same terms as the self-actor had once been—as a curb on "obstreperous" workers—only this time these were the minders of self-acting mules![77]

In part, the failure of capitalists to achieve their goals can be put down to workers' resistance; to the extent that it can be explained in this way, it offers no fundamental challenge to Marx's account. Workers are not passive clay in capital's hands; quite the opposite. Even highly automated factories with close, harsh labor supervision offer major opportunities both for individual acts of noncompliance and for collective action to change conditions.[78] Further, the very fact that the labor process, however much it is affected by the valorization process, remains a material process of production constrains what capital can achieve. In his work on automatically controlled machine tools, David Noble found that, despite all their efforts, managements were unable to do without skilled machinists. As one machinist put it: "Cutting metals to critical tolerances means maintaining constant control of a continually changing set of stubborn, elusive details. Drills run. End mills walk. Machines creep. Seemingly rigid metal castings become elastic when clamped to be cut, and spring back when released so that a flat cut becomes curved, and holes bored precisely on location move somewhere else. Tungsten carbide cutters imperceptibly wear down, making the size of a critical slot half a thousandth too small." Experienced machinists were needed to make sure that "automatic" machines did not produce junk parts or have expensive "smashups."[79]

The intractability of both workers and the material world is, however, not fully sufficient to explain the type of development described by Lazonick. Here we come to an area where Marx's account clearly requires modification. The social relations of production within which technology develops are not simply between worker and capitalist, but also between worker and worker. Crucially, they include relations between male workers and female workers, between older workers and younger workers, and, sometimes at least, between workers of different ethnic groups.

Marx was of course aware of the division of labor by age and sex, but he slid far too readily into a facile description of it as "natural."[80] Lazonick's account of the history of the self-acting mule, for example, shows that adult male minders in Britain retained their position not through any "natural" attributes, nor because of their power to resist capital, but because British employers found useful, indeed indispensable, the hierarchical division in the work force between minders and "piecers," whose job it was to join the inevitable broken threads. And this relation within the work force conditioned technical change. It made it rational for capitalists to work with slightly less automated mules than were technically possible, so that failures of attention by operatives led not to "snarls" that could be hidden in the middle of spun "cops" but to the obvious disaster of "sawney," where all of the several hundred threads being spun broke simultaneously, with consequent loss of piecework earnings for the minder.[81]

Of the divisions within the work force that affect the development of technology, that between women and men is perhaps the most pervasively important. Marx's account captures only one of the (at least) three ways in which this division interacts with change in the technology of production. He focuses on the very common use of machinery plus low-paid, less unionized women workers to replace skilled men. Ruth Schwartz Cowan, in her review of "women and technology in American life," shows this process at work in American cigar making.

But she also points to the very different situation of the garment industry, arguing that there the sewing process had not been automated
(beyond the use of the sewing machine) in large part because of the
availability of "successive waves" of immigrant women. Their undoubted
skills cost employers nothing extra. Those skills were learned largely in
the home, rather than at the employers' expense. And because sewing
is "women's work," it is defined as unskilled (Phillips and Taylor argue
that this, not the opposite as commonly assumed, is the real direction of
causation) and thus is poorly paid.[82]

A third form of the interaction between gender divisions and workplace technology is that identified by Cynthia Cockburn in her study of
the history of typesetting technology in Britain. Up to a point, the
process was exactly parallel to that described by Marx. Employers
sought to invent a machine that could "bypass the labor-intensive
process of hand typesetting," thus undermining the well-paid, well-
unionized male hand compositors. By the end of the nineteenth century several such mechanized typesetters had become available, and the
compositors and their employers struggled over their introduction. But
here the story diverges from Marx's archetype. The male compositors
(like the mule spinners) were able to retain a degree of control over the
new technology, and the machine that became the dominant means of
mechanizing typesetting, the Linotype, was the one that offered least
threat to their position. Unlike its less successful predecessor, the
Hattersley typesetter, the Linotype did not split the process of typesetting into separate parts. As the men's union, the London Society of
Compositors, put it, by not splitting up the process "the Linotype
answers to one of the essential conditions of trade unionism, in that it
does not depend for its success on the employment of boy or girl labor."
The choice of the Linotype, backed up by vigorous campaigning by the
union to exclude women, eventually left the composing room still "an
all-male preserve." Technology, according to Cockburn, can thus reflect
male power as well as capitalist power.[83]

The Politics of Design and the History of Technology

Perhaps the most intriguing question of all those that are raised by
Marx's account of the machine is one that Marx neither put clearly nor
answered unequivocally: Does the design of machinery reflect the social
relations within which it develops? Do capitalists (or men) merely abuse
machinery for their own purposes, or do those purposes somehow
shape the machine?

At this point, of course, the issues raised by Marx's theory converge with a central question—perhaps the central question—of the history of technology. George Daniels posed it when he organized his essay "The Big Questions in the History of American Technology" around the "nature and the direction of causation" in the relationship between technology and society, asserting his belief that "the direction of the society determines the nature of its technological innovations." "The influence of economics, politics, and social structure on technology" is among the topics mentioned by Thomas Hughes in his survey "Emerging Themes in the History of Technology." According to Carroll Pursell, arguments about the neutrality of technology—whether "the purposes (ethics and values) of our society are built into the very form and fabric of our technology"—have "grave implications . . . for the way in which the history of technology is studied and taught." If the history of technology needs to be rescued, as David Hounshell believes, from becoming "increasingly internalistic" in its approach, then pursuit of this question offers a way of combining attention to technical detail with concern for broader issues of social history.[84]

Replying to Hounshell, Darwin Stapleton notes that Karl Marx "has always been in the background" of the history of technology.[85] Unfortunately, Marx himself equivocated on this crucial question. Sometimes he appears to treat machines as subject to abuse by capital but not in their design inherently capitalist: "It took both time and experience before the workers learnt to distinguish between machinery and its employment by capital, and therefore to transfer their attacks from the material instruments of production to the form of society which utilizes those instruments." Marx also writes, however, that a "specifically capitalist form of production comes into being (at the technological level too)."[86] While it seems to me that extending Marx's theory to the level of detailed technical design would be a natural step, we have no unequivocal evidence that Marx took it. *A priori*, it would not be unreasonable (indeed, as outlined above, it would be orthodox) to accept that the *pace* of technical change was affected by social relations—that mechanization was hastened by valorization-imposed needs to undermine the power of skilled workers, for example—while denying that those relations affected the actual design of technical artifacts. Without clear information about what Marx believed, we can but turn to the more important question of what actually is the case.

Fortunately, historians have found it possible to obtain at least partial, tentative answers to the question of the effect of social relations on technical design. Perhaps the most straightforward way of doing this hinges on documenting the *contingency* of design, identifying instances where "things could have been different," where, for example, the same artifact could have been made in different ways, or differently designed artifacts could have been constructed. Having identified contingency, the historian can then ask why one way, or one design, was chosen rather than another. In that way the question of the effect of social relations becomes a matter for empirical inquiry as well as for theory.[87]

Langdon Winner's stimulating essay "Do Artifacts Have Politics?" provides a rudimentary but clear example. Robert Moses could have had the bridges over Long Island's parkways constructed with a wide range of clearances. He chose to build them low, with "as little as nine feet of clearance at the curb." The reason, Winner argues, was that the buses which might otherwise take poor people along the parkways to Moses's "widely acclaimed public park" at Jones Beach were 12 feet high![88] (Why contingency is important is obvious here. If it had not been clearly possible for Moses to choose to build higher overpasses, we would have no way of assessing the relevance of his social prejudices to his bridge design.)

There is of course nothing new about the approach of identifying contingency,[89] nor is identifying contingency in itself enough.[90] An explanation of the causes of the choices actually made is necessary too. But here Marx's theory *is* useful, because it does suggest where to look for such an explanation—in the area of the technology of production, at least. In any society, the design of production technology will reflect the need for that technology to be part of a labor process that is a functioning whole. This implies obvious physical constraints: the instruments of production must be compatible with the raw materials available. But it also implies social constraints. The labor process in a capitalist society must function effectively not simply as a material process of production but also as a valorization process. Production technology will thus be designed with a view to ensuring successful valorization, and valorization will typically not simply be a matter of "profit maximizing" but will involve the creation and maintenance of desired social relations.

David Noble's analysis of the automation of machine tools can be seen as an attempt to apply this perspective to technical design. Noble identifies contingency in that development. There were *two* ways to automate—record-playback and numerical control—and it is far from clear that only numerical control was *a priori* viable. Noble also identifies a problem of valorization: the capacity of skilled machinists to control the pace of production, or indeed to disrupt it completely. He suggests that the choice of numerical control reflected its perceived superiority as a solution to this problem of valorization. As one engineer central to the development of both systems put it: "Look, with record-playback, the control of the machine remains with the machinist—control of feeds, speeds, number of cuts, output; with N[umerical] C[ontrol] there is a shift of control to management. Management is no longer dependent upon the operator and can thus optimize the use of their machines. With N.C., control over the process is placed firmly in the hands of management—and why shouldn't we have it?"[91]

Contingency and the Politics of Technology

There is of course one major objection to making contingency the way into the study of the social relations embodied in the actual design of artifacts and of the technologies of production: we may not be able to identify contingency. The most obvious way to legitimate any particular design decision or choice of technique is to say it is "technically necessary." A vested interest thus typically arises in disguising the actual extent of contingency. Even more serious, particular ways of designing things and making things can become so routine and habitual that our minds may be closed to the possibility of doing things otherwise. Though Seymour Melman may be right that choice in production techniques and the consciousness of choice among engineers and designers are pervasive, the parameters within which choice operates may well be much narrower than those within which it could operate.[92]

Several attempts have been made to reveal the extent of contingency by designing "alternative technologies." Best known are the efforts to embody in technology the virtues of small scale, decentralization, and ecological awareness. But there have also been attempts from within high-technology industry to alter in fundamental ways both what is produced and how it is produced. In Britain this was best exemplified by the "alternative plans" put forward by the work force at Lucas Aerospace. These plans involved attempts to shift production from military to "socially useful" products, and also to change the nature of production—to reverse deskilling and the separation of head and hand. The Lucas employees' work in this latter sphere seems to have been informed explicitly by Marx's analysis of the machine.[93]

Whatever the eventual success or failure of efforts to alter the nature of technology, our understanding of how technology changes can only profit from them. By making contingency and choice actual rather than merely hypothetical, they throw into ever-sharper light the ways in which social relations shape technical development. Perhaps, too, the process can be dialectical rather than one-way. Perhaps understanding how existing technology has been and is being socially shaped can help in reconstructing it. If that can be so, and if Marx's account of the machine is useful to that understanding, then the shade of Marx will surely be happy, for it was of the essence of the man that he believed not simply in understanding the world but also in changing it.[94]

3

Economic and Sociological Explanations of Technological Change

This chapter seeks to identify tools to overcome the cleavage between economic and sociological analyses of technological change. It draws on the tradition of "alternative economics" deriving from Herbert Simon. A more implicit debt is to Marx's critique of political economy, and an explicit, but of necessity highly tentative, attempt is made to argue that the sociology of scientific knowledge might be brought to bear on the economist's discussion of the unmeasurable uncertainty (rather than quantifiable risk) of technological change.

I am painfully aware of many places where I shall stray into areas where I am ignorant. There may well be answers to the questions I ask and a relevant literature of which I am unaware. It may be that, as a sociologist, I have misunderstood what economists mean. In some places I suspect, though I am not certain, that I am calling for the bringing of coals to Newcastle. If any of this is true, I would be most grateful for both pardon and enlightenment. Unless we take the risk of revealing our ignorance, interdisciplinary bridges will not be built.

In studies of technology, the gap between economic and sociological explanations is pervasive. Economic analyses are often based upon assumptions sociologists regard as absurd, while sociological writing often almost ignores the dimension of cost and profit in its subject matter. Though there are thinkers who provide rich resources for transcending the gap (despite their considerable differences, Karl Marx and Herbert Simon are the two central ones), it is far more common to find economic and sociological studies, even of the same topic, existing in separate conceptual universes.[1]

In the first section of the chapter I contrast neoclassical economics, particularly its assumption of profit maximization, with the alternative economics associated with Simon and more recently developed by Richard Nelson and Sidney Winter. I then go on to discuss possible applications of that alternative view to a false dichotomy sometimes found in labor-process studies, to pricing behavior in the computing industry, and to the setting of research and development budgets.

Next I examine the idea of a "technological trajectory" or a "natural trajectory" of technology, found in the work of Nelson and Winter and other recent contributors to the economics of technology. I argue that, although persistent patterns of technological change do exist, there is a crucial ambiguity in their description as "natural," and that a different understanding of them would help bridge the gap between economic and sociological explanations.

In the next section I discuss another way of bridging the gap, one again loosely in the tradition of Simon but in practice little pursued: the "ethnoaccountancy" of technological change (that is, the empirical study of how people actually reckon financially about technology, as distinct from how economic theory suggests they should reckon).

Finally, I turn to the topic of uncertainty and the construction of the economic. Despite their "thing-like" character, economic relations are never wholly self-sustaining and self-explaining. Whereas this point is normally argued in the large (Marx justifies it by an examination of the evolution of capitalism), technological innovation demonstrates it on a smaller scale. As is well known, the inherent uncertainty of radical innovation makes economic calculation applicable only *ex post*, not *ex ante*—that is, once networks have stabilized, not before. This makes radical innovation a problem for orthodox economics, but it points, I argue, to the relevance here of the sociology of scientific knowledge.

Neoclassical and Alternative Economics

It is convenient to begin with our feet firmly on the economic side of the gap. The neoclassical economics of production technology is crystalline in its explanations. Although the full neoclassical structure is dauntingly complex, its central pivot is simple and clear: firms choose production technology so as to maximize their rate of profit.

Unfortunately, that clarity is purchased at too high a price. The notion of maximization at the heart of the neoclassical structure is incoherent, at least as a description of how firms do, or even could, behave. Perhaps the most cogent statement of why this is so comes from Sidney Winter:

It does not pay, in terms of viability or of realized profits, to pay a price for information on unchanging aspects of the environment. It does not pay to review constantly decisions which require no review. These precepts do not imply merely that information costs must be considered in the definition of profits. For without observing the environment, or reviewing the decision, there is no way of knowing whether the environment is changing or the decision requires review. It might be argued that a determined profit maximizer would adopt the organization form which calls for observing those things that it is profitable to observe at the times when it is profitable to observe them: the simple reply is that this choice of a profit maximizing information structure itself requires information, and it is not apparent how the aspiring profit maximizer acquires this information, or what guarantees that he does not pay an excessive price for it.[2]

This critique of neoclassical economics draws most importantly upon the work of Herbert Simon. It has been elaborated by Winter, by his collaborator Richard Nelson, and by a goodly number of other economists. Its logic seems inescapable.[3] Furthermore, Simon and his intellectual descendants do not simply highlight the central incoherence haunting neoclassical economics' formidable apparatus of production functions, isoquants, and the like. They provide a different vision of economic activity. In this alternative economics, actors follow routines, recipes, and rules of thumb while monitoring a small number of feedback variables. As long as the values of these variables are satisfactory ("satisficing" is Simon's famous replacement for "maximizing"), the routines continue to be followed. Only if they become unsatisfactory will they be reviewed. But the review will not be an unconstrained evaluation of the full universe of alternatives in search of the best; it will be a local search, given direction by the perceived problem in need of remedy and using heuristics (which are rather like routines for searching).

This intellectual tool kit offers a bridge toward sociological analysis as it is conventionally understood. Routines can be entrenched for a variety of organizational reasons, and different parts of a firm typically follow different routines and different heuristics of search. Since in this perspective there is no longer any ultimate arbiter of routines (such as profit maximization), firms become political coalitions rather than unitary rational decision makers. The actual behavior of a firm may represent a compromise between different and potentially contending courses of action.[4]

Intrafirm processes are not, of course, ultimately insulated from what goes on outside the firm. That outside is a "selection environment" favoring certain routines over others. Nelson and Winter, especially, draw an explicit parallel with evolutionary biology, seeing routines as akin to genes, being selected for or against by their environment. This environment is not just "the market"; it includes other institutional structures as well. It is not necessarily or even generally stable, nor is it simply external and "given." One particular firm may be able to alter its environment only slightly (although some firms patently alter it more than slightly), but the behavior of the ensemble of firms is in large part what constitutes the environment.[5]

This "alternative economics" promotes a subtle change in ways of thinking, even in areas where its relevance is not apparent. Take, for example, David Noble's justifiably celebrated, empirically rich study of the automation of machine tools in the United States. Noble frames his most general conclusion in terms of a dichotomy between profit and capitalists' control over the work force:

> It is a common confusion, especially on the part of those trained in or unduly influenced by formal economics (liberal and Marxist alike), that capitalism is a system of profit-motivated, efficient production. This is not true, nor has it ever been. If the drive to maximize profits, through private ownership and control over the process of production, has served historically as the primary means of capitalist development, it has never been the end of that development. The goal has always been domination (and the power and privileges that go with it) and the preservation of domination.[6]

This analytical prioritization of the sociological[7] over the economic cannot be correct: a firm or an industrial sector that pursued control at the expense of profit would, unless protected from competition, shrink or die. Much of the American industrial sector studied by Noble did indeed suffer this fate, in the period subsequent to the one he examined, at the hands of the Japanese machine-tool manufacturers, who were equally capitalist but who, in their organizational and technological choices, were less concerned with control over the work force. Arguably it was only the protection offered by military funding (a factor to which Noble rightly gives considerable emphasis) that allowed American machine-tool manufacturers to follow the technological strategy they did.

The temptation to counterpose profit and domination, or economics and sociology, arises, I would suggest, from the way our image of economics is permeated by neoclassical assumptions. The alternative economics associated with Simon allows us to make analytical sense of capitalists who are profit oriented (as any sensible view of capitalists must surely see them) without being profit maximizers. The urge to achieve and maintain control over the work force is not an overarching imperative of domination, overriding the profit motive; it is a "heuristic"[8] with deep roots in the antagonistic social relations of capitalist society. When facing technological choices, American engineers and managers, in the period studied by Noble, often simplified production technology decisions by relying on an entrenched preference for technological solutions that undercut the position of manual labor. Noble quotes a 1968 article by Michael Piore that was based on an extensive survey of engineers: "Virtually without exception, the engineers distrusted hourly labor and admitted a tendency to substitute capital whenever they had the discretion to do so. As one engineer explained, 'if the cost comparison favored labor but we were close, I would mechanize anyway.'"[9]

Any significant technological change (such as the automation of machine tools) involves deep uncertainty as to future costs and therefore profits—uncertainty far more profound than the quotation from Piore's work implies. Relying on simple heuristics to make decisions under such circumstances is perfectly compatible with giving high priority to profit: there is simply no completely rational, assuredly profit-maximizing way of proceeding open to those involved. Analyzing the decisions taken under such circumstances in terms of heuristics rather than imperatives opens up a subtly different set of research questions about the interaction of engineers' culture with the social relations (including the economic relations) of the workplace, and about the different heuristics found under different circumstances (including different national circumstances).

Existing attempts to give empirical content to the ideas of the alternative economics have, however, naturally been more traditionally "economic" than that sort of investigation. Pricing behavior is perhaps the most obvious example.[10] Prices do typically seem to be set according to simple, predictable rules of thumb. Even in the sophisticated U.S. high-performance computer industry, what appears to have been for many years the basic rule is startlingly simple: set the selling price at three times the manufacturing cost.[11] Of course, much more elaborate sets of procedures have evolved (along with the specialist function of the pricing manager). These procedures, however, still seem likely to be comprehensible in the terms of the alternative economics, and indeed open to research (although, perhaps through ignorance, I know of no published study of them). Cray Research, for example, traditionally set its supercomputer prices according to a well-defined financial model whose relevant rule is that from 35 to 40 percent of the proceeds of a sale should cover manufacturing cost plus some parts of field maintenance, leaving a 60 or 65 percent overhead.[12] Discounting and different ways of determining manufacturing cost make such rules, even if simple in form, flexible in application; I would speculate, however, that understanding them is an essential part of understanding the computer industry, and that they are by no means accidental, but (like the control heuristic) have deep roots. It would, for example, be fascinating to compare pricing in the Japanese and American computer industries. There is certainly some reason to think that, in general, Japanese prices may be set according to heuristics quite different from those that appear prevalent in the United States.[13] If this is correct for computing, it is unlikely to be an accidental difference; it is probably related to the considerable differences in the organizational, financial, and cultural circumstances of the two computer industries.

Similarly, it has often been asserted that large firms determine their total research and development (R&D) budgets by relatively straightforward rules of thumb.[14] At Cray Research, for example, the R&D budget is set at 15 percent of total revenue.[15] On the other hand, some recent British evidence suggests that matters are not always that straightforward,[16] and there seem likely to be many other complications, such as the significance of the definition of expenditure as R&D for taxation and for perception of a firm's future prospects. Here too, however, empirical investigation inspired by the alternative economics might be most interesting.[17]

Trajectories

What, however, of the *content* of R&D, rather than its quantity? Perhaps the most distinctive contribution in this area of recent work within the tradition of alternative economics is the notion of the technological trajectory, or the "natural trajectory" of technology.[18]

That there is a real phenomenon to be addressed is clear. Technological change does show persistent patterns, such as the increasing mechanization of manual operations, the growing miniaturization of microelectronic components, and the increasing speed of computer calculations. Some of these patterns are indeed so precise as to take regular quantitative form. For example, "Moore's Law" concerning the annual doubling of the number of components on state-of-the-art microchips, formulated in 1964, has held remarkably well (with at most a gradual increase in doubling time in recent years) from the first planar-process transistor in 1959 to the present day.[19]

The problem, of course, is how such persistent patterns of techno-logical change are to be explained. "Natural" is a dangerously ambiguous term here. One meaning of "natural" is "what is taken to follow as a matter of course"—what people unselfconsciously set out to do, without external prompting. That is the sense of "natural" in the following passage from Nelson and Winter: "The result of today's searches is both a successful new technology and a natural starting place for the searches of tomorrow. There is a 'neighborhood' concept of a quite natural variety. It makes sense to look for a new drug 'similar to' but possibly better than the one that was discovered yesterday. One can think of varying a few elements in the design of yesterday's successful new aircraft, trying to solve problems that still exist in the design or that were evaded through compromise."[20] The trouble is that "natural" has quite another meaning, connoting what is produced by, or according to, nature. That other meaning might not be troublesome did it not resonate with a possible interpretation of the mechanical[21] metaphor of "trajectory." If I throw a stone, I as human agent give it initial direction. Thereafter, its trajectory is influenced by physical forces alone. The notion of "technological trajectory" can thus very easily be taken to mean that once technological change is initially set on a given path (for example, by the selection of a particular paradigm) its development is then determined by technical forces.

If Nelson and Winter incline to the first meaning of "natural," Giovanni Dosi—whose adoption of the notion of trajectory has been at least equally influential—can sometimes[22] be read as embracing the second. To take two examples:

"Normal" technical progress maintains a momentum of its own which defines the broad orientation of the innovative activities.

Once a path has been selected and established, it shows a momentum of its own.[23]

A persistent pattern of technological change does indeed possess momentum, but never momentum *of its own*. Historical case-study evidence (such as Tom Hughes's study, rich in insights, of the trajectory of hydrogenation chemistry) can be brought to bear to show this, as can the actor-network theory of Michel Callon, Bruno Latour, John Law, and their colleagues.[24] I shall argue the point rather differently, drawing on an aspect of trajectories that is obvious but which, surprisingly, seems to not to have been developed in the literature on the concept[25]—namely, that a technological trajectory can be seen as a self-fulfilling prophecy.

Persistent patterns of technological change are persistent in part because technologists and others believe they will be persistent.

Take, for example, the persistent increase in the speed of computer calculation. At any point in time from the mid 1960s to the early 1980s there seems to have been a reasonably consensual estimate of the likely rate of increase in supercomputer speed: that it would, for example, increase by a factor of 10 every five years.[26] Supercomputer designers drew on such estimates to help them judge how fast their next machine had to be in order to compete with those of their competitors, and thus the estimates were important in shaping supercomputer design. The designer of the ETA[10] supercomputer told me that he determined the degree of parallelism of this machine's architecture by deciding that it must be 10 times as fast as its Cyber 205 predecessor. Consulting an expert on microchip technology, he found that the likely speedup in basic chips was of the order of fourfold. The degree of parallelism was then determined by the need to obtain the remaining factor of 2.5 by using multiple processors.[27]

Although I have not yet been able to interview Seymour Cray or the designers of the Japanese supercomputers, the evidence suggests similar processes of reasoning in the rest of mainstream supercomputing (excluding massively parallel architectures and minisupercomputers).[28] Where possible, speed has been increased by the amount assumed necessary by using faster components, while preserving the same architecture and thus diminishing risks and reducing problems of compatibility with existing machines. When sufficiently faster components have not been seen as likely to be available, architectures have been altered to gain increased speed through various forms of parallelism.

The prophecy of a specific rate of increase has thus been self-fulfilling. It has clearly served as an incentive to technological ambition; it has also, albeit less obviously, served to limit such ambition. Why, the reader may ask, did designers satisfice rather than seek to optimize? Why did they not design the fastest possible computer (which is what they, and particularly Seymour Cray, have often been portrayed as doing)? The general difficulties of the concept of optimization aside, the specific reasons were risk and cost. By general consensus, the greater the speed goal, the greater the risk of technological failure and the greater the ultimate cost of the machine. Though supercomputer customers are well heeled, there has traditionally been assumed to be a band of "plausible" supercomputer cost, with few machines costing more than $20 million. If designers did not moderate their ambitions to take risk and cost into account, their managers and financiers would.[29] The assumed rate of speed helps as a yardstick for what is an appropriately realistic level of ambition.

In the case of supercomputers, all those involved are agreed that increased speed is desirable. Similarly, all those involved with chip design seem to assume that, other things being equal, increased component counts are desirable. Trajectories are self-fulfilling prophecies, however, even when that is not so. Take the "mechanization of processes previously done by hand." Though analyzed as a natural trajectory by Nelson and Winter,[30] it has of course often seemed neither natural nor desirable to those involved—particularly to workers fearing for their jobs or skills, but sometimes also to managements disliking change, investment, and uncertainty. A powerful argument for mechanization, however, has been the assumption that other firms and other countries will mechanize, and that a firm that does not will go out of business. Increasing missile accuracy is a similar, if simpler, case: those who have felt it undesirable (because it might make attractive a nuclear first strike on an opponent's forces) have often felt unable to oppose it because they have assumed it to be inevitable and, specifically, not stoppable by arms control agreements. Their consequent failure to oppose it has been one factor making it possible.

The nature of the technological trajectory as self-fulfilling prophecy can be expressed in the languages of both economics and sociology. As an economist would put it, *expectations* are an irreducible aspect of patterns of technological change. The work of Brian Arthur and Paul David is relevant here, although it has, to my knowledge, largely concerned either/or choices of technique or standard rather than the cumulative, sequential decisions that make up a trajectory. In an amusing and insightful discussion of the almost universal adoption of the inferior QWERTY keyboard, David writes:

Intuition suggests that if choices were made in a forward-looking way, rather than myopically on the basis of comparisons among currently prevailing costs of different systems, the final outcome could be influenced strongly by the expectations that investors in system components—whether specific touch-typing skills or typewriters—came to hold regarding the decisions that would be made by the other agents. A particular system could triumph over rivals merely because the purchasers of the software (and/or the hardware) expected that it would do so. This intuition seems to be supported by recent formal analyses of markets where purchasers of rival products benefit from externalities conditional upon the size of the compatible system or "network" with which they thereby become joined.[31]

Actors' expectations of the technological future are part of what make a particular future, rather than other possible futures, real. With hindsight, the path actually taken may indeed look natural, indicated by the very nature of the physical world. But Brian Arthur's "nonergodic," path-dependent models of adoption processes are vitally helpful in reminding us of ways in which technologies devoid of clear-cut, initial, intrinsic superiority can rapidly become irreversibly superior in practice through the very process of adoption.[32]

The sociological way of expressing essentially the same point is to say that a technological trajectory is an *institution*. Like any institution, it is sustained not through any internal logic or through intrinsic superiority to other institutions, but because of the interests that develop in its continuance and the belief that it will continue. Its continuance becomes embedded in actors' frameworks of calculation and routine behavior, and it continues because it is thus embedded. It is intensely problematic to see social institutions as natural in the sense of corresponding to nature (although that is how they are often legitimated), but institutions do of course often become natural in the sense of being unselfconsciously taken for granted. The sociological work most relevant here is that of Barry Barnes, who has argued that self-fulfilling prophecy should be seen not as a pathological form of inference (as it often was in earlier sociological discussions), but as the basis of all social institutions, including the pervasive phenomenon of power.[33]

My claim is not the idealist one that all prophecies are self-fulfilling. Many widely held technological predictions prove false. Not all patterns of technological change can be institutionalized, and it would be foolish to deny that the characteristics of the material world, of Callon and Latour's "nonhuman actors," play a part in determining the patterns that do become institutionalized. One reason for the attractiveness of the notion of a natural trajectory to alternative economics is that the latter field has been reacting not against technological determinism (as has much of the sociology of technology), but against a view of technology as an entirely plastic entity shaped at will by the all-knowing hands of market forces.[34]

I entirely sympathize with the instinct that technology cannot be shaped at will, whether by markets or by societies. The risk, however, of expressing that valid instinct in the notion of natural trajectory is that it may actually deaden intellectual curiosity about the causes of persistence in patterns of technological change. Although I am certain this is not intended by its proponents, the term has an unhappy resonance with widespread (if implicit) prejudices about the proper sphere of social-scientific analysis of technology—prejudices that shut off particular lines of inquiry. Let me give just one example. There is wide agreement that we are witnessing an information-technology "revolution," or a change of "technoeconomic paradigm" based on information and communication technologies. Of key importance to that revolution, or new paradigm, is, by general agreement, microchip technology and its Moore's Law pattern of development: "clearly perceived low and rapidly falling relative cost"; "apparently almost unlimited availability of supply over long periods"; "clear potential for . . . use or incorporation . . . in many products and processes throughout the economy."[35] Yet in all the many economic and sociological studies of information technology there is scarcely a single piece of published research—and I hope I do not write from ignorance here—on the determinants of the Moore's Law pattern.[36] Explicitly or implicitly, it is taken to be a natural trajectory whose effects economists and sociologists may study but whose causes lie outside their ambit. In Dosi's work on semiconductors, for example, Moore's Law is described as "almost a 'natural law' of the industry," a factor shaping technical progress, but not one whose shaping is itself to be investigated.[37] Until such a study of Moore's Law is done, we cannot say precisely what intellectual opportunities are being missed, but it is unlikely that they are negligible.

Ethnoaccountancy

A revised understanding of persistent patterns of technological change offers one potential bridge over the gap between economic and sociological explanations of technical change. Another potential bridge I would call "ethnoaccountancy." I intend the term as analogous to ethnomusicology, ethnobotany, or ethnomethodology. Just as ethnobotany is the study of the way societies classify plants, a study that should not be structured by our perceptions of the validity of these classifications, ethnoaccountancy should be the study of how people do their financial reckoning, irrespective of our perceptions of the adequacy of that reckoning and of the occupational labels attached to those involved.

Ethnoaccountancy has not been a traditional concern of writers within the discipline of accounting. Their natural concern was with how accountancy ought to be practiced, rather than with how it actually is practiced.[38] Although studies of the latter have been become much more common over the past decade (see, for example, the pages of the journal *Accounting, Organizations and Society*), there has still been little systematic study by accountancy researchers of the ethnoaccountancy of technological change. Sociologists, generally, have not been interested in ethnoaccountancy, again at least until very recently.[39] Compare, for example, the enormous bulk of the sociology of medicine with the almost nonexistent sociology of accountancy.[40] Since the latter profession could be argued to be as important to the modern world as the former, it is difficult not to suspect that sociologists have been influenced by accountancy's general image as a field that may be remunerative but is also deeply boring.

It is somewhat more surprising that economists have ignored the actual practices of accounting, but this appears to be the case. Nelson and Winter suggest a reason that, though tendentiously expressed, may be essentially correct: "For orthodoxy, accounting procedures (along with all other aspects of actual decision processes) are a veil over the true phenomena of firm decision making, which are always rationally oriented to the data of the unknowable future. . . . Thanks to orthodoxy's almost unqualified disdain for what it views as the epiphenomena of accounting practice, it may be possible to make great advances in the theoretical representation of firm behavior without any direct empirical research at all—all one needs is an elementary accounting book."[41]

Ethnoaccountancy most centrally concerns the category of "profit." As noted above, even if firms cannot maximize profit, it certainly makes sense to see them as oriented to it. But they can know their profits only through accounting practices. As these change, so does the meaning, for those involved, of profit. Alfred Chandler's *The Visible Hand*, for example, traces how accounting practices and the definition of profit changed as an inseparable part of the emergence of the modern business enterprise.[42] Unfortunately, Chandler clothes his insightful analysis in teleological language—he describes an evolution toward correct accounting practice and a "precise" definition of profit[43]—and he does not directly tie the changes he documents to changing evaluations of technology.

The teleology has largely been corrected and the connection to technological change forged, albeit in a much more limited domain, by the historian of technology Judith McGaw.[44] Though adequate for the purposes of those involved, accounting practice in early-nineteenth-century U.S. papermaking, she notes, "hid capitalization" and highlighted labor costs, facilitating the mechanization of manual tasks. Though others have not made the same connections McGaw has, it is clear that the practices she documents were not restricted to the particular industry she discusses.[45]

The general issue of whether accounting practice highlights one particular class of cost, thus channeling innovation toward the reduction of that cost, is of considerable significance. Accounting practices that highlight labor costs might generally be expected to accelerate mechanization. They may, however, be a barrier to the introduction of capital-saving or energy-saving technologies, and many current information-technology systems are regarded as having these advantages.

There is also fragmentary but intriguing evidence that the techniques of financial assessment of new technologies used in the United Kingdom and the United States may differ from those used in Japan. In effect, "profit" is defined differently. In the United Kingdom and the United States there is typically great reliance (for decision-making purposes, and also in rewarding managers) on what one critic calls "financial performance measures, such as divisional profit, [which] give an illusion of objectivity and precision [but which] are relatively easy to manipulate in ways that do not enhance the long-term competitive position of the firm, and [which] become the focus of opportunistic behavior by divisional managers."[46] Japanese management accounting, by contrast, is less concerned with financial measurement in this short-term sense. While Japanese firms are patently not indifferent to profit, and are of course legally constrained in how profit is calculated for purposes such as taxation, they seem much more flexible in the internal allocation of costs and the definition of profit. Japanese firms "seem to use [management] accounting systems more to motivate employees to act in accordance with long-term manufacturing strategies than to provide senior management with precise data on costs, variances, and profits."[47]

Uncertainty and Closure

Ethnoaccountancy is one aspect of the much larger topic we might call the construction of the economic. Economic phenomena such as prices, profits, and markets are not just "there"—self-sustaining, self-explaining—but exist only to the extent that certain kinds of relations between people exist. This insight, simultaneously obvious and easy to forget, is perhaps Marx's most central contribution to our topic.[48] Marx devoted the final part of volume 1 of *Capital* to an analysis of the historical emergence of capital as a way of mediating relations between persons. Implicit, too, in Marx's account is the reason why the insight is forgettable. It is not just that capitalism gives rise to a particular type of economic life. Under capitalism, aspects of social relations inseparable in previous forms of society (such as political power and economic relations) achieve a unique degree of separation, giving rise to the "thing-like" appearance of the economic.

One of the fascinations of technological change is that it turns the question of the construction of the economic from a general question about capitalist society into a specific and unavoidable concern. The oft-noted unquantifiable uncertainty of technological change defies the calculative frameworks of economics. Chris Freeman, for example, compares attempts at formal evaluation of R&D projects to "tribal war-dances."[49] He is referring to participants' practices, but it is worth noting that the economists of technological change, in their search for an ancestor to whom to appeal, have often turned to Joseph Schumpeter, with his emphasis on the noncalculative aspects of economic activity, rather than to any more orthodox predecessor.

The issue can usefully be rephrased in the terms of actor-network theory. Radical technological innovation requires the construction of a new actor-network.[50] Indeed, that is perhaps the best way of differentiating radical innovation from more incremental change. Only once a new network has successfully been stabilized does reliable economic calculation become possible.[51] Before it is established, other forms of action, and other forms of understanding, are needed.

Unstabilized networks are thus a problem for economics, at least for orthodox economics. By comparison, their study has been the very lifeblood of the sociology of scientific knowledge.[52] Scientific controversy, where the "interpretative flexibility" of scientific findings is made evident, has been the latter field's most fruitful area of empirical study, and interpretative flexibility is the analogue of what the economists refer to as "uncertainty."[53] The weakness of the sociology of scientific knowledge has, rather, been in the study of "closure"—the reduction of (in principle, endless) interpretative flexibility, the resolution of controversy, the establishment of stable networks.

The economics of technological change and the sociology of scientific knowledge thus approach essentially the same topic—the creation of stable networks—from directly opposite points of view. I confess to what is perhaps a disciplinary bias as to how to proceed in this situation: using tools honed for stable networks to study instability seems to me likely to be less fruitful than using tools honed for instability to study stability.[54] Indeed, attempting the former is where, I would argue, the alternative economists have gone wrong in the concept of technological trajectory.

The latter path, using the tools developed in the study of instability, does, however, require a step back in research on technological change: a return to the "natural history"[55] of innovation of the 1960s and the 1970s, but a return with a different focus, highlighting the empirical study of heuristics, the role of the self-fulfilling prophecy in persistent patterns of technological change, and the ethnoaccountancy of technological change. We need to know more about the structure of the interpretative flexibility inherent in technological change, and about the ways that interpretative flexibility is reduced in practice. How, in the economists' terminology, is uncertainty converted into risk?[56] How, for example, do participants judge whether they are attempting incremental or radical innovation?[57] What is the role of the testing of technologies (and of analogues such as prototyping and benchmarking)?[58] How is technological change "packaged" for the purposes of management— in other words, how is a process that from one perspective can be seen as inherently uncertain presented as subject to rational control? What are the roles here of project proposals, project reviews, and milestones—of the different components of Freeman's "war-dances"? How is the boundary between the "technical" and the "nontechnical" negotiated? What are the determinants of the credibility of technical, and of nontechnical, knowledge claims?

Even if we set aside the fact that technological change is not substantively the same as scientific change, we cannot look to the sociology of scientific knowledge for theories or models that could be applied directly in seeking to answer questions such as these. That is not the way the field has developed. It is more a question of sensitivities, analogies, and vocabularies. Nevertheless, the parallels between closure in science and successful innovation in technology, and between interpretative flexibility and uncertainty, are strong enough to suggest that exploring those parallels may be an important way forward for the study of technological change. In the closure of scientific controversies and in successful technological innovation, an apparently self-sustaining realm (of objective knowledge, of economic processes) emerges, but only as the end product of a process involving much more than either natural reality or economic calculation. Understanding of the one should surely help develop understanding of the other.

Conclusion

I have argued that the alternative economics associated with Simon, Nelson, Winter, and others is more plausible than neoclassical economics, with its incoherent notion of profit maximization. Ideas from the former tradition could help bridge the gap between the economic and the sociological in fields where those ideas have not (to my knowledge) been widely drawn upon, such as labor-process studies. This alternative economics can also fairly straightforwardly be applied to pricing and to firms' overall R&D budgets, although recent empirical work in these areas seems surprisingly sparse.

Applying the alternative economics to the content of R&D is more difficult. The metaphor of "technological trajectory" can mislead. Persistent patterns of technological change do exist, but they should not been seen as "natural" in the sense of corresponding to nature. Nor do they have a momentum of their own. Expectations about the technological future are central to them: they have the form of self-fulfilling prophecies, or social institutions. Conceiving of persistent patterns in this way offers one way of bridging the gap between economic and sociological explanations of technological change.

Another way of bridging the gap is what I have called ethnoaccountancy. Studying how people actually do the financial reckoning of technological change would bring together the economist's essential concern for the financial aspects of innovation with the sociologist's equally justified empiricism. I have suggested that ethnoaccountancy would not be a marginal enterprise, rummaging though the boring details of economic activity, but ought to throw light on central questions such as the practical definition of profit and the relative rate of technological change in different historical and national contexts.

Finally, I have argued that, because of the centrality of uncertainty (or nonstabilized networks) to technological change, the sociology of scientific knowledge, with its experience in the study of the essentially equivalent matter of interpretative flexibility, ought to be of relevance here. Scientists construct stable, irreversible developments in knowledge in a world where no knowledge possesses absolute warrant; out of potential chaos, they construct established truth. Technologists, workers, users, and managers construct successful innovations in a world where technological change involves inherent uncertainty; out of potential chaos, they construct a world in which economics is applicable.

Acknowledgments

The research reported here was supported by the Economic and Social Research Council's Programme on Information and Communication

Technologies. Earlier versions of this chapter were read to the conference on Firm Strategy and Technical Change: Micro Economics or Micro Sociology? (Manchester, 1990) and to a meeting of the ESRC New Technologies and the Firm Initiative (Stirling, 1991). Thanks to Tom Burns, Chris Freeman, John Law, Ian Miles, Albert Richards, Steve Woolgar, and both the above audiences for comments, and particularly to the neoclassical economists for their toleration of a sociologist's rudeness.

4

From the Luminiferous Ether to the Boeing 757

Inertial navigation systems are central to modern navigation. They permit wholly self-contained navigation of remarkable accuracy. They are now standard in long-range civil aircraft and most modern military aircraft, as well as in ballistic missiles, cruise missiles, space boosters, and submarines. They are increasingly to be found in shorter-range tactical missiles, in tanks and self-propelled artillery, and in some surveying applications.

At the heart of inertial navigation are the inertial sensors themselves: gyroscopes, which sense rotation, and accelerometers, which measure acceleration. During the last twenty years, the former have undergone what those involved see as a technological revolution. Since the beginnings of inertial navigation in the 1930s, the gyroscopes used had remained analogues—however sophisticated—of the child's spinning toy, reliant in their detection of rotation on the mechanics of a rapidly revolving rotor. But they have now been challenged by inertial sensors in which the detection of rotation is achieved by optical rather than mechanical means: laser gyroscopes. All but one of the major corporate suppliers of inertial technology are heavily committed to laser gyroscope technology. A basic shift has thus taken place in this key modern technology.

This chapter begins with the conceptual origins of the laser gyroscope, which are remote from the "high-tech" world of the modern device. They lie in experiments probing the controversial question of the existence of the ether, the massless substance that pre-Einsteinian physics took to be the medium of the transmission of light. In particular, the physicist Georges Sagnac (1869–1928) believed that his work on the optical detection of rotation refuted Einstein. The second section of the chapter describes the move of what became known as the "Sagnac effect" from science to technology, a move that took place between 1959 and 1963. The invention of the laser was fundamental to this move, but more was involved than just a new light source. As quantum electronics flowered, the optical detection of rotation was reconceptualized.

On January 7, 1963, a prototype laser gyroscope first detected rotation, and that date can be taken as indicating the end of the process of "inventing" the laser gyroscope and the beginning of the "development" phase of the device's history. That development phase is the subject of the third section. It stretched from 1963 to the first unequivocally successful tests of a practical laser gyro in 1975, and it proved as crucial and as troublesome in the case of the laser gyro as elsewhere in the history of technology.[1] The fourth section describes the growing acceptance of the laser gyro after 1975. It highlights the single most crucial event in that process of acceptance: the decision to adopt the new device as the core of the standard navigation and attitude reference system for Boeing's new civil air transports, the 757 and the 767.

The chapter ends with a discussion of what can be learned from this episode about the nature of technological change. The history of the laser gyroscope underlines the significance of the fusion of scientific and technological concerns in the new field of quantum electronics. It supports those who have noted the pervasiveness of military involvement in quantum electronics, while showing that the resultant technology may not bear the stamp of any specifically military need. The history of the laser gyroscope is one in which economic considerations, market processes, and corporate structures are central, yet it is a history that does not correspond to orthodox economic theory, with its assumption of profit maximizing by unitary firms. Perhaps most interesting of all, the process of the acceptance of the laser gyroscope reveals the role of self-fulfilling prophecy in technological revolutions.[2]

Searching for the Ether

The ether was a paradoxical substance. It was believed to pervade the universe and to be the medium for such phenomena as electromagnetism, gravitation, and nervous impulses. Yet it was also thought to be devoid of the qualities that made the grosser forms of matter easily perceptible. It could not be seen, felt, or touched. It played a crucial role in orthodox physics, chemistry, and even biology; it was of theological significance too. The physicist Sir Oliver Lodge was not alone in seeing the ether as "the primary instrument of Mind, the vehicle of Soul, the habitation of Spirit." "Truly," he wrote, "it may be called the living garment of God."[3]

The most famous attempt to demonstrate the existence of the ether was the series of experiments conducted in the 1880s by the physicist Albert A. Michelson and the chemist Edward W. Morley.[4] If the ether was at rest in absolute space, as most assumed, then as the Earth moved it would be moving relative to the ether. From the point of view of an observer on the Earth, an "ether wind" would thus exist. It would not be directly perceptible to the senses, but it would affect the speed of transmission of light, since light was a wave in the ether. Michelson and Morley's apparatus split a beam of light into two, one part traveling parallel to the Earth's motion and one at right angles to it, and sought to detect the predicted effect of the ether wind in the interference pattern when the two beams were recombined in an interferometer.[5] Michelson and Morley were unable to find that effect.[6] The fame of their experiments lies in this null result. Later, the null result was taken as proof of the nonexistence of the ether and as leading to Einstein's Special Theory of Relativity, a key postulate of which is that the velocity of light is the same for all observers and therefore no difference is to be expected between "looking" along the direction of the Earth's motion through space and "looking" at right angles to it.

Matters were not, however, quite as clear as this simple hindsight history suggests.[7] When Morley's colleague Dayton C. Miller repeated the experiments, he believed he did find at least some significant effect.[8] Furthermore, a null result by no means compelled rejection of the ether. It could, for example, be taken as showing simply that the moving Earth dragged the ether along with it, so that no "ether wind" would be found at the Earth's surface.[9]

So the search for the ether did not end with the Michelson-Morley experiments, and here Georges Sagnac enters the story. Sagnac was a professor of physics, first at Lille and then at the University of Paris. His early work had been on the recently discovered x rays. In his ether experiment, he sought to create an ether wind in the laboratory by mounting an interferometer on a rotating platform. A beam from an electric light was split, and the two resulting beams, R and T, were sent in opposite directions around a path formed by four mirrors, M_1, M_2, M_3, and M_4 (figure 1). Sagnac used a camera to observe the interference patterns when the two half beams were recombined.[10] As Sagnac's apparatus rotated, first in one direction and then in the other, the camera did indeed record a shift in the interference fringes. He reported his results in a brief, exuberant paper to the Académie des Sciences in 1913. The fringe shift occurred, he claimed, because his apparatus was rotating in

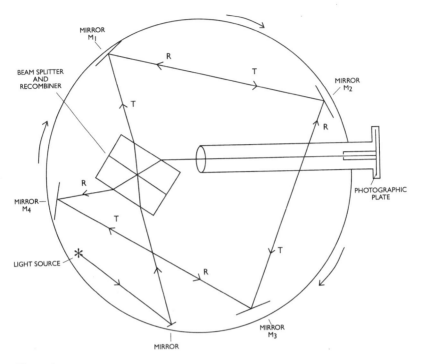

Figure 1
Sagnac's interferometer on its turntable. Simplified from diagram in G. Sagnac, "Effet tourbillonnaire optique: la circulation de l'éther lumineux dans un interférographe tournant," *Journal de Physique*, fifth series, 4 (March 1914), p. 187.

the ether. Relative to the turntable, one beam was retarded, and the other accelerated, according to the direction of the turntable's rotation in the ether. Sagnac calculated what the effect of this on the interference pattern ought to be and found that the measured shift was as predicted. His experiment, he concluded, was "a proof of the ether"; the interferometric effect "directly manifested the existence of the ether."[11]

Though Einstein's name was not mentioned, the challenge could not have been clearer; and it was made within a French scientific community predominantly hostile to relativity. (Even as late as the 1950s, "with rare exceptions, teaching, textbooks, and university programs" did not allow detailed attention to relativity to disturb an image of "Science . . . as a fully realized achievement, encased in certainty, organized around Newtonian categories."[12]) The relativist Paul Langevin vigorously disputed Sagnac's interpretation of his results.[13] Nevertheless, the Sagnac effect seems to have counted in France as evidence for the ether. Thus, when Sagnac was awarded the Pierson-Perrin Prize of the Académie des Sciences in 1919, his experiment was described as having verified the theory of the ether. It was repeated in a different form (with the "observer" fixed in the laboratory, rather than rotating) in 1937, and again the results were found to confirm "classical theory" and to violate the predictions of relativity.[14]

In the Anglo-Saxon world matters were different. Sagnac had his defenders there too, notably the anti-relativist Herbert E. Ives. But mainstream opinion was firmly in favor of Einstein, and to the extent that Sagnac's work was considered at all it was dismissed. There were doubts about the reliability of Sagnac's results.[15] But, more important, the conclusion became accepted that the theory of relativity could explain them just as well as ether theory. With a rotating system, the relevant aspect was argued to be general, not special, relativity. According to the former, "two observers, traveling around a closed path that is rotating in inertial space, will find that their clocks are not in synchronization when they return to the starting point (traveling once around the path but in opposite directions). The observer traveling in the direction of rotation will experience a small increase, and the observer traveling in the opposite direction a corresponding small decrease in clock time." If the two "observers" are photons, each traveling at the speed of light, "the time difference appears as an apparent length change in the two paths," causing the shift in the interference fringes reported by Sagnac.[16]

Therefore, it did not help the case against Einstein when, in 1925, Michelson and his colleague Henry Gale also reported a change in interference pattern as a result of rotation. They employed the Earth itself as the turntable. Using a rectangular system of pipes in which they created a vacuum, they constructed an optical circuit a mile in circumference (figure 2). A smaller rectangular circuit provided a "fiducial mark from which to measure the displacement" of the interference fringes formed by the clockwise and counterclockwise beams in the larger circuit.[17]

Michelson and Gale's results were in agreement with "the calculated value of the displacement on the assumption of a stationary ether," just as Sagnac's had been. However, concluded Michelson and Gale, they were "in accordance with relativity too." There was little doubt where Michelson's heart lay—in 1927 he wrote of "the beloved old ether (which is now abandoned, though I personally still cling a little to it)"—but the ambiguous experiment did not help bring the ether back to life.[18]

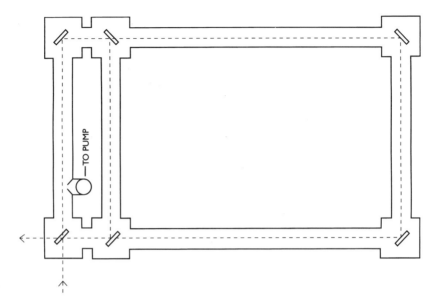

Figure 2
Ground plan and arrangement of mirrors in Michelson-Gale experiment. Based
on diagram in A. A. Michelson and Henry G. Gale, "The effect of the earth's rotation on the velocity of light: Part II," *Astrophysical Journal* 61 (April 1925), p. 141.

As late as 1965 there were still those who claimed that Sagnac had
indeed "discovered the existence of a luminiferous ether" and denied
that relativity theory explained his results. By then, though, this was
a distinctly unusual opinion to hold. True, the author of this claim
could point out that, using the novel technology of the laser, "the
Sagnac experiment has been repeated, with the same but more refined
outcome."[19] The meaning of that replication had, however, shifted
decisively. There was indeed widespread interest in it, but the question of the existence of the luminiferous ether was certainly not the
source.

From Science to Technology

Sagnac had speculated that it might be possible to use his effect to measure rotation in a practical context:

I hope that it will be possible to repeat these measurements of the optical whirlwind effect [*l'effet tourbillonnaire optique*] with an optical circuit at least some tens
of meters square, fastened to the rigid sides of a ship. If the circuit is horizontal,

the displacement of the central [interference] fringe will make known at each instant the speed of rotation of the ship about a vertical axis; slow rotations could thus be revealed without any external benchmark. . . . A circuit installed parallel to one of the vertical planes of the ship would permit similar observation or photographic recording of the speed of oscillatory rotation in roll and pitch.[20]

This 1914 speculation is, however, as far as the practical application of the Sagnac effect went for many years. Yet when interest in the optical detection of rotation revived around 1960, theoretical issues (though not absent) quickly became less salient than technological ones.

In the intervening half-century, the measurement of rotation had become a central technical activity. When Sagnac was conducting his experiments on the eve of the First World War, the practical application of the mechanical gyroscope was a relatively new field: the first successful trials of a marine gyrocompass, for example, had taken place in 1908.[21] Between then and the late 1950s, the marine and aircraft uses of the gyroscope had grown in importance and sophistication and had been joined by the new and uniquely demanding field of inertial guidance and navigation. Inertial systems were seen as having one decisive advantage over other forms of navigation: being wholly self-contained, they could not be disrupted by either hostile action or bad weather. Though inertial navigation had yet to find significant civilian applications, by the late 1950s it was a crucial military technology.[22]

That did not mean, however, that the place of the mechanical gyroscope was secure. The dominant variety in inertial navigation in the United States—the fluid-floated gyro—could be made highly accurate, but it was difficult to produce and therefore expensive. The mechanical gyros of the 1950s also suffered from reliability problems. There was thus a conscious search for alternative means of detecting rotation.

That search led at least one military organization in the United States back to the ether experiments. The Navigation and Guidance Laboratory of the Air Force Systems Command at Wright-Patterson Air Force Base had been "interested for several years in an angular rate sensing device without moving parts for the obvious reason of reliability," its chief wrote in 1962. Since an optical circuit a mile in circumference was patently too large for a practical navigation system, the laboratory had sought to "miniaturize the Michelson-Gale experiment."[23] Its attempts, however, were "notably unsuccessful at both optical and gamma ray wavelengths."[24] Success was to require the transformation, and not merely the miniaturization, of the Sagnac and Michelson-Gale experiments.

That transformation was wrought by quantum electronics. This new field fused science, notably quantum theory, with the technological concerns of radar and radio engineering. Like inertial navigation, it emerged in large part under military tutelage. The U.S. military supported the field financially, organized key conferences, and actively sought defense applications for its products.[25]

A key element in quantum electronics was experience in the use of resonant cavities, in which large quantities of electromagnetic radiation are generated at a frequency such that the wave "fits" the cavity exactly (in other words, the length of the cavity is an integral number of wavelengths). An example crucial to radar was the resonant cavity magnetron, a powerful new microwave generator developed at the University of Birmingham (England) in 1940.[26] Another element in quantum electronics was the physics of quantum transitions, in which electrons move from higher to lower energy orbits or vice versa. These two elements were brought together in the development in the 1950s of the maser (an acronym for microwave amplification by stimulated emission of radiation). In this device, electrons in an appropriate material are "pumped" by an input of energy to higher energy orbits. If then properly stimulated in a suitable resonant cavity, they will return to lower-energy orbits in unison, producing a powerful output of coherent microwave radiation. By 1954 the first maser was working, and by 1956–57 there was already interest in moving to light frequencies, and thus to an optical maser or laser (for light amplification by stimulated emission of radiation). T. H. Maiman of the Hughes Aircraft Company demonstrated the first such device, a solid-state ruby laser, in July 1960. In February 1961 a gas laser, using as its material a mixture of helium and neon, was announced.[27]

Between 1959 and 1961, three people independently saw that it was possible to transform the Sagnac and Michelson-Gale experiments, which they probably knew about primarily through the account in an optics textbook of the day, R. W. Ditchburn's *Light*.[28] Not only did they see that the electric light of the earlier experiments could be replaced by a laser; a conceptual shift was involved. The first hint of this shift came in the autumn of 1959, before the operation of the first laser. There was no reference to either masers or lasers, but the source was a man with considerable experience of the general field of quantum electronics. Ohio State University physicist Clifford V. Heer was working as a consultant for Space Technology Laboratories, an offshoot of Ramo-Woolridge (later TRW) set up to manage the intercontinental ballistic missile program of the U.S. Air Force. In September 1959, Heer proposed to the firm's Guidance Research Laboratory a "system for measuring the angular velocity of a platform [that] depends on the interference of electromagnetic radiation in a rotating frame." He noted that in experiments such as Sagnac's a path enclosing a large area was necessary to achieve sensitivity, and this would clearly be a limitation on their technological use. He suggested investigating four areas in the light of this problem, including "the use of resonant structures in a rotating frame."[29] A month later, in a patent disclosure, he added a further new element to the idea of using resonance: that frequency differences, as well as the interference effects used by Sagnac and Michelson, could be used to measure rotation. As a resonant structure rotated, there would be a shift in resonant frequencies.[30]

Those two elements—using a resonant structure and detecting rotation by frequency differences rather than changes in interference patterns—were central in the conceptual shift that led to the laser gyroscope. In 1959, however, Heer was not necessarily thinking of light as the appropriate form of electromagnetic radiation to use. He was at least equally interested in employing radiation of "lower frequencies such as radio and microwave frequencies" confined in a "coaxial cable or waveguide," with "N turns of cable or guide . . . used to increase the phase difference over that for one traversal."[31] In the version of his ideas presented for the first time in public, at the January 1961 meeting of the American Physical Society, he even suggested that the interference of matter waves in a rotating system could be studied.[32]

Heer's first proposal to study the use of masers (including optical masers) in the measurement of rotation came in March 1961, but only as nonhighlighted aspects on the third and fourth pages of a proposal for research on "measurement of angular rotation by either electromagnetic or matter waves."[33] Though copies were sent to NASA, the Air Force Office of Scientific Research, and the Office of Naval Research, funds were not forthcoming. Heer's interest in the use of the laser rapidly grew, however, in part as a result of his attending the Second International Conference on Quantum Electronics at Berkeley, at which Ali Javan of the Bell Laboratories described the first gas laser, in late March 1961. In October 1961, Heer forwarded his original proposal to the Chief Scientist of the Aeronautical Systems Division of the Air Force Systems Command, along with a cover letter stating: "The experiments in the microwave region remain of considerable interest, but in view of the recent development of the optical masers I feel a study of the feasibility of the use of optical masers and the eventual use of optical masers must be given consideration." In January 1962, Heer sent potential sponsors a further paper containing a description of a square resonant structure with "laser amplification along the path." Such a structure a meter square, he noted, would make possible the measurement of "angular rotation rates as small as 10^{-6} radians/sec."[34]

By October 1961 a second researcher, Adolph H. Rosenthal of the Kollsman Instrument Corporation, had also become convinced that, in the words of a paper he read to the Optical Society of America, "interferometry methods making use of optical maser oscillations . . . permit [us] to increase considerably the accuracy of the historical relativistic experiments of Michelson, Sagnac, and others, and have also potential applications to studies of other radiation propagation effects."[35] Before Rosenthal died in July 1962, he had developed his ideas sufficiently that a posthumous patent application using them in a "optical interferometric navigation instrument" could be submitted.[36]

One member of Rosenthal's audience at the Optical Society had already been thinking along the same lines. He was Warren Macek, a young physics-and-mathematics major working for the Sperry Rand Corporation. Much of the original strength of that company had been built around Elmer Sperry's use of the mechanical gyroscope for navigation, stabilization, and aircraft instruments.[37] However, Macek worked not on gyroscopes but in a new optics group Sperry Rand had set up in 1957. After the announcement of the ruby and gas lasers, the optics group built its own versions of each, with help from specialists on microwave resonant cavity devices.

Macek had read Ditchburn's *Light* for a course in physical optics he had taken as part of his Ph.D. work at the Brooklyn Polytechnic Institute, and through that he knew of the Sagnac and Michelson-Gale experiments. In October 1961, when he heard Rosenthal's paper, Macek was already working on a proposal to Sperry management which included, among other novel rotation sensor techniques, the idea of building an interferometer, analogous to that used in the ether experiments, using a laser as its light source.[38]

In early 1962, Macek and colleagues at Sperry set to work to construct a device in which lasers would be used to measure rotation, adapting resources they already had on hand.[39] They used gas laser tubes the optics group had built. Sufficiently good mirrors were hard to find, so one mirror used by Macek was coated in gold by a relative of his who worked for a gold-plating firm. An old radar pedestal was modified to form the

turntable on which the apparatus was placed. One of the group's technicians who was a radio "ham" tuned the device to achieve resonance.

On January 7, 1963, their device worked successfully.[40] Four helium-
neon lasers were arranged in a square a meter on each side (figure 3).
These lasers were modified so that, unlike conventional lasers, they radi-
ated light from both ends. Mirrors at the corners of the square reflected the light from one laser tube into the next. In this way, laser
oscillations were sustained in both directions around the ring, clockwise
and counterclockwise (until this was achieved in the Sperry work, it was
not clear that oscillations could be sustained in both directions). One
of the four mirrors was only partially coated. Some light from both
beams passed through it, and, with use of a further reflector, light from
both beams fell on a photomultiplier tube used as a detector.

Although the paper reporting the Sperry work cited Sagnac and
Michelson and Gale, it made clear that what was being detected was not
the conventional optical interference fringes they had used, and here the
input from quantum electronics was clearest. Like all lasers, the device
was a resonant cavity, with resonant frequencies "determined by the condition that the cavity optical path length must equal an integral number
of wavelengths."[41] When the system was not rotating, the clockwise and

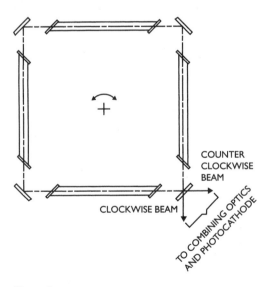

Figure 3
Schematic diagram of the Sperry ring laser. Based upon diagram in W. M.
Macek and D. T. M. Davis, Jr., "Rotation rate sensing with traveling-wave ring
lasers," *Applied Physics Letters* 2 (February 1, 1963), p. 67.

counterclockwise path lengths were identical, so the frequencies of clockwise and counterclockwise waves were the same. When the system was rotating, however, the path lengths became unequal.[42] The frequencies of the two waves were no longer exactly the same, so, when they were superimposed, the combined wave oscillated in amplitude with a "beat" frequency proportional to the difference in their frequencies, and thus to the rotation rate of the platform. It was those beats that formed the device's output. Such a use of the beats resulting from the superimposition of waves of slightly different frequencies—"heterodyne" action—was a radio engineering method already widely used in laser work. As the platform was rotated at between 20 and 80 degrees per minute, the beat frequencies changed in a satisfactorily linear fashion.

The technological meaning of what they had done was clear to the members of the Sperry team: "The principle demonstrated in this experiment may be utilized for rotation rate measurement with high sensitivity over an extremely wide range of angular velocities. Such sensors would be self-contained, requiring no external references."[43] Along with the conceptual work of Heer (who, together with a doctoral student, P. K. Cheo, had his own device working by August 1963, with funding finally obtained from the National Science Foundation),[44] and that of Rosenthal, the construction of this prototype can be said to constitute the invention of the laser gyroscope.

Developing the Laser Gyro

What had been achieved by January 1963 needs to be put in perspective. At the time, an "inertial grade" mechanical gyroscope was one with a drift rate of a hundredth of a degree per hour, corresponding roughly to an average error of a nautical mile per hour's flying time in an aircraft inertial navigator. The 20°/minute threshold of the Sperry device meant a sensitivity several orders of magnitude poorer. Both Heer and Macek were predicting much better future performance, but that remained a prediction. Furthermore, the meter-square prototype was much larger than the small mechanical gyros (2 inches in diameter, or thereabouts) then available, and the theory of the laser device indicated that its sensitivity would decrease in proportion to any reduction in the area enclosed in the path. Finally, the laser device had many competitors as a potential replacement for the conventional mechanical gyroscope. The gamut of physical phenomena was being searched for new ways to detect rotation. One review listed 29 candidate technologies, several of which—dynamically tuned, electrostatically supported, fluid sphere, nuclear magnetic resonance, and superconductive, as well as laser—were being pursued actively.[45]

So the invention of the laser gyro need not necessarily have led anywhere. Macek and the Sperry group realized this clearly, and what they did once they had their prototype working is of some interest. Instead of keeping their work confidential within the company, they immediately and effectively sought the maximum publicity for it—even though this might be expected to generate competition, and indeed did so. Within a week of its first successful operation, Macek and a colleague had dispatched a paper describing their device to *Applied Physics Letters*; the paper was published within $2\frac{1}{2}$ weeks. They rigged up an impressive audio-visual display, with glowing lasers and beat frequencies relayed through a loudspeaker. Among those they invited to see their device was an influential technical journalist, Philip J. Klass. A mere month after their laser gyro first worked, he rewarded them with an article describing their work (in which the term "laser gyro," which Klass may have coined, was used for the first time) in the widely read *Aviation Week and Space Technology*, and with a color picture on the cover.[46]

Publicity was necessary because the most immediate problem facing Macek and his colleagues was their own company's management. Their original proposal had been rejected on the grounds of infeasibility, and in the company that had pioneered the mechanical gyroscope in the United States the commitment to the existing technology was strong. Even the name "laser gyro" was taboo at Sperry: "the company shuns the use of the word 'gyro' because the device lacks the familiar spinning mass."[47] Competition arguably turned out to be harmful to the long-term interests of the company as a whole: Sperry's laser gyroscopes had less market success than those of the company's competitors. However, competition was in the immediate interest of the team developing the device—that others took it to be feasible was a powerful argument to use with a skeptical management—and certainly was to the benefit of the overall development of the laser gyro.[48]

Several different research and development teams in the United States—and groups in the Soviet Union, the United Kingdom, and France—began laser gyro work soon after the device's invention and the success of the Sperry prototype became known.[49] The American researchers included groups at the Kearfott Division of General Precision, the Autonetics Division of North American Aviation, the Hamilton Standard Division of United Aircraft, and the MIT

Instrumentation Laboratory.[50] Most consequential, however, was a team at Honeywell, members of which freely admit to having learned of the laser gyro from Klass's article in *Aviation Week.*[51]

Like quantum electronics more generally, this R&D effort was strongly supported by the armed services—particularly in the United States, where there was keen appreciation of the military importance of inertial guidance and navigation and of the deficiencies of existing systems. Much of the work within corporations received military funding, and the Bureau of Naval Weapons and the Air Force Systems Command sponsored an annual series of classified symposia on "unconventional inertial sensors" at which work on the laser gyro—and on its competitors—was presented and discussed.[52]

Military support was not, on its own, sufficient to move the laser gyro from prototype to product. At Autonetics, for example, "every year we [the laser gyro developers] told them [higher management] that ring lasers were going to take over everything, and every year they kept us on the back burner. . . . They wanted to stay up with the technology but weren't willing to commit. It costs lots and lots of money to go into production. Because their [Autonetics's] marketplace was strategic vehicles and high accuracy devices, and the devices they were manufacturing were successful, there was no real reason to develop a new product." The founder of MIT's Instrumentation Laboratory, Charles Stark Draper, considered the laser gyro a diversion from the pursuit of ultimate accuracy through the evolutionary refinement of floated mechanical gyros.[53]

The long-term significance of the Honeywell team was thus that they, more than any other group, were able to sustain the development of the laser gyro through the extended period it took to turn the invention into a navigational instrument able to compete on the market. The team, the most central members of which were Joseph E. Killpatrick, Theodore J. Podgorski, and Frederick Aronowitz,[54] possessed not only theoretical and technological expertise but also a capacity to persuade Honeywell's management of the need to do more than keep the laser gyro work on a risk-free, military-funded "back burner." Defense Department support was crucial, especially when the project ran into difficulties within Honeywell. Over the years, however, government funding was matched by a roughly equal volume of internal funding. Honeywell was also prepared to develop a laser gyro production facility in the absence of any firm military orders.[55]

Honeywell's unique position with respect to the inertial navigation business helped make it possible for the laser gyro team to extract this level of commitment from corporate management. Important mechanical gyroscope development work had been done at Honeywell in the 1950s and the early 1960s. Whole navigation systems had been built, too, but they were largely for small-volume and highly classified programs.[56] As a wider military market and then a civil-aviation market for inertial navigation opened up in the 1960s and the early 1970s, Honeywell was largely excluded. It was successful in producing inertial components to others' designs, especially those of the MIT Instrumentation Laboratory, but not in designing and selling its own inertial systems. This meant that at Honeywell (in contrast with Autonetics, for example) there was no existing, successful product line that was threatened by novel inertial sensor technologies, and indeed the latter were seen as providing an opportunity to move Honeywell from the margins to the center of the inertial market. The first technology with which Honeywell attempted this was the electrostatic gyro— a mechanical gyroscope, without conventional bearings, in which the spinning mass is a sphere suspended in an electrostatic field. This device temporarily brought Honeywell an important share of the high-accuracy strategic bomber navigation market, but it was defeated in its primary intended niche, ballistic missile submarine navigation, by a similar gyro produced by the niche's established occupant, Autonetics.[57] Furthermore, the electrostatic gyro never became accepted in the largest market of all: the market for medium-accuracy (around 1 nautical mile per hour error) military and civil aircraft navigators.

Success in this last market was what Honeywell sought with the laser gyro. The potential advantages of the device had been listed in Klass's *Aviation Week* article: it "has no moving parts and, in theory, should be long-lived, sensitive and stable," and, because it measures discrete beats, "its output is available in digital form, for use by digital guidance computers." But to turn this promise into practice clearly required replacement of what those involved would certainly have admitted were "bulky and unwieldy" experimental configurations.[58] This could have been done by modification of these configurations—that in essence was the strategy adopted in further ring laser gyro development at Sperry—but the Honeywell team chose instead to simplify the design radically.[59] They moved from a square to a triangular path drilled in a single solid quartz block (figure 4). In their "monolithic" design, there is no distinction between the path and the laser. Lasing in the entire triangular

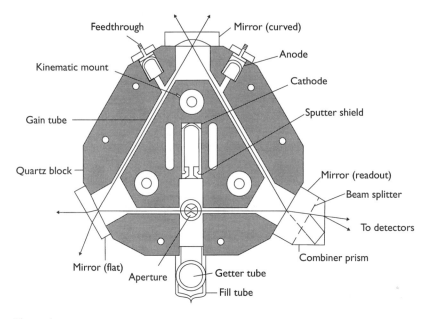

Feedthrough
Mirror (curved)
Kinematic mount
Anode
Cathode
Gain tube
Sputter shield
Quartz block
Mirror (readout)
Beam splitter
To detectors
Mirror (flat)
Combiner prism
Aperture
Getter tube
Fill tube

Figure 4
Monolithic solid block ring laser gyro as developed at Honeywell. Based on diagram provided by Theodore J. Podgorski, Military Avionics Division, Honeywell, Inc.

path is sustained by energy supplied by a high voltage difference between a cathode and two anodes.

A second change from the early prototype laser gyros was perhaps even more consequential, because it differentiated the approach taken at Honeywell from those of the other development efforts. All the developers quickly identified a major problem in developing a laser gyro that would be competitive with mechanical gyros: at low rotation rates the laser gyro's output vanished (figure 5). Below a certain threshold (which could be as high as 200°/hour), rotation could not be measured. If uncorrected, this would be a fatal flaw in a device whose mechanical competitors were by the 1960s sensitive to rotations of 0.01°/hour or less.

The cause of the phenomenon now seems obvious, but it was not immediately so to the early investigators. The scattering of light from imperfect mirrors and various other causes meant that the two beams were not in practice wholly independent. They acted like coupled oscillators in radio engineering, "pulling" each other's frequencies toward convergence, and therefore toward zero output and the phenomenon those involved call "lock-in."[60]

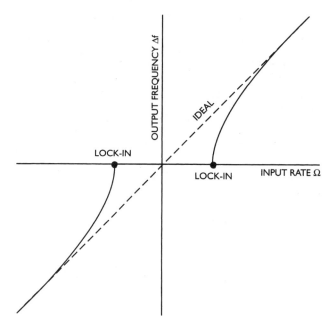

Figure 5
The input-output function for an "ideal" laser gyro and for the actual device. Based
on diagram in "Presentation of the Elmer A. Sperry Award for 1984 to Frederick
Aronowitz, Joseph E.. Killpatrick, Warren M. Macek, Theodore J. Podgorski."

One approach to solving the problem of lock-in was to seek an electro-optical means of preventing the beams from coupling at low rotation rates. The team at Sperry introduced a "Faraday cell" into the cavity
(figure 6). This increased the effective travel path of one of the beams
more than the other; the device was thus "biased" so that the region
where lock-in would occur was no longer within the gyro's normal operating range. Later the Sperry workers substituted an alternative electro-
optical biasing technique, the "magnetic mirror."

For the laser gyro to measure rotation rates accurately, however, the
bias had to be dauntingly stable, according to calculations at Honeywell.
Joseph Killpatrick, the most prominent champion of the laser gyro at
Honeywell, had an alternative solution to the problem of lock-in. This
was, in effect, to shake the laser gyro rapidly so that it would never settle into lock-in. The idea flew in the face of the "no moving parts" image
of the laser gyro that had been created by the publicity for it, such as
Klass's article; thus it met considerable resistance: "Shaking it was just

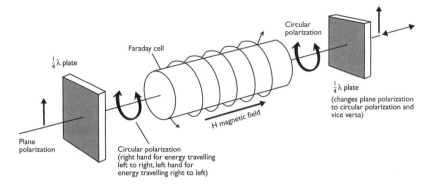

Figure 6
Use of a Faraday cell to bias a laser gyro. Based on diagram in Joseph Killpatrick, "The laser gyro," *IEEE Spectrum* 4 (October 1967), p. 53.

repugnant to people, and so the natural thing was to build in a Faraday cell, but [that] was wrong." Mechanical dither, as it is known, triumphed nevertheless, first at Honeywell and then more widely, even though its incompatibility with the laser gyro's "image" meant that research funds continued to be readily available for investigations of other ways of circumventing lock-in.[61]

Crucial in persuading Killpatrick's colleagues at Honeywell of the virtues of dither were experiments conducted there in 1964. A prototype laser gyro was placed on a large spring-mounted granite block in the Honeywell laboratories, and the block was set oscillating. The results were a remarkable improvement: the device detected the Earth's relatively slow rotation with considerable accuracy. Paradoxically, though, Killpatrick found that too regular a dither motion would lead to considerable errors as a result of the cumulative effect of the short periods of lock-in when the device was at rest at the extremities of its dither motion. "Noise"—a random element in the dither motion—prevented this cumulation.[62]

During this period, the laser gyro was increasingly connected to a hoped-for major redesign of inertial systems. Traditionally, accelerometers and gyroscopes had been mounted on a platform supported by a complex gimbal structure that gave it freedom to change its orientation with respect to the vehicle carrying it. Any rotation of the platform with respect to the fixed stars (or, in some systems, with respect to the local direction of gravity) would be detected by the gyroscopes, and a feedback system would cancel out the rotation, thus maintaining the platform in the desired orientation irrespective of the twists and turns of the vehicle carrying it. The configuration was, therefore, called a "stable platform."

During the 1960s, there was growing interest in the mechanically much simpler "strapdown" configuration, in which the gyroscopes and accelerometers would simply be attached to the body of the vehicle carrying them. There were two barriers to implementing this. One was that a powerful onboard computer would be needed. Because the instruments were no longer in a fixed orientation, more complex mathematical processing of their output was needed to permit velocity and position to be calculated. With digital computers growing more powerful, smaller, and more robust, this first barrier was rapidly eroding by the late 1960s. The laser gyroscope promised to remove the second barrier. In a stable platform the gyroscopes had to be highly accurate, but only over a limited range of rotations. Strapdown gyroscopes had to maintain that accuracy over a much wider range. This was acknowledged as hard to achieve with most forms of mechanical gyroscope, and one of the most crucial claims for the laser gyroscope was that "excellent linearity" had been achieved in the measurement of rotation rates as high as 1000°/second.[63]

Simultaneous with the attempts to improve the laser gyro practically and to make it the centerpiece of a reconfigured inertial system, a more sophisticated theoretical understanding of it was developing. Though many contributed, including Heer, the theoretical effort at Honeywell was led by Frederick Aronowitz, a physics graduate student hired by Killpatrick from New York University. Drawing on both classical electromagnetic theory and quantum mechanics, Aronowitz had by 1965 developed an elaborate mathematical theory of the operation of the laser gyro, a theory he continued to develop over the following years.[64]

By 1966, then, the laser gyroscope had been considerably refined from the earliest prototypes, a role for it and a solution to the main development problem had been found, and it was well understood theoretically. It was no longer restricted to the laboratory. Honeywell had a military contract with the Naval Ordnance Test Station at China Lake, California, to develop not a full inertial navigator but a prototype attitude reference system (orientation indicator) for launching missiles from ships. The laser gyro attitude reference system constructed by Honeywell was small and rugged enough to be operated while being transported by air to China Lake in September 1966, allowing Honeywell to claim the first flight test of a laser gyro system. The

Honeywell group's confidence was high: they were already able to mea-sure rotation rates of 0.1°/hour, and they believed that "within a year" they would achieve the goal of measuring 0.01°/hour.[65]

That "year," however, stretched into almost a decade. At issue was not merely achieving the final order-of-magnitude increase in accuracy but increasing reliability (the working lifetimes of the early devices were typ-ically less than 200 hours) and reducing size (though considerably smaller than the laboratory prototype, laser gyros were still typically larger than their mechanical competitors). Achieving these goals required ingenuity, considerable resources, and far more time than had been forecast: "the late sixties–early seventies were trying times." Even within Honeywell, the patience of higher management began to run out—"internal funding went almost to zero because one vice-president had something bad to eat or something"—and military funding, espe-cially a contract from the Naval Weapons Center, was crucial in keeping development going.[66]

Almost every element in the laser gyro was refined and changed in the continuing Honeywell development effort: the material of the block (which was changed from quartz, through which the helium leaked, to the new glass ceramic Cer-Vit), the mirrors, the seals, the cathode, the quantum transition employed (which was shifted from 1.15 microns, in the infrared spectrum, to 0.63 microns, in the visible spectrum), the dither motor, and the output optics.

Slowly, these efforts bore fruit. By 1972, Cer-Vit, improved seals, and a new "hard coating" mirror fabrication process led to laser gyros that finally began to live up to the original promise of high reliability. This enabled Honeywell, rather than its competitors Sperry and Autonetics, to win the key contract from the Naval Weapons Center that helped per-mit resolution of the device's other problems. Worth $2.5 million, that contract was again not for a full inertial navigator but for prototypes of a more modest system for the guidance of tactical missiles. As these became more sophisticated, there was increasing interest in providing them with inertial guidance systems. The simplicity of strapdown, the fast reaction of the laser gyroscope (with no fluid to be heated or rotor to "spin up"), and the apparent insensitivity of the laser gyro to accel-eration-induced errors all made laser systems seem an attractive option for such applications. At a time when pessimists had begun to doubt whether the laser gyro would ever achieve the "magic" figure of a 0.01°/hour error, its application to tactical missiles had the advantage of permitting drift rates much worse than that.[67]

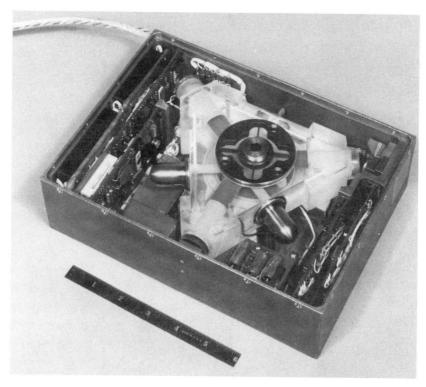

Figure 7
Early version of Honeywell GG 1300 laser gyro. The ruler (marked in inches) gives an indication of the device's size. Courtesy of Theodore J. Podgorski, Military Avionics Division, Honeywell Inc.

Yet 0.01°/hour, and with it the mainstream aircraft navigation market, remained the goal of the Honeywell team, particularly Killpatrick, and they continued to seek higher accuracy. In 1974, laser gyros finally began to demonstrate the 0.01°/hour error level in Honeywell's own laboratory tests. In February and March 1975, laboratory tests of prototype inertial systems delivered to the U.S. Navy under the tactical missile contract yielded an accuracy figure of 0.64 nautical miles per hour—far beyond the demands of that contract, and indeed better than the one-nautical-mile criterion for an aircraft navigator.[68]

In May 1975, Honeywell submitted an inertial navigation system based around its new GG1300 laser gyro (figure 7) for flight testing at the most authoritative military test center, the Central Inertial Guidance Test Facility at Holloman Air Force Base in New Mexico. Its accuracy was assessed there in flight as well as in laboratory tests. The official report on the tests concluded that they "demonstrated the successful application of ring laser gyros to strapdown inertial navigation system technology," and that the Honeywell system "appears to be better than a 1 nautical mile per hour navigator."[69]

The Laser Gyro Revolution

It was a turning point. Quiescent laser gyro programs at other inertial suppliers were infused with resources even before the successful tests—whose likely significance was underlined in January 1975 by Philip J. Klass in *Aviation Week*. Several firms outside the traditional inertial business also began laser gyroscope development, seeing an opportunity to break into the market.[70] After the excitement of the early 1960s and the long struggle of the late 1960s and the early 1970s, the laser gyro had finally proved itself a competitor to its established mechanical rivals.

Yet even this success was not, on its own, sufficient to ensure the laser gyro's future. Its test accuracy and reliability, though now acceptable, by no means surpassed those of contemporary mechanical gyroscopes, and its cost advantages were "projected."[71] Only prototypes had been built.

Military interest in the United States was nevertheless keen. A Ring Laser Gyro Navigator Advanced Development Program was set up within the Naval Air Systems Command to further refine and evaluate the Honeywell system. Funding increased sharply as the technological focus began to shift from performance to production. A tri-service (Army, Navy, Air Force) laser gyro manufacturing and producibility program provided Honeywell with $8 million. Honeywell's competitors benefited too, as the armed services, fearing future dependence on a single supplier, also funded work at Sperry, Litton, and elsewhere.[72]

Despite this support, however, a military market for the laser gyroscope opened up only in the mid 1980s, several years behind the civil market. The delay was due in part to remaining performance difficulties. By the late 1970s, the U.S. Air Force was demanding from fighter aircraft inertial navigators an error rate of 0.8 nautical miles per hour. Given the often violent maneuvers of military aircraft, which impose a greater strain on a strapdown system than the gentle flight path of an airliner, this remained a demanding goal when combined with strict limits on the size and weight of operational (rather than test) inertial systems. The accuracy specifications for bomber navigation were tighter still. Furthermore, a military aircraft navigator must provide information not just on position but also on velocity for accurate bombing or missile launches. In 1980, after the device's breakthrough into the civil market, Major General Marc Reynolds told the Joint Services Data Exchange Group for Inertial Systems that, in the Air Force's opinion, the laser gyro "does not yet have the velocity accuracy required for fighter aircraft." Another problem (at least as seen from Honeywell) was that the U.S. military was less centralized in its decision making than the civil aviation world: "If you deal with Boeing, at some point you're going to find a . . . man who is empowered to make a decision. If you go to the Air Force, you can never find a guy who is going to make a decision. You can find advocates . . . but you can't find a decision maker."[73]

Boeing was, in fact, central to the most crucial decision in the laser gyro revolution. In the late 1970s, Boeing was designing two new airliners: the 757 and the 767. Mechanical gyro inertial navigation systems had proved their worth on the long-range 747 "jumbo jet." Though the 757 and the 767 were to be smaller, medium-range planes, Boeing engineers believed that there was a role for strapdown inertial systems on them, especially if the orientation information they provided was used to eliminate the previously separate attitude and heading reference system.

These engineers became enthusiasts for the laser gyro. The 757 and the 767 were to be the most highly computerized civil aircraft yet built by Boeing, and the laser gyro's digital output would fit in well with this vision. The laser system's fast reaction reduced the risk that a takeoff would be delayed because the inertial navigator was not ready for use. Its promise of high reliability was attractive in an airline environment that was conscious not only of the initial cost of buying a system but also of the cost of maintaining and repairing it over its lifetime. Finally, the sheer glamour of the laser gyro was appropriate to the "high-tech" image that Boeing was cultivating for the new planes.

An informal alliance developed between proponents of the laser gyro within Honeywell and Boeing. Both groups knew that winning a commitment from Boeing to the laser gyro required an equally visible prior commitment from Honeywell. Specifically, Honeywell had to build a laser gyro production facility, in advance of any contract to sell the device, and this would require a large and apparently risky corporate investment. (The military funding, though helpful, fell far short of what was needed to build such a facility.) The night before a crucial meeting with Honeywell's top managers, Boeing and Honeywell engineers met at the house of a Honeywell engineer to prepare. Next day, as planned, the Boeing engineers emphasized the need for Honeywell investment: "Honeywell had got to put some money into that laser stuff or we're never going to put it on the airplane."[74]

This informal alliance succeeded in its twin tasks. Honeywell's top management was persuaded that the risk of investment in a laser gyro production facility was worthwhile, and Boeing's top management was persuaded of the virtues of a laser system for the 757 and the 767. More than the two managements needed convincing, however. New-generation avionics specifications are decided not by the manufacturer alone but by a wider semiformal body, which includes representatives of all the main aircraft manufacturers, the avionics companies, and the airlines. The Airlines Electronic Engineering Committee, as it is known, is a section of ARINC (Aeronautical Radio, Incorporated), created in December 1929 by the U.S. airlines to provide radio communications with aircraft. Despite the apparently *ad hoc* nature of the arrangement and the considerable potential for conflict of interest, the system works remarkably smoothly to define "Characteristics"—agreed understandings of the function, performance, physical dimensions, and interfaces of avionics equipment.[75] To seek to market a new system in advance of a Characteristic, or in violation of it, would be self-defeating.

The laser gyroscope was able to meet any plausible accuracy requirement. Extremely high accuracy has never been demanded in civil air inertial navigation; average error as great as 2 nautical miles per hour is acceptable. Rather, the crucial aspect of the Characteristic was physical size. (The weight of laser systems was also an issue, but it was around size that debate crystallized.) State-of-the-art mechanical systems, using sophisticated "tuned rotor" designs, were substantially smaller than the Honeywell laser gyroscope system, despite the continuing efforts to make the latter smaller. If the manufacturers and the airlines opted to save physical space by adopting a small box size, the laser gyro would be ruled out and the new mechanical systems would triumph by default.

"We met individually with every guy on the committee," recalls Ron Raymond of Honeywell. The crucial 1978 meeting was held in Minneapolis, where Honeywell is based. Some 300 delegates were present. Honeywell bought advertising space at airline gates throughout the country, "getting our message to the guys coming out on the planes."[76]

Honeywell carried the day on size, obtaining in the key specification, ARINC Characteristic 704, a box size 25 percent larger than what was needed to accommodate the mechanical systems. Because nothing prevented manufacturers and airlines from opting for mechanical systems, a pricing battle had also to be won. Bolstered by what turned out, for the reasons outlined above, to be a grossly optimistic (or at least premature) forecast of a market for 12,000 laser gyro systems in military aircraft, Honeywell priced its civil laser gyro system very keenly.

Honeywell's laser gyro system was selected for the 757 and the 767. With the predicted military market slow to appear and the production costs higher than anticipated, quick profits were not to be found. The financial details are confidential, but the industry's consensus in the mid 1980s was that Honeywell had yet to recoup its investment in the laser gyro. (U.S. law permits such an investment to be set against corporate taxes, which reduces the effect of any loss on a large, diversified corporation such as Honeywell.)

Although profits were slow in coming, market share was not. Despite fierce competition from Litton Industries, including legal battles over alleged patent and antitrust violations, Honeywell has secured a dominant share of the world's market for inertial navigation systems in civil aircraft (around 50 percent by the mid 1980s, and perhaps 90 percent by 1990).[77]

During the latter part of the 1980s, the laser gyro also established Honeywell firmly in the military market for inertial navigation. In 1985 the U.S. Air Force began to make large purchases of laser gyro systems, selecting Honeywell and Litton as competitive suppliers of laser inertial navigation units for the C-130, the RF-4, the F-4, the EF-111, and the F-15.[78] International military sales climbed rapidly as laser systems became standard on new military aircraft and as the retrofitting of older planes increased. In the United States, Honeywell, Litton (the previously dominant supplier of mechanical gyro systems for military aircraft), and Kearfott (now a division of the Astronautics Corporation of America) competed vigorously for the military market.

The form taken by competition in the market for inertial systems, both civil and military, changed during the 1980s. At the beginning of the decade, laser systems were striving to establish a foothold in a market dominated by mechanical systems. By the end of the decade, competition was almost always between laser systems offered by different companies. Although Sperry developed and sold several laser devices, it never successfully entered the air navigation market, and in 1986 the Sperry Aerospace Group was bought by Honeywell. Litton began a low-level laser gyro effort in 1973. In mid 1974, under the leadership of Tom Hutchings, the program was expanded. By the end of 1980 Litton had achieved satisfactory flight test results with its laser gyro system. Though its work lagged behind that of Honeywell, the desire of airlines to avoid dependence on a single supplier helped a Litton laser system win the next major civil air transport contract, for the Airbus Industrie A310.[79] Kearfott also developed laser systems, as did all but one of the other U.S. suppliers of inertial systems, the European firms, and Japan Aviation Electronics Industry, Limited.

With the exception of Sperry, which continued to use electro-optical biasing, the laser systems developed by these other firms generally followed the main features of Honeywell's design. There were differences, such as Litton's use of a square path with four mirrors rather than a triangular path with three, but the monolithic solid-block design and the use of dither supplemented by noise predominated. Honeywell's patents on these features did not prevent their use by other firms. Honeywell sued Litton for alleged patent infringement, but the action was settled out of court, and other firms seem to have been able to employ these features with impunity.[80]

The success of the laser gyro during the 1980s cannot be attributed to its exceeding its mechanical competitors in accuracy, although by the end of the decade the accuracy advantage of mechanical systems was eroding as substantial U.S. military research and development funds were devoted to improving the laser gyro and development money for mechanical gyros diminished. In 1984 Honeywell received $60.9 million, and Litton $74.8 million, to develop laser gyro guidance systems for a proposed new U.S. missile, the Small ICBM. Success in this would have been an enormous step toward acceptance of the laser gyro, since self-contained prelaunch alignment of a ballistic missile guidance system to the accuracy required of the Small ICBM is extraordinarily demanding of gyroscope performance. Error rates between $0.0001°$ and $0.00001°$ per hour are needed, rather than the $0.01°/\text{hour}$ of aircraft navigation. The former figures are close to what is held to be a physical limit on the performance of laser gyroscopes roughly comparable in size to mechanical gyros—a limit arising ultimately from quantum effects. In the end, though, the Air Force, advised by the Draper Laboratory (formerly the MIT Instrumentation Laboratory), concluded that the laser system could not provide the requisite accuracies and opted to modify the existing mechanical gyro guidance system of the MX.[81]

Nor did the laser gyro turn out (at least in the short term) to possess the clear advantage over mechanical gyros in cost of production that had been hoped for.[82] Rather, reliability has been the major claimed (and widely accepted) advantage of the laser gyro. A typical Honeywell advertisement contrasted the 8000 hours mean time between failures achieved by its laser system on the Boeing 757 and 767 with the much lower mean times between failures achieved by its competitors' previous-generation mechanical systems in military aircraft.[83]

There are still skeptics, however, even on the question of reliability. They argue that it is unfair to contrast civil systems with traditionally less reliable military ones; that the large box size won by Honeywell meant that the laser system worked at a lower temperature than mechanical ones, and temperature was the crucial determinant of failure; that Honeywell engaged in extensive preventive maintenance, especially mirror replacement, to keep the mean time between failures high; that modern mechanical gyros are as reliable as laser gyros; and that the main determinant of a system's reliability is the electronic components (which were more modern and thus more reliable in the Honeywell system than in its older competitors), not the gyros.[84] These counterarguments counted for little, however, as the laser gyro revolution became irreversible. The skeptics worked for firms that had seen no alternative to heavy investment in laser gyroscopes, and even they did not disagree with that decision. As one proponent of the laser gyro put it: "Anyone who wants to play in the future has got to have a laser gyro. Spinning iron won't do any more. Even if spinning iron was truly better, you can't do it—it doesn't have the technology charisma."[85]

Often the decision seems to have been an either/or one: commitment to the laser gyro meant a reduction in support for continued development of mechanical devices. At Kearfott, for example, research was focused in the early 1970s on a sophisticated new mechanical design, the Virex gyro. Its development was going well, but when Kearfott's vice-president of engineering heard of Honeywell's success with the laser gyro he insisted that the Virex work be stopped and that the resources be devoted to the laser gyro instead.[86]

The one major firm to stand aside from the laser gyro revolution has been the Delco Division of General Motors. As AC Spark Plug, Delco pioneered inertial navigation for civil aviation. Its Carousel system, based on traditional spinning-wheel gyros and used in the 747, was the first successful system of its kind. During the mid 1960s, Delco researchers had become interested in the idea of a "hemispherical resonator gyro" (figure 8). (The device is analogous to a ringing wine glass; it senses rotation through changes in vibration patterns.) When other firms set up or revived their laser programs in the mid 1970s, Delco instead devoted resources to the resonator gyro. Delco believes the

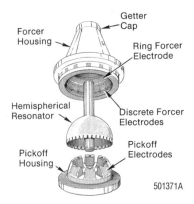

Figure 8
Hemispherical Resonator Gyro. Courtesy David Lynch, Delco Systems Operations, General Motors Corporation.

hemispherical resonator gyro to have even greater reliability than the laser gyro, together with an important military advantage: lack of susceptibility to the electromagnetic pulse from a nuclear explosion.[87]

Like Warren Macek with the first laser gyro 30 years before, Delco's researchers understand why it can be better for technologists to have competitors also seeking to develop the same device: that makes it easier to "keep management on board."[88] Unlike Macek, however, they have not succeeded in generating competitors. The fate of their solitary dissent from the laser gyroscope revolution remains to be seen.

Conclusion

Several issues concerning the relationships among science, technology, and society emerge from the history of the laser gyroscope. There was no direct path from "science" (the ether experiments of Sagnac and Michelson) to "technology" (the laser gyroscope). The crucial intermediary was the development of quantum electronics, a field that involved fundamental physics but did not fit the traditional stereotype of "pure science." The "greater and rapidly growing part of quantum electronics owed its very existence to wartime radar work,"[89] and its postwar direction was still shaped by technological concerns and at least to some extent by military interests. The development of the laser gyroscope (and quantum electronics more generally) may best be seen as what Bruno Latour calls "technoscience"—the construction of an interconnected network of elements of "science," "technology," and "social processes" or "social interests."[90]

No single element of this network was able to write the script of the story of the laser gyroscope. "Science" did not determine "technology": the meaning of the "Sagnac effect," for example, was radically transformed in its passage from being a claimed proof of the existence of the ether to being the oft-cited foundation of the laser gyroscope. Neither, however, was there any internal logic of technological change that led of necessity from the mechanical to the optical sensing of rotation. Inertial navigation's "founding father," Charles Stark Draper, and the researchers at Delco saw the path of technical evolution quite differently, and it would be rash to assert that either was definitely wrong.

Nor did social processes and interests have free rein: they had to interact with an only partially tractable material world. The members of the Honeywell team were adroit engineers of social support (from their management and the military) as well as of cavities and mirrors, yet what is most impressive about what they did is their persistence in the face of obstacles they could shift only slowly. The successful development of the laser gyroscope (and perhaps even its invention) is hard to imagine without the U.S. military, yet the resultant technology was not shaped (initially, at least) by specifically military needs. Indeed, where those needs are most specific—in the guidance of strategic ballistic missiles, with its extreme demands for accuracy—the laser gyroscope has not met with success, and it was accepted in military aviation only after its triumph in the civil sphere.

Similarly, despite the central importance of economic phenomena—markets, profits, and the like—to the history of the laser gyroscope, the history cannot be told in the terms of orthodox neoclassical economics, with its all-seeing, unitary, rationally maximizing firms. Honeywell, the central firm in the story, was not all-seeing: the laser gyroscope proponents within Honeywell had to work to keep their vision of the future in front of the eyes of senior management. Neither was Honeywell (or Sperry, or other firms) unitary: the story of the laser gyroscope cannot be understood without understanding the tensions between engineers and their senior managers, or the informal alliances that can develop between staff members of different firms (notably Honeywell and Boeing). Nor was Honeywell in any demonstrable sense a rational maximizer. Profit calculations were certainly prominent in the decisions of senior managers, but the data on which the crucial early calculations were based (particularly the estimates of production costs and the size of the market for the laser gyroscope) appear in retrospect to have been little better than guesses (brave and consequential guesses though they were).

If an economic theory of the laser gyroscope revolution is sought, then the neoclassical economists, with their assumption of finely tuned optimization, are less relevant than Joseph Schumpeter, who emphasized product-based rather than price-based competition, "gales of creative destruction," and what John Maynard Keynes called the "animal spirits" of entrepreneurs. Although they were corporate rather than individual entrepreneurs, the Honeywell staffers possessed those "spirits" in good measure. They aimed high, they took risks, and they knew that to achieve their goal they had to shape the market as well as meet its demands (as is demonstrated by their intensive lobbying to secure a Characteristic that the laser gyro could meet).[91]

The history of the acceptance of the laser gyroscope reveals at least one interesting facet of the dynamics of "technological revolutions."[92] It is difficult to attribute the device's success to any unambiguously inherent technological superiority over its rivals. It has not yet succeeded in ousting mechanical systems in applications that demand the greatest accuracy; the hopes that it would be much cheaper to make were unfulfilled for a long time; and its claims to intrinsically superior reliability, though highly influential, are not universally accepted. Until recently, laser systems have been bulkier and heavier than mechanical systems of comparable accuracy. The laser gyro's digital output and its compatibility with the simpler strapdown configuration of inertial systems gave it a certain "systemic" advantage, but even that is not unique. The analog output of other devices can be digitized. Compatibility with strapdown was one of the main initial attractions of the electrostatically suspended gyro; dynamically tuned mechanical gyros have been developed for strapdown configurations, and the hemispherical resonator gyro has been used in a strapdown system. Other varieties of gyro also offer quick startup.

There is a sense, however, in which the intrinsic characteristics of different gyroscope technologies are irrelevant. What matters in practice are the *actual* characteristics of such technologies and the systems built around them, and these reflect to a considerable degree the extent of the development efforts devoted to them.

There is thus an element of self-fulfilling prophecy in the success of the laser gyroscope. In the pivotal years of the revolution (from 1975 to the early 1980s), firms in the business of inertial navigation had to make a hard decision on the allocation of development funds. Was a technological revolution about to occur? Would they be able to compete in the mid or late 1980s without a laser gyroscope? All but Delco decided that the revolution was likely and that the risk of not having a laser gyroscope was too great. Accordingly, they invested heavily in the development of laser gyroscopes and systems incorporating them while cutting back or even stopping development work on mechanical gyroscopes and systems. And some firms without mechanical gyroscope experience began laser programs in anticipation of the revolution.

The result was a rapid shift in the balance of technological effort—even by 1978, "optical rotation sensor . . . technology [was] being pursued more broadly for inertial reference systems applications than any other sensor technology"[93]—that helped make the laser gyroscope revolution a reality. By the end of the 1980s, laser gyro systems were beginning to seem unequivocally superior to their traditional mechanical rivals, at least in aircraft navigation. Proponents of traditional mechanical systems claim that with equivalent development funds they could still match or outstrip laser systems; however, the argument has become untestable, as no one is now prepared to invest the necessary sums (tens of millions of dollars) in further development work on traditional systems.

There is nothing pathological in this aspect of the laser gyro revolution. The outcome of a political revolution, after all, depends in part on people's beliefs about whether the revolutionaries or the established order will be victorious, and on the support the different parties enjoy as a consequence. Indeed, it has been argued, convincingly, that all social institutions have the character of self-fulfilling prophecies.[94] Technology is no exception, and the role of prediction and self-fulfilling prophecy in technological change, especially technological revolution, is surely worthy of particular attention.

Acknowledgments

The interviews drawn on here were made possible by a grant from the Nuffield Foundation for research on "the development of strategic missile guidance technology" (SOC442). Their further analysis was part of work supported by the Economic and Social Research Council under the Programme on Information and Communication Technologies (A35253006) and the Science Policy Support Group Programme of Science Policy Research in the Field of Defence Science and Technology (Y307253006). I am grateful to Wolfgang Rüdig for assistance in the research.

5

Nuclear Weapons Laboratories and the Development of Supercomputing

One theme of recent social studies of technology has been the need to look "inside the black box"—to look at technology's content, not just at its effects on society.[1] This chapter seeks to do this for one particular technology: high-performance digital computers (or "supercomputers," as they have come to be called). I shall examine the influence on super-computing of two powerful organizations: the national laboratories at Los Alamos, New Mexico, and Livermore, California. These labs have been heavily involved in supercomputing ever since the supercomputer began to emerge as a distinct class of machine, in the latter part of the 1950s. What has their influence been? What demands does their key task—designing nuclear weapons—place upon computing? How far have those demands shaped supercomputing? How deep into the black box—into the internal configuration and structure, or "architecture," of supercomputers—does that shaping process go?

I begin by reviewing the history of high-performance computing and the nature of the computational tasks involved in designing nuclear weapons. I then describe Los Alamos's influence in the early years of digital computing, and the role of Los Alamos and of Livermore as sponsors and customers for supercomputing. Being a customer and sponsor, even a major one, does not, however, automatically translate into the capacity to shape the product being bought or supported. In an attempt to specify the sense in which the laboratories have influenced (and also the sense in which they have failed to influence) the develop-ment of supercomputing, I address the effect of the laboratories on the evolution of supercomputer architectures.

Reprinted, with permission, from *Annals of the History of Computing* 13 (1991). ©1991 AFIPS (now IEEE).

Supercomputing: A Brief History

The terms "high-performance computing" and "supercomputing" are relative. The level of performance required to make a computer a high-performance computer or a supercomputer has changed through time. The criterion of performance has been stable, though, at least since the latter part of the 1950s: it has been speed at arithmetic with "floating-point" number representation—the representation most suitable for scientific calculations.[2] This speed, now conventionally expressed as the number of floating-point operations ("flops") carried out per second, has increased from the thousands (kiloflops) in the 1950s to the millions (megaflops) in the 1960s to thousand millions (gigaflops) in the 1980s, and may increase to million millions (teraflops) by the end of the 1990s.

The category "supercomputer" (though not the word, which came later) emerged toward the end of the 1950s out of the earlier distinction between "scientific" and "business" computers.[3] IBM's early digital computers, most notably, were divided along these lines. The 1952 IBM 701 and the 1954 IBM 704 were seen as scientific computers, whereas the 1953 IBM 702 and the 1954 IBM 705 were business data processors.[4]

Two partially contradictory efforts emerged in the latter part of the 1950s. One was the effort to transcend the scientific/business distinction by designing a family of architecturally compatible computers. First finding expression in the 1954 "Datatron" proposal by Stephen Dunwell and Werner Buchholz of IBM,[5] this effort came to fruition in the IBM System/360 of the 1960s. The other was the effort to develop a computer that would be substantially faster than the IBM 704 at floating-point arithmetic. The most immediate expressions of this were the Univac LARC and IBM Stretch computers designed in the second half of the 1950s.

Though I have not found an example of the use of the term to describe them at the time, LARC and Stretch were supercomputer projects in the above sense. They were certainly perceived as such in Britain, where they prompted a national "fast computer project" that eventually, after many vagaries, gave birth to the 1962 Atlas computer.[6] Supercomputer projects were also begun in France and in the USSR. The French project led to the Bull Gamma 60,[7] and the Soviet project to the BESM-6. (BESM is the acronym for Bystrodeystvuyushchaya Elektronnaya Schotnaya Mashina, meaning High-Speed Electronic Computing Machine.)

With the exception of the BESM-6 (many examples of which, remarkably, were still in operation in the late 1980s[8]), none of these early projects were unequivocally successful on both commercial and technical criteria. The first "supercomputer" that achieved success in both these senses was the 1964 Control Data Corporation 6600, the chief designer of which was Seymour Cray.

The 6600, which won a significant part of the scientific computing market away from IBM, was followed in 1969 by the Control Data 7600. Thereafter the mainstream of U.S. supercomputing divided.[9] Seymour Cray left Control Data to form Cray Research, which produced the Cray 1, the Cray X-MP, the Cray Y-MP, and the Cray 2, while Control Data developed the STAR-100, the Cyber 205, and eventually the ETA[10]. In April 1989, however, Control Data closed its ETA Systems supercomputing subsidiary and left the supercomputer market. IBM remained uneasily placed on the margins of supercomputing, producing some very fast machines but concentrating on high-speed versions of its mainframes rather than on producing a specific supercomputer range.[10]

In the second half of the 1960s and in the 1970s, American supercomputing faced no real overseas competition. Indeed, access to the technology was used, in the words of one Control Data executive, as "the carrot or the stick in the U.S. government's effort to reward or punish other governments in the realm of foreign policy."[11] Neither the Gamma 60 nor the Atlas was in any full sense followed up in France or Britain,[12] and the Soviet supercomputer designers proved unable to build on their success with the BESM-6.[13] In the 1980s, however, Japanese firms began to compete with Cray Research and Control Data. In 1982 Fujitsu announced the FACOM VP-100 and VP-200—supercomputers "clearly designed to combine the best features of the CRAY-1 and CYBER 205."[14] Hitachi and NEC launched supercomputers soon afterward, and the Ministry of International Trade and Industry began a national supercomputing project aimed at a 10-gigaflop-per-second machine.

In all these efforts—from the 1950s to the 1980s, in America and elsewhere—speed has been sought by two means: improving component technology and changing computer architecture. In regard to component technology, "improvement" means lower gate delays—reduction in "the time taken for a signal to travel from the input of one logic gate to the input of the next logic gate."[15] The "first generation" electronic valve computers of the early 1950s had gate delays of around a microsecond; the fastest integrated circuits of the mid 1970s permitted that to be reduced to around a nanosecond. That three-orders-of-magnitude improvement cannot, however, on its own account for the increase by roughly five orders of magnitude in processing speed over the same period.[16] The other two orders of magnitude can be attributed to changes in computer architecture—the "organization and interconnection of components of computer systems."[17]

These changes can, loosely, be described as the gradual introduction of various forms of parallelism or concurrency. Six of these forms deserve special mention: concurrent input/output operations, pipelining, memory interleaving and hierarchy, parallel functional units, vector processing, and multiple central processors.

Providing specialized hardware and software, so that input of data and programs and output of results can go on concurrently with processing, both predates and is more widespread than supercomputing. In the search to eliminate all barriers to speed, it was nevertheless developed to a considerable degree in supercomputing. The central processor of a Control Data 6600, for example, was never slowed by having to communicate directly with any peripheral device. Ten small computers arranged in parallel could communicate through any of twelve channels with peripheral equipment (such as printers and card readers) and with the 6600's memory (figure 1).

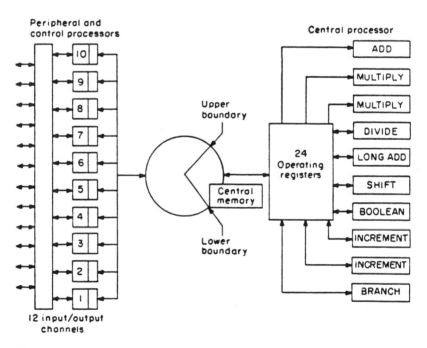

Figure 1
Block diagram of the Control Data 6600. Source: Thornton 1980, p. 346.

Pipelining is a technique rather more specific to supercomputing, at least originally. It was introduced in the earliest of the machines listed above, LARC and Stretch.[18] In a nonpipelined computer the different phases of the execution of a single instruction—accessing and interpreting the instruction, accessing the operands, performing the operation, returning the result to memory—are performed one after the other. In a pipelined computer they are overlapped, so that while one instruction is being interpreted another is being accessed and so on. In Stretch, up to eleven instructions could be in the pipeline simultaneously.[19]

Memory interleaving and hierarchy—also early and widespread techniques—are designed to keep the low speed of memory relative to the central processor from becoming a bottleneck. In interleaving, memory is arranged so as to allow simultaneous access to different segments of memory. In memory hierarchy, small amounts of ultrafast (and expensive) memory are provided in addition to the slower (and cheaper) main memory, the aim being that as many transfers as possible involve the small, fast memory rather than main memory.

The provision of separate specialized units for addition, multiplication, division, and so on that can operate independently and in parallel was a particular feature of the Control Data 6600, which contained ten parallel functional units: a Boolean unit, a shift unit, a fixed-point adder, a floating-point adder, two multiply units, a divide unit, two increment units, and a branch unit.

Vector processing means hardware and software provision for a single instruction to be executed on all the members of an ordered set of data items. The first pipelined vector computer to be proposed was the Control Data STAR-100, which, though conceived in the mid 1960s, was not operational until 1973.[20] The first pipelined vector computer to be an unequivocal success, however, was the 1976 Cray 1.

During the 1980s, the last of the aforementioned six forms of parallelism was introduced. It involved constructing supercomputers with multiple central processing units. Two, four, eight, and sixteen units have been the most common choices, but in the near future we will likely see larger numbers of units configured into a single supercomputer. Though this is a potentially major step in the direction of parallelism, these multiple processors still share a common main memory, and in practice they are often run primarily as a collection of separate processors. Rather than the components of a single task being distributed over all of them, each processor individually runs unrelated tasks, such as different programs for different users.

With the partial exception of the last two, all these six forms of parallelism represent incremental alterations of the fundamental sequential computer architecture that has become associated with the name of John von Neumann. Seymour Cray put matters succinctly: his Control Data 6600, he said, attempted to "explore parallelism in electrical structure without abandoning the serial structure of the computer programs. Yet to be explored are parallel machines with wholly new programming philosophies in which serial execution of a single program is abandoned."[21] Even vector processors and multiple central processing units, while allowing considerably greater degree of parallelism in program execution, did not wholly abandon this.

The evolutionary addition of parallel features to an originally sequential computer architecture, especially as exemplified in the development of the Control Data and Cray Research machines, constitutes what we might call "mainstream supercomputing." The overall pattern of technical change in mainstream supercomputing resembles that found in a range of other technologies (notably electricity-supply networks) by the historian of technology Thomas P. Hughes.[22] In the case of supercomputers, there is a single dominant objective: speed at floating-point arithmetic. At each stage of development the predominant barriers to progress toward the goal—Hughes calls them "reverse salients"—are sought, and innovation focuses on removing them. For example, Cray's Control Data 6600 and 7600, despite their pipelining and their multiplicity of parallel functional units, could not perform floating-point arithmetic at a rate faster than one instruction per clock period.[23] This "operation issue bottleneck" was "overcome in the CRAY-1 processor by the use of vector orders, which cause streams of up to 64 data elements to be processed as a result of one instruction issue."[24]

Some of the developments in mainstream supercomputing—notably the move to vector processing—have been daring steps. However, other computer designers, outside the mainstream of supercomputing, have not found them radical enough. They have not agreed among themselves on the best alternative architecture, but all the alternatives they have proposed have involved parallelism greater in degree than and different in kind from the parallelism used in the mainstream supercomputing of the corresponding period.[25] Perhaps the most important example is the processor array.

The central figure in the development of the processor array, Daniel Slotnick, dated his interest in parallel computers to his work on the von Neumann/Goldstine computer at Princeton's Institute for Advanced

Study in the early 1950s.[26] The architecture of that computer was paradigmatic for a generation and more of computer development. It was a "word-serial, bit-parallel" machine: though "words," or units of data, were processed sequentially, all the bits in a word were processed concurrently. The Institute for Advanced Study machine was definitely a "scientific" rather than "business" computer, and "bit parallelism" was seen as the most immediate route to increased arithmetic speed.

Slotnick's inspiration came not from abstract argument but from contemplation of a material object[27]: the magnetic drum being built to supplement the machine's main memory. His idea was to invert the word-serial, bit-parallel design by building a computer that would perform the same operation or sequence of operations concurrently on many words. Such a machine might be particularly useful for the large class of problems where an equation has to be solved for every point in a large mesh of points.[28]

Slotnick was not the only person to whom such a notion occurred,[29] and his idea did not take hold at the Institute for Advanced Study (where von Neumann dismissed it as requiring "too many tubes").[30] Yet in the 1960s Slotnick became the key proponent of the array processor, first at the Air Arm Division of the Westinghouse Corporation and then at the University of Illinois.

The first concrete form of Slotnick's scheme was called SOLOMON, "because of the connotation both of King Solomon's wiseness and his 1000 wives."[31] It was to have 1024 separate bit-serial processing elements, each performing the same fixed-point operation concurrently on different data.[32] In the later terminology of Michael Flynn, it was to be a SIMD (single instruction stream, multiple data stream) parallel computer.[33]

After Slotnick's move to the University of Illinois, SOLOMON evolved into an even more ambitious scheme, ILLIAC IV.[34] The number of processing elements decreased to 256, arranged in four quadrants of 64. But the processing elements were no longer the original simple bit-serial fixed-point processors. Each would now be capable of concurrent operation on all the bits of a 64-bit floating-point number. The overall performance goal was a gigaflop per second, and "Illiac IV ultimately included more than a million logic gates—by far the biggest assemblage of hardware ever [at the time] in a single machine."[35]

The failure of suppliers to produce the required integrated circuits, the antiwar demonstrations, the sit-ins, and the firebombing on the campus of the University of Illinois in 1970, and other circumstances prevented the smooth development of ILLIAC IV.[36] It was never built in full, though eventually one 64-element quadrant was installed at NASA's Ames Research Center in California in 1972.

ILLIAC, however, was only a harbinger of the next decade's wave of highly parallel challengers to mainstream supercomputing—MIMD (multiple instruction stream, multiple data stream) machines as well as SIMD ones. Although other factors, such as relative ease of access to venture capital, were important, what was most important in giving force to that new wave was the emergence of a viable alternative to the bipolar technology that dominated mainframe computer microcircuitry.

The challenge was from field-effect chips, in which, in principle, current flows only in the surface plane of the microchip; in a bipolar chip the current flows perpendicular to the chip as well as along it (figure 2). Field-effect technology is relatively amenable to mass production, but

(a)

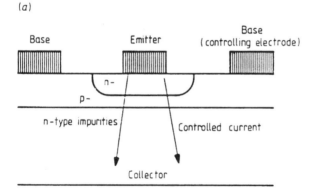

(b)

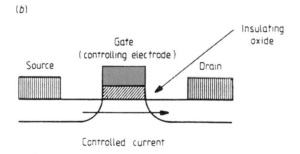

Figure 2
Schematic cross section of (a) bipolar transistor, (b) metal-oxide-semiconductor field-effect transistor. Source: Hockney and Jesshope 1988, p. 555.

for many years it was believed to be too slow for use in mainframe computers. In the 1960s it was used mainly in electronic calculators and digital watches, but the steady growth of the number of components in a given chip area (famously summarized by Moore's Law),[37] together with the intimately linked development of the microcomputer market in the 1970s, made the field-effect chip a competitor of the bipolar chip.

With the exception of the ETA[10], which used field-effect chips (cooled in liquid nitrogen to reduce gate delays), mainstream supercomputing eschewed field-effect technology, remaining with bipolar or, in the case of Cray 3, moving to the faster but even harder to fabricate gallium arsenide. However, the wider dominance of the less demanding, more highly integrated field-effect technology opened up an opportunity that during the 1980s was taken up by dozens of projects. Why not adopt field-effect technology, benefit from its maturity, ease of fabrication, and economies of scale, and try to compensate for the relative slowness of individual field-effect chips by configuring large numbers of them in highly parallel architectures? As one important early paper put it: "The premise is that current LSI [large-scale integration] technology would allow a computational facility to be built around a large-scale array of microprocessors. . . . We anticipate that individual microprocessors would use a technology with intermediate values of gate speed and gate density to keep costs low. Therefore, the individual microprocessor circuitry is likely to be of only moderate speed. Total processing speed and throughput for the entire system would be obtained through parallelism. The network itself might contain as many as $2^{14} = 16,384$ microprocessors to obtain a very high degree of parallelism."[38]

Hence the 1980s explosion of parallel architectures, such as the various "hypercubes" and the Connection Machine.[39] Until the very end of the 1980s, these did not claim to rival mainstream supercomputing in absolute floating-point performance, promising instead a superior price-performance ratio. However, by the start of the 1990s, with the most advanced field-effect chips (such as the million-transistor Intel i860) being claimed to offer on a single chip a floating-point processing performance approaching that of a 1976 Cray 1, rivalry in absolute performance was growing.

The Computational Demands of Nuclear Weapons Design

Before we turn to the impact of the Los Alamos and Livermore National Laboratories on these processes of technical development, it is necessary to examine the computational demands of what has been their central task: the designing of nuclear weapons. Although Los Alamos, more than Livermore, has diversified into civil science and technology, at Los Alamos the "weapons people" still used "60% of supercomputer cycles" at the end of the 1980s, according to head of computing Norman Morse.[40]

The dominant feature of the computational demands of nuclear weapons design is their sheer magnitude. The explosion of an atomic or a hydrogen bomb is a complex event. Nuclear, thermodynamic, and hydrodynamic processes interact within a physical structure that may have a far-from-simple in shape and which may contain as many as 4000 components[41]—a structure that, moreover, is subject to catastrophic destruction as the processes continue. The processes unfold very rapidly. The scientists at Los Alamos invented their own unit of time during the original Manhattan Project: the "shake," a hundred millionth of a second. It was "supposedly given this name because it was 'faster than a shake of a lamb's tail.'"[42]

The temperatures (several hundred million degrees) and pressures (10^{12} atmospheres) involved in nuclear explosions are obviously hard to reproduce by any other means. Thus, knowledge of the processes of a nuclear explosion has been seen as obtainable in essentially only two ways: by constructing and exploding a nuclear device and attempting as far as possible to measure what goes on, or by constructing from physical first principles a model of the processes.

The first path cannot yield knowledge of an as-yet-unconstructed device and thus cannot resolve, ahead of time, the sort of disputes about feasibility that took place over the hydrogen bomb.[43] Also, the speed and the destructive power of a nuclear explosion plainly limit the amount of monitoring of the processes that is possible. Furthermore, nuclear testing is expensive and is increasingly subject to legal and political constraints.

The difficulties in the way of the second path are primarily computational. First-principles theoretical knowledge of the physical processes involved is held to be reasonably good, certainly in recent decades. But the resultant equations are susceptible of no analytical solution; interactions and nonlinearities abound. Computation, in massive quantities, is needed to move from the basic equations to a model that can inform design and can be calibrated against the results of nuclear testing.

It is therefore necessary to "compute" as well as to "shoot," as those involved put it.[44] This does not imply an absolute requirement for a digital computer. The first atomic bombs were designed and built without any such assistance, though both desk calculators and then IBM punched card machines were used[45]; the first Soviet and Chinese[46] bombs were likewise designed without digital computers.

But the demands upon nuclear weapons designers inexorably grew. The shift from the atomic to the hydrogen bomb brought a great increase in the complexity of the physical processes involved. And whereas with the first bombs little mattered other than that a substantial nuclear explosion took place, soon that was not good enough. "Improvement" was necessary—in yield-to-weight ratio, in yield-to-diameter ratio, in proportion of energy released as prompt radiation, and in safety and security, to name but five parameters of particular importance.

The search for change has not been due entirely to demands from the military, at least in the United States. Indeed, in several cases it seems as if the military have needed to be persuaded that developments were necessary and/or feasible.[47] As one person involved put it, "in most cases it's technology push rather than employment demand" that generates innovation.[48] Furthermore, the existence in the United States of *two* organizations responsible for designing nuclear weapons—Los Alamos and Livermore—generates competition. Neither laboratory can afford not to press the state of the art, for fear of being overtaken by the other.

Continuous pressure over more than four decades to enhance what is in a sense "the same technology" has led to what many of the individuals involved clearly perceive as diminishing returns. "Improvements" are still seen as possible, but their size has diminished. The weapons of the 1970s improved on their 1960s predecessors by a factor of about 2. The improvement from the 1970s to the 1980s was smaller. By the later 1980s a 10 percent improvement was hard to come by, though that percentage still was still significant and would have given an edge in interlaboratory competition.[49]

The laboratories were unable to "shoot" their way to more sophisticated nuclear weapons, since as time went on they were able to conduct fewer and fewer tests explosions. Numbers of nuclear weapons tests have fallen. In the 1960s the United States routinely conducted about 40 test explosions per year; however, after 1971 the annual total never exceeded 20, and in the early 1990s U.S. testing ceased altogether.[50] While weapons designers at Los Alamos and Livermore defended testing as necessary, they increasingly saw it as a means of validating computer models rather than an independent, self-sufficient source of knowledge. Enormous effort went into the development of what those involved refer to as the "codes": computer programs to assist in the designing of weapons.

What the codes mean in terms of the labs' demand for computer power can be seen by considering one of the two main types of computational process found in them: the mesh problem. This involves modeling the evolution through time of a physical quantity or a set of interrelated physical quantities in a region of space. The behavior of the quantity or quantities is understood to be governed by a partial differential equation or equations, but nonlinearities prevent these being solved analytically. So a numerical solution is attempted by superimposing a mesh of subdivisions in the relevant space, transforming the relevant partial differential equations into finite difference equations, and calculating for a series of time steps the changing values of the physical quantities for all the points in the mesh. This method predated the atomic bomb project (the classical discussion of it dates from 1928)[51]; however, it was not "put to use in practical problems" until the Second World War, particularly in problems at Los Alamos involving "the calculation of certain time-dependent fluid flows."[52]

Even 50 years ago, adequate precision in the solution of physical problems in this way was seen as requiring the breaking up of a linear dimension into 50–100 subdivisions.[53] Moving to two dimensions implies a mesh with at least $50 \times 50 = 2500$ cells, and possibly as many as $100 \times 100 = 10,000$ cells. Three dimensions takes us to a mesh of 125,000–1,000,000 cells.

This problem will be found in any field where equations have to be solved numerically over a two-dimensional or, worse, a three-dimensional space. Nuclear weapons design adds a further twist in the number of physical variables that have to be solved for simultaneously. An early-1960s weapons-design code sought to compute around 15 quantities per cell; a modern one seeks to compute 200 or 300. The requisite calculations for one time step for one cell might amount to 200 floating-point operations for the early-1960s code and 20,000 for the modern one.[54]

An early-1960s code, if it employed a 2500-cell two-dimensional mesh, would thus require a memory size of at least 37,500 words to store all the values of the variables. A single time step of the model would require half a million floating-point operations. A current code, if used with a 125,000-cell three-dimensional mesh, would require at least 25 million words of memory, and a single time step would require 2500 million floating-point operations. Even a late-1980s supercomputer, operating at around a gigaflop, would take $2\frac{1}{2}$ seconds to advance such a code through a single time step. Increasing sensitivity by going from 50 to 100 subdivisions of each linear dimension would increase the demands on memory size and processing speed by a factor of 8.

The scaling properties of this sort of computational problem show how easily increased computer speed can be absorbed. In the 1940s and the 1950s most of the hydrodynamics modeling done at Los Alamos used "only a single space variable, either spherical symmetry or the symmetry of an infinite cylinder."[55] Not until the late 1980s did Livermore computer specialists feel that enough computer power was becoming available for a move from two-dimensional to three-dimensional modeling.[56] Even on the supercomputers of the late 1980s, a single run of a weapons-design code could take 2 or 3 hours,[57] and 100 hours was not unheard of.[58] Memory size, as well as arithmetic speed, has also been a persistent constraint. Given that no one wanted data moving between main memory and peripheral storage with every iteration of a model, limited memory capacity was an issue even with the million-word memory of the Cray 1.[59]

Though mesh computation vividly illuminates the roots of the laboratories' apparent insatiable demand for computer power, it at least has the characteristic of computational predictability, with relatively few data-dependent branches in the program. Quite the opposite is true of the other major type of computational problem of nuclear weapons design: Monte Carlo simulation. In contrast with the deterministic mesh model, this is a probabilistic technique, developed at Los Alamos by John von Neumann on the basis of a suggestion by Stanislaw Ulam, for the analysis of problems such as the development of a nuclear chain reaction.[60]

High-precision Monte Carlo modeling makes heavy computational demands. Three-hour supercomputer runs are common.[61] It is, however, the *nature* of the computation, with its large number of conditional branches, that is particularly important. Up to 30 percent of the instructions in a Monte Carlo program may be branches.[62]

The magnitude of the computational demands of nuclear weapons design is not a clear imperative. Among exponential changes in computational demands and capacities, one parameter has remained close to constant: run time.[63] The mundane and familiar effects of the rhythms of the working week, and the demands of several different research and development groups sharing a central computer resource, are presumably the cause. Even in the late 1980s, large-scale simulations were executed as batch jobs overnight or on weekends, the working day being reserved for short jobs, code development, and so on.[64] So there is pressure to choose computational complexities such that, with the hardware available, a simulation can be run in a convenient time slot, such as overnight.

There are also some differences in style between the two laboratories. Livermore codes are more computationally intensive than Los Alamos ones.[65] At Los Alamos, in the late 1980s, the design of a single nuclear weapon was reckoned to consume about 1000 hours of Cray CPU time; the figure at Livermore would have been significantly larger.[66] But these differences pale in comparison with the two labs' similarity in posing the most extreme demands on computer speed, and it is to the consequences of this that I now turn.

The Early Years: Los Alamos and the Beginnings of the Computer Age

The early years of computing at Los Alamos (the 1940s and the early 1950s) have been relatively well documented in the literature[67] and need be recapped only briefly here. The Livermore Laboratory, established in September 1952, becomes relevant only at the very end of this period.

The first program run on the ENIAC was run for Los Alamos scientists.[68] Even before the atomic bomb was successfully constructed, Edward Teller was pushing research work on the hydrogen bomb. "The more complex calculations of hydrogen-bomb simulation exceeded the capabilities of the punched-card machine operation" used for atomic bomb design,[69] and in 1945 von Neumann arranged for Stanley Frankel and Nicholas Metropolis of Los Alamos to use the new electronic computer to run the hydrogen bomb simulation. One million IBM cards carried the requisite initial values, one card for each point in the computational mesh, and "the computations to be performed required the punching of intermediate output cards which were then resubmitted as input."[70]

The Los Alamos scientists used the ENIAC again for the computationally complex "liquid drop fission model" and other work. They also used the IBM SSEC in New York, the SEAC at the National Bureau of Standards in Washington, and the UNIVAC 1 machines at New York University and in Philadelphia.[71] Metropolis even significantly modified the ENIAC, contributing a key idea to the attempt to convert it into "a limited stored-program mode of operation instead of its gigantic plugboard mode."[72] The first major program run on the Institute for Advanced Study machine at Princeton in 1952 was a hydrogen- bomb simulation.[73]

It is not surprising that Los Alamos wanted its own digital computer and, with all this experience, felt confident enough to design and build one. The MANIAC (Mathematical Analyzer, Numerical Integrator, and Computer) was begun in 1948 and completed in 1952. Though modeled on the Princeton machine, MANIAC diverged in detail, notably to avoid problems encountered in developing the Princeton machine's memory.[74] In 1957 it was succeeded by MANIAC II, chiefly designed, like the original MANIAC, by Nicholas Metropolis.[75] MANIAC II is perhaps most noteworthy for an associated software development: the 1958 Madcap programming language. Unusually, the symbols in a line of Madcap code did not need all to be on the line. Subscripts and binary coefficients were permitted in code that closely resembled ordinary mathematics.[76]

The wider influence of Los Alamos was perhaps of greater significance than the machines used and built by the scientists who worked there. The Manhattan Project involved an unprecedented scale of the use of numerical modeling as a research and development tool. It also demonstrated the time and effort needed to do that modeling with existing technology. As scientists and engineers from the project "dispersed to laboratories, universities, companies, and government agencies after the war . . . they provided . . . a receptive climate for the introduction of electronic computing."[77] Here the key individual was John von Neumann, who moved between Los Alamos, the early computer projects, the Institute for Advanced Study, and IBM. His Los Alamos experience may have led von Neumann to doubt the practicality, with then-existing technology, of parallelism (other than in the limited form of bit-parallelism) in computer design:

In March or April 1944, [von Neumann] spent two weeks working in the punched-card machine operation [at Los Alamos], pushing cards through the various machines, learning how to wire plugboards and design card layouts, and becoming thoroughly familiar with the machine operations. He found wiring the tabulator plugboards particularly frustrating; the tabulator could perform parallel operations on separate counters, and wiring the tabulator plugboard to carry out parallel computation involved taking into account the relative timing of the parallel operations. He later told us this experience led him to reject parallel computations in electronic computers and in his design of the single-address instruction code where parallel handling of operands was guaranteed not to occur.[78]

Also, in this period Los Alamos facilitated IBM's move into digital computing. IBM entered the digital computer business slowly, and enthusiasts within the corporation for the new technology had actively to seek grounds for making the move. They turned to defense research and development, rather than commercial computing, for evidence of a market. The case for the move into stored-program digital computing was framed as a "special undertaking in support of the [Korean] war effort, an interpretation artfully emphasized in the name chosen soon afterward for the [IBM 701] machine: the Defense Calculator."[79] Los Alamos was only one of several defense R&D organizations whose demand for digital computing legitimated this epoch-making decision, but it was the first external organization to receive a 701, at the end of March 1953.

The situation of the Livermore Laboratory in relation to digital computing in the early years was of course quite different from that of Los Alamos. By 1952, when Livermore was established, it was becoming possible to buy, rather than have to build, a digital computer. The new laboratory bought a UNIVAC 1 from Remington Rand. The machine that was installed in April 1953 already had a place in computer history, having been used to predict on television the outcome of the 1952 presidential election.[80]

Though Livermore continued to purchase computers,[81] buying an IBM 701, four IBM 704s, and so on,[82] it was not content simply to buy what computer manufacturers chose to produce for sale. Livermore's first major active intervention in the process of computer development can, indeed, be seen as the beginning of supercomputing.

The Laboratories as Sponsors and Customers for Supercomputing

Livermore's role is enshrined in the very name of the first computer I am defining as a supercomputer. LARC was the acronym of the Livermore Automatic Research Computer.[83] The project was initiated from the highest levels at Livermore, by the lab's founder Edward Teller and by its Director of Computing, Sidney Fernbach. One inspiration was von Neumann, who at the end of 1954 had decided that it was desirable to push the computer industry toward speed by writing "specifications simply calling for the most advanced machine that is possible in the present state of the art."[84] The weapons designers at Livermore estimated that "they would need a system having one hundred times the computing power of any existing system."[85]

Teller and Fernbach sought bids from both IBM and Remington Rand for such a machine, requiring that it employ transistors, not tubes. "Teller was convinced that future machines should use transistors instead of vacuum tubes, so the use of transistors became an important requirement of the proposal."[86] Fierce conflict between Remington Rand's two computer operations (one in Philadelphia and one in St. Paul) was resolved with the decision that the former should bid, and there followed ten days of "heroic and frenzied effort to get a proposal together."[87]

The Remington Rand bid was accepted, and there followed intensive negotiations between Livermore and the company on the detailed specifications. These the machine ultimately met,[88] but the process of designing and building it was protracted and painful, and the final development cost of $19 million far exceeded the bid price of $2,850,000.[89] Nor was it, by the time it was ready, clearly a supercomputer in the sense of standing out in terms of speed from the other machines of the day. It was only around twice as fast as IBM's new transistorized 7090.[90] So the LARC had only a marginal speed advantage over a machine that was a commercial product, and while Livermore bought only the one LARC it had contracted for, it bought three IBM 7090s.[91] Only two LARCs were ever built; the other went to the U.S. Navy's ship and reactor designers at the David Taylor Model Basin.[92]

IBM had also bid on the LARC specifications but had simultaneously indicated its desire to renegotiate the specification to a more ambitious design with a clock speed of a 100 nanoseconds rather than the 500 nanoseconds envisaged for LARC. That plan became the Stretch project, whose goal was "a computer system operating 100 times faster than today's fastest machines."[93]

Stretch embodied at least three tensions. One, reflected in the ambivalent bid for LARC, was between the IBM tradition of technical conservatism (as reflected in its avoidance of publicly taking on dangerously overambitious tasks) and the fear that unless IBM "stretched" the technology of semiconductor components it might be left behind by those who did. Another tension arose from the desire to transcend the business/scientific dichotomy in computer design. In the case of Stretch, this took the form of attempting simultaneously to meet the demands of the nuclear weapons laboratories and those of the cryptanalysts at the National Security Agency. Finally, there was reportedly an internal divide in the large team that designed Stretch, with separate groups that did not communicate well responsible for designing the hardware and constructing the instruction set.[94]

With Livermore committed to Sperry Rand for the LARC, IBM's obvious target for Stretch was Los Alamos. The company offered the laboratory a "good deal"[95]: a supercomputer significantly faster than LARC at below cost. It also offered partnership in the design, not just in order to gain access to the technical expertise at Los Alamos but also as a way of avoiding possible antitrust legal difficulties involved in selling a machine at a price that was known in advance to be below cost. Eight members of the Los Alamos staff worked full time on the Stretch design.[96]

Like LARC, Stretch was a financial disaster. Unlike LARC, it did not meet its ambitious performance specifications, even though it was later seen as successful in the sense that many of the technical innovations used in the IBM System/360 flowed from it.[97] Livermore sought, unsuccessfully, to cancel the order it had placed for a Stretch.[98] A particular problem, from Livermore's point of view, was that Monte Carlo code, with its many conditional branches, defeated Stretch's instruction pipelining: while Stretch ought to have been 240 times as fast as the IBM 704 on such code, it was actually only 11 times as fast.[99]

On the other hand, Stretch's large memory permitted the data required for a two-dimensional weapon-design code to be held in main memory for the first time.[100] The overall increase in efficiency in handling weapons codes meant that weapons could be redesigned and retested during the last series of American atmospheric nuclear tests (Operation Dominic, April–November 1962).[101] Stretch was also the preferred machine of the French Atomic Energy Commission laboratories at Saclay and the British Atomic Weapons Research Establishment at Aldermaston.

In all, four of the eight Stretch computers that were sold went for nuclear research and development. (This figure that indicates how important nuclear weapons design was in the market for supercomputing at the time.) Two more went for other forms of military research and development (at the MITRE Corporation and Dahlgren Naval Proving Ground), one to the National Security Agency, and one to the U.S. Weather Bureau.[102]

The Livermore Laboratory played a crucial role in making possible the next-generation supercomputer, Seymour Cray's Control Data 6600. Control Data was a relatively young company, and the development costs of the 6600 threatened to overtax its financial resources. Desiring the machine, and "unwilling to see the excellent team of people dispersed," the laboratory stepped in and purchased a 6600 "while it was still a small bit of junk." It was, according to the individual just quoted, "a year and a half before the computer became semi-alive." Carl Hausmann, a leading Livermore weapons designer who enjoyed good contacts on the powerful Congressional Joint Committee on Atomic Energy, "bankrolled the operation," while Sidney Fernbach "made it happen" and the influential Edward Teller mobilized high-level political support.[103]

The direct sponsorship of supercomputing by the national laboratories became more difficult when more restrictive government purchasing regulations were introduced in the early 1970s. The replacement of the Atomic Energy Commission by the Department of Energy in 1974, and the demise of Joint Committee on Atomic Energy in 1976 diluted the laboratories' access to power in Washington. Simultaneously, the laboratories' share in the supercomputer market declined. Their purchases increased in absolute number but decreased relative to expanding the overall market. Jointly, the Los Alamos and Livermore Laboratories have bought around ten of their chosen machine from each generation of supercomputer: the Control Data 6600 and 7600, the Cray 1, and the Cray X-MP.[104] Additional purchases have been made by Aldermaston and by the Sandia National Laboratory (an American lab whose primary task is to integrate nuclear weapons and delivery systems). But by mid 1988 no fewer than 147 of the various variants of the Cray X-MP had been installed.[105] Worldwide annual revenues from sales of supercomputers were $967 million in 1988—up from $323 million in 1984.[106]

Thus, the nuclear weapons laboratories are no longer anything like half of the supercomputer market, as they had been for Stretch. Nevertheless, they remain in an important sense prime customers, and they are understood as such by vendors. Cray Research lent the first Cray 1 to Los Alamos for six months free of charge, because the firm was anxious to have the machine accredited there, but by then government regulations prohibited its purchase prior to accreditation.[107] In the late 1980s IBM had a team working at Los Alamos developing an understanding of supercomputing needs there,[108] and Los Alamos explicitly saw its role as being to "encourage the development of the next-generation supercomputer."[109]

The laboratories are discriminating purchasers of supercomputers; they do not simply buy "the latest machine." Neither Los Alamos nor Livermore used a Cray 2 in weapons-design work, preferring Cray Research's X-MP and then its Y-MP. Neither bought a Cyber 205 or an ETA[10], and this may well have contributed to Control Data's demise as a supercomputer supplier. Neither bought one of the Japanese super-computers, though here questions of nationalism and protectionism come into play as well as the questions of the suitability of particular machines for the laboratories' computational tasks.

Thus, the labs, as large, highly visible, and discriminating purchasers, retain an influence on the development of mainstream supercomputing. Computer staffs at Livermore and Los Alamos agree, however, that their influence is declining as the supercomputer market expands and as vendors must concern themselves with a wider range of customers.[110] Furthermore, the laboratories only slowly became important customers for the more massively parallel architectures described above. In the late 1980s Los Alamos bought a hypercube from Intel and one from Floating Point Systems, and also a Connection Machine,[111] but these systems were seen as experimental devices rather than computational workhorses.

While the laboratories' primary role in supercomputing has been that of "customers" since LARC and Stretch, they have also continued to commission supercomputing technology and even to seek to develop it themselves. As far as entire supercomputers rather than system components are concerned, the two main episodes were those of the STAR-100 and the S-1.

The STAR-100 episode was pivotal because it secured the commitment of the laboratories to mainstream supercomputing rather than to the more massively parallel alternatives. In 1964, Daniel Slotnick, his Department of Defense funding coming to an end, offered to build his highly parallel SOLOMON machine for Livermore.[112] Computer specialists there were enthusiastic. The SOLOMON structure was designed explicitly to handle iterative mesh problems of the kind that are so important to the laboratories. Though there were misgivings about the programmability of the novel architecture, Livermore staff members encouraged Slotnick to move from the original fixed-point SOLOMON design to a floating-point SOLOMON 2 design.[113]

Sidney Fernbach was, however, unable to persuade the Atomic Energy Commission to fund SOLOMON development.[114] Proposals were instead sought for a machine with a "new and somewhat radical structure,"[115] and this was done on the basis of the Commission's agreeing to lease the machine once it was developed rather than directly supporting its development.[116] Three proposals were entered, and lengthy negotiations ensued. One proposal, from IBM, is described by a Livermore interviewee as a "non-bid"[117]; it perhaps signaled IBM's retreat from the supercomputer market. Slotnick's employer, Westinghouse, as envisaged, entered the SOLOMON 2 design.

Although "everyone at the Lab felt that [Westinghouse] was the best bid,"[118] Slotnick could not persuade Westinghouse to take on the financial commitment demanded by the terms of the competition. That was understandable, since the design was a novel, technologically radical one, and Westinghouse would have had to commit itself to an R&D investment that might well not have been recoupable by leasing the final machine. Slotnick's development group was disbanded by Westinghouse. He resigned and sought venture capital to continue with the project, but he was not successful.[119]

So the competition was won by a design from Control Data for the machine that became known as the STAR-100. As was noted above, this was the first proposal for a vector computer. "STAR" referred to the STrings of binary digits used to carry information about, and sometimes to manipulate, ARrays of data[120]; the "100" referred to the 100-megaflop-per-second performance goal.

The STAR-100 was unquestionably an influential machine. Its architecture was the basis of Control Data's later Cyber 205 vector supercomputer. It was the first supercomputer to use integrated circuits and the first to have a million-word memory.[121] It was also an intensely problematic machine. Learning how to use its novel architecture proved traumatic. "For seven goddamn years we didn't do any physics while we worked out how to get that machine to work," said one exasperated member of the Livermore Laboratory.[122] Los Alamos refused to buy a STAR-100, and the decision that Livermore would purchase a second one had seriously detrimental consequences for Sidney Fernbach's career there.

The S-1 was more of an "in-house" development at Livermore. The originator of the S-1 project was Lowell Wood, the head of a special section at Livermore (known as the O Group) that was not tied to "routine" weapons design. In the early 1970s, Wood's widening involvement in the Department of Defense made him aware that other defense systems made far less use of computer technology than did nuclear weapons design. For example, massive arrays of hydrophones had been placed on the continental shelves around the United States, and in other strategic areas such as between the United Kingdom and Greenland, to detect hostile submarines.[123] But data analysis was lagging badly behind this data-collection effort, and with defense spending in a slump and supercomputers costing around $10 million apiece it was likely to continue to do so.

Wood had unique resources that enabled him to embark on the ambitious project of setting out to design from scratch a series of supercomputers intended to be significantly cheaper than conventional machines. A close associate of Teller, he knew how to win high-level political support in Washington, even after the abolition of the Joint Committee on Atomic Energy. Through the Hertz Foundation he had access to a stream of exceptionally talented graduate students.[124] Two such Hertz Fellows, Thomas M. McWilliams and L. Curtis Widdoes, Jr., who arrived at Livermore in 1975, were assigned by Wood the task of designing the supercomputer system, christened S-1.[125]

Their design was extremely ambitious by mid-1970s standards. It was for a MIMD architecture with 16 pipelined vector supercomputer central processors, each equivalent in power to a Cray 1, connected to 16 memory banks through a crossbar switch.[126] The plan was to retain this architecture through several generations of S-1 while making use of developing semiconductor component technology to miniaturize it, ending with an S-1 Mark V—a "supercomputer on a wafer."[127]

Though a considerable amount of prototype hardware was built, the project never fully realized its ambitious goals. This did not surprise staffers at the Livermore Computer Center, who were skeptical to the point of hostility to the project.[128] Its main product was the computerized design method, developed by McWilliams and Widdoes, that enabled them to design the original S-1 Mark I with remarkable speed: SCALD (Structured Computer-Aided Logic Design). McWilliams and Widdoes left Livermore to set up their own company, Valid Logic Systems, Inc., to market SCALD. By 1984 the firm was worth $150 million.[129]

The Influence of the Laboratories on the Development of Supercomputer Architecture

Given this very considerable involvement of the National Laboratories as sponsors and customers for supercomputers, can we go on to conclude that their particular computational requirements have shaped computer architecture? In one sense, of course, this is a wholly meaningless question. All the computers we have been discussing are general-purpose machines, and, in the famous words of Alan Turing, "This special property of digital computers, that they can mimic any discrete state machine, is described by saying that they are *universal* machines. The existence of machines with this property has the important consequence that, considerations of speed apart, it is unnecessary to design various new machines to do various computing processes. They can all be done with one digital computer, suitably programmed for each case.

It will be seen that as a consequence of this all digital computers are in a sense equivalent."[130]

The catch is in Turing's qualification, "considerations of speed apart." In supercomputing, where speed is of the essence, architectures can be shaped with particular computational tasks in mind. An example is the architecture of the Texas Instruments Advanced Scientific Computer (TI ASC), a vector supercomputer almost contemporaneous with the STAR-100. Texas Instruments, originally a supplier of instrumentation to the oil industry, had designed the TI ASC with the computation needs of oil exploration geophysics directly in mind: "A significant feature of this type of processing is the frequent use of triple-nested indexing loops, and an important characteristic of the ASC is the provision of three levels of indexing within a single vector instruction."[131]

Even when architecture is not shaped by explicit goals (as it was with the TI ASC), the institutional circumstances of computer design can leave their mark on it. Although the accuracy of the imputation is uncertain, the following quotation from Tracy Kidder's book *The Soul of a New Machine* captures what I mean: "Looking into the [architecture] of the VAX, [Data General Corporation computer designer Tom] West had imagined he saw a diagram of DEC's corporate organization. He felt that VAX was too complicated. He did not like, for instance, the system by which various parts of the machine communicated with each other; for his taste, there was too much protocol involved. He decided that VAX embodied flaws in DEC's corporate organization. The machine expressed that phenomenally successful company's cautious, bureaucratic style."[132]

Stretch exemplifies how the circumstances of a project can have unintended effects on its technical design. The project was formulated in an "almost pathological atmosphere of optimism—and its corollary, fear of being left behind"[133]; as outlined above, the designers were also trying to satisfy the needs of quite different kinds of users. The result was an extraordinarily complex instruction set: "The 'stretch' principle that infected planners made it easier to accept than reject ideas, perhaps especially so because they were in no position to assess accurately the direct and indirect costs of each embellishment."[134]

My main concern here, however, is with the more deliberate kind of influence. The laboratories have perceived themselves as having particular needs, and have been perceived the same way by suppliers. Thus, the IBM proposal to the Atomic Energy Commission for Stretch stated that "the general design criteria for this computer include: suitability and ease of use for atomic energy computing problems."[135] As mentioned above, IBM and Los Alamos staffers collaborated in the detailed design of Stretch, while Livermore and Remington Rand staff worked together to draw up detailed specifications for LARC. Such explicit channels for influence did not exist for Seymour Cray's supercomputers, but informal channels did. Livermore's Sidney Fernbach "was reportedly one of the few people in the world from whom Seymour Cray would accept suggestions,"[136] and Cray took care to become aware of the laboratories' computational needs. He visited Livermore to ask architecturally key questions such as the frequency of branches in the code used there.[137]

What are the architectural consequences of this kind of influence? One is the Stretch machine's "noisy mode" facility.[138] The inspiration for this facility came from Los Alamos's Nicholas Metropolis, who in the 1950s developed what he called "significance arithmetic": the attempt to determine the consequences, for the reliability of results, of errors caused by the need to represent numbers by words of finite length. In "noisy mode" the effects of truncation were handled differently than in normal operation so as to allow errors caused by truncation to be detected.[139]

Los Alamos was also able to make sure that the Stretch instruction set contained "great debugging tools" and "lots of great instructions useful for the guy coding in machine language." Los Alamos computer specialists were worried about what they saw as the inefficiency of the new high-level languages (such as Fortran), and in the late 1950s much Los Alamos code was still machine code. Even in the 1980s the computationally intensive inner loops in weapons codes were still sometimes "hand-tailored."[140]

However, Stretch's instruction set was particularly open to influence, and other instances of successful, specific intervention by the laboratories in the details of design are harder to find. In the 1980s, for example, Los Alamos was unable to persuade Cray Research to provide as a design feature in Cray supercomputers what Los Alamos would like to see in the way of hardware devices to assist debugging.[141] On the other hand, Livermore computer specialists influenced the handling of zeros in the Control Data 6600. They pressed successfully for a "normalized zero," in which a register is cleared completely if the significand of the number represented in it consists only of zeros, even though there are ones in the exponent. Their view was that without this feature, which Cray was not originally going to provide, significant errors would be introduced in hydrodynamic calculations important to their work.[142]

Livermore staffers also believe that it was the laboratories' needs for fast Monte Carlo simulation that led Cray Research to provide special facilities in the Cray X-MP/4 for the vector processing operations known as "gather" and "scatter."[143] Frank McMahon and other computer scientists at Livermore persuaded Seymour Cray to add to the instruction set of the Cray 2 a related instruction called "compress iota" to assist the vectorization of loops containing IF statements.[144]

Matters such as "gather/scatter" and "compress iota" concern specific, detailed modifications to preexisting architectures. Of the six major developments in supercomputer architecture reviewed above, the most plausible candidate for identification as a case of direct influence from the computational needs of the laboratories is vector processing. The STAR-100, the central machine in the early evolution of vector processing, came into being in response to a Livermore request, and its design was optimized for the handling of long vectors, which were "common to many scientific problems at the Lawrence Livermore Laboratory."[145]

There is, however, a striking paradox here. The STAR-100 in this sense represents the peak of the laboratories' influence on the development of supercomputer architecture. Yet the outcome, as we have seen, was a machine perceived at Livermore and Los Alamos as ill suited to their computational needs. Its offspring—the Cyber 205 and the ETA[10]—retained its distinctive optimization for long vectors, and were spurned by the laboratories.

How did this paradox—the disowning by the laboratories of their major legacy to supercomputer architecture—come about? The answer is that "computational needs" are neither simple nor self-evident. The STAR-100 was designed according to a particular vision of these "needs," a vision that ultimately could not be sustained.

That much of the laboratories' computational work is highly classified is relevant here. Without security clearance the individuals responsible for supercomputer designs (even those designs directly commissioned by the laboratories) cannot have access to actual weapons design codes, so they lack immediate contact with the "need" which they should be trying to satisfy.

The solution to this attempted with the STAR-100 was to declassify and pass to the designers segments of code that occupied a large proportion of run time. The segment chosen as a contract benchmark (figure 3) was a fragment of Livermore's main nuclear weapons design code of the 1960s, Coronet. This segment of code became Kernel 18 of the Livermore Loops (see below). However, this declassified sample was later

```
C
C
C
C
C
C
C********************************************************************
C*** KERNEL 18     2-D EXPLICIT HYDRODYNAMICS FRAGMENT
C********************************************************************
C
          DO 75  L= 1,Loop
              T= 0.0037
              S= 0.0041
             KN= 6
             JN= n
          DO 70  k= 2,KN
          DO 70  j= 2,JN
           ZA(j,k)= (ZP(j-1,k+1)+ZQ(j-1,k+1)-ZP(j-1,k)-ZQ(j-1,k))
       .            *(ZR(j,k)+ZR(j-1,k))/(ZM(j-1,k)+ZM(j-1,k+1))
           ZB(j,k)= (ZP(j-1,k)+ZQ(j-1,k)-ZP(j,k)-ZQ(j,k))
       .            *(ZR(j,k)+ZR(j,k-1))/(ZM(j,k)+ZM(j-1,k))
   70     CONTINUE
C
          DO 72  k= 2,KN
          DO 72  j= 2,JN
           ZU(j,k)= ZU(j,k)+S*(ZA(j,k)*(ZZ(j,k)-ZZ(j+1,k))
       .            -ZA(j-1,k) *(ZZ(j,k)-ZZ(j-1,k))
       .            -ZB(j,k)   *(ZZ(j,k)-ZZ(j,k-1))
       .            +ZB(j,k+1) *(ZZ(j,k)-ZZ(j,k+1)))
           ZV(j,k)= ZV(j,k)+S*(ZA(j,k)*(ZR(j,k)-ZR(j+1,k))
       .            -ZA(j-1,k) *(ZR(j,k)-ZR(j-1,k))
       .            -ZB(j,k)   *(ZR(j,k)-ZR(j,k-1))
       .            +ZB(j,k+1) *(ZR(j,k)-ZR(j,k+1)))
   72     CONTINUE
C
          DO 75  k= 2,KN
          DO 75  j= 2,JN
           ZR(j,k)= ZR(j,k)+T*ZU(j,k)
           ZZ(j,k)= ZZ(j,k)+T*ZV(j,k)
   75     CONTINUE
C
C.................
          CALL TEST(18)
C
C
C
C
C
C
C
C
C
```

Figure 3
Kernel 18 of the Livermore Loops, earlier the contract benchmark for the
STAR-100. Source: McMahon 1986, p. 44.

judged untypical of Livermore weapons code because it contained no
data-dependent branches. That made it too "easy" a test for a pipelined
vector computer such as the STAR-100. The STAR's designers could satisfy Livermore's "need" expressed in the declassified code (on the benchmark it was 7 times as fast as the CDC 7600, when the specification called
for it to be only 5 times as fast[146]), and yet the machine they produced was
successfully run on only twelve of the several hundred Livermore weapons
codes. "The STAR met our specifications, but not our expectations," was
how Livermore staff put it. "Everybody, especially at the Lab, was slow to
recognize the effect of branches on STAR performance. If there's a
branch, the [vector] pipeline has to be drained . . . but the only time STAR
was fast was when the vector units were running."[147]

There was, of course, nothing absolute about this "failure." Algorithms could have been redesigned, and codes rewritten, so as to make them more suitable to the architecture of the STAR-100. To a limited extent this did happen. As the years have gone by and vector machines have became the norm, the laboratories have learned how to vectorize even seemingly intractable problems of Monte Carlo simulation.[148]

But the central task of the laboratories introduces a specific difficulty in making the algorithm fit the architecture, just as classification causes problems in making the architecture fit the algorithm. Any weapons simulation involves approximations. Those embodied in existing weapons design codes have been "calibrated on"[149]—their empirical validity has been checked in nuclear weapons tests. To change algorithms radically would involve making use of new approximations, which would, in the opinion of Livermore interviewees, require test validation.

As we have seen, there are powerful constraints on numbers of nuclear tests. Thus, wholly new weapons-design codes are now rare. Designers have preferred to modify and improve mature codes rather than start again from scratch. We have here a specific reason for the reluctance to shift to radically new computer architectures, a reason over and above the pervasive "dusty deck" problem of heavy investment in existing codes.[150] There has thus been a strong source of architectural inertia in the laboratories' weapons design work, an inertia that may help to explain why the laboratories were not in the lead in pioneering or sponsoring new massively parallel computer architectures in the 1980s. Whereas evolutionary developments in mainstream supercomputing, such as the Cray X-MP and Y-MP series, were adopted readily, more radically parallel architectures were much harder to integrate into the labs' work of designing nuclear weapons.

Conclusion

What does it mean for an institution to have "influenced" the development of an area of technology? Perhaps the clearest way of thinking about this is to ask what would be different if the institution had not existed. Would the area of technology still exist? Would it have developed more slowly, or more rapidly? Would it have developed in a qualitatively different technical direction? Ultimately, of course, these questions are beyond empirical resolution. There is no alternative world, similar in all respects other than the absence of nuclear weapons laboratories, for us to examine.[151] At best, judgment is all we can bring to bear.

My judgment, based on the evidence I have reviewed here, is that without the weapons laboratories there would have been significantly less emphasis on floating-point-arithmetic speed as a criterion (in certain circumstances *the* criterion) of computer performance. Business users typically cared relatively little, at least until quite recently, for megaflops. Cryptanalysts (practitioners of an activity that tied computing almost as closely to state power as did nuclear weapons design) also wanted different things: the National Security Agency's emphasis, writes one of its chief computer specialists, "was on manipulation of large volumes of data and great flexibility and variety in non-numerical logical processes."[152] There were other people—particularly weather forecasters and some engineers and academic scientists—for whom floating-point speed was key, but they lacked the sheer concentrated purchasing clout, and perhaps the sense of direct connection to a mission of prime national importance, that the weapons laboratories possessed. Only since the early 1980s has a supercomputer market fully independent of its original core—Los Alamos and Livermore—come into being.

Without Los Alamos and Livermore we would doubtless have had a category of supercomputing—a class of high-performance computers— but the criterion of performance that would have evolved would have been much less clear cut. What we would mean by "supercomputer" would thus be subtly different.

Developments at the Livermore Laboratory were central to popularizing, from 1969 on, the megaflop as the appropriate measure of supercomputer performance. In the wider computer world, instructions performed per second was a widely quoted metric. But, especially with the advent of vector machines, that metric was of little use at Livermore—one STAR instruction could correspond to many floating-point operations. Francis H. McMahon, a member of the compiler group at Livermore in the late 1960s and the early 1970s, often addressed weapons designers at Livermore on the constraints placed on optimizing compilers by the way they formulated source code. He would give examples from Livermore of the vast differences in speed between "clean" code and "messy" code full of IF statements and the like. McMahon came to realize that it was possible to predict the speedup on full weapons codes gained from the introduction of a new-generation supercomputer by examining only speedup on these samples. Gradually the samples were codified as the Livermore Fortran Kernels, or Livermore Loops, and a statistical average of performance over them was used to define the megaflop rate of a given machine.[153]

Using the megaflop per second as a performance metric, and the Livermore Loops as the way of determining that rate, diffused well beyond Livermore in the 1970s and the 1980s. To the extent that computer designers shaped architectures to optimize their machine's performance on the Loops, an indirect Livermore influence on computer architecture thus continued even when Livermore's direct influence was declining. However, the megaflop escaped its creators' control. A variety of other means of determining megaflops emerged, and even when the Livermore Loops were used manufacturers tended to quote simply arithmetic mean speed (which was strongly influenced by the kernels on which the machine ran fast). In McMahon's opinion, machine performance would have been characterized better by reporting the megaflop rate between the harmonic mean (strongly influenced by the kernels on which a machine runs slowly) and the arithmetic mean.[154] "Lies, Damned Lies, and Benchmarks," wrote two exasperated technologists.[155]

The existence of the national laboratories played a major part in establishing floating-point performance as the criterion of supercomputer status, and Livermore, in particular, influenced how that floating-point performance was measured. Beyond this, however, is hard to specify any precise, major effect of the laboratories on supercomputer architecture. One reason for this is that the computational task of the laboratories, though it certainly falls within the general field of high-speed numerical computation, is diverse. If the laboratories did only large-scale mesh computations, with few conditional branches, then their impact would have been clear-cut. They would have fostered either the array processor (e.g. SOLOMON) or the long-vector supercomputer (e.g. STAR-100). But, as we have seen, the algorithms used in weapons design are by no means all of the mesh-computation kind. In particular, Monte Carlo code is quite differently structured, full of conditional branches.

Thus, it seems to have been impossible straightforwardly to optimize supercomputer architecture for the laboratories' computational task. The nearest attempt to do so, with the commissioning of the STAR-100, foundered on the diversity of this computational task and on the difficulty, in part caused by security classification, of formulating precisely what the laboratories' needs were.

The successful supplier to the laboratories, at least until recently, and the dominant force in the evolution of supercomputer architecture, has thus been Seymour Cray, who kept himself at some distance from the needs of particular users. He has listened to the laboratories, but he has also listened to the quite different demands of the National Security Agency, in whose original computer supplier, Engineering Research Associates,[156] he began his career. He is also, of course, a single-minded technical visionary.

Visionaries succeed, however, only to the extent to which they tailor their vision to the world, or tailor the world to their vision. If the above analysis is correct, Cray's success was based on the design of rather robust supercomputer architectures. Machines might exist that were seen as better than his for one particular type of algorithm (as the CYBER 205 was seen as surpassing the Cray 1 in the processing of long vectors). But none existed that met so well the perceived needs of both of the two major types of computational task at the weapons laboratories *and* the different tasks of the National Security Agency. And, in a fashion familiar in other areas of technology, success bred success. A wider set of users meant longer production runs, a more solid financial base, economies of scale, progress along the "learning curve," and, perhaps crucially, the development of a relatively large body of applications software. The last issue, as we have seen, is of particular importance at the weapons laboratories, because of the specific difficulty of radically rewriting weapons-design codes.

The robust Cray strategy for supercomputer development minimized the influence of particular users' needs on supercomputer architecture. Its success, and the gradual growth in the number and variety of supercomputer customers, has intensified this effect.[157] Cray Research, or Seymour Cray's new spinoff, the Cray Computer Corporation, now could not satisfy the particular needs of Los Alamos and Livermore if those needs conflicted with the needs of other users and if significant costs (financial or technological) were involved in meeting them. The anonymous—though not asocial—logic of the market has come to shape supercomputing.

The question why the evident general influence of the laboratories on supercomputing has not translated into major, durable, particular influence can thus be answered simply in two words: Seymour Cray. As we have seen, however, this answer was possible—Cray could appear to have a demiurgic role—only because there were limits to the extent to which the laboratories, or, more generally, the users of supercomputers, could define what they *specifically* needed.

It is perhaps appropriate to end with a speculation about the future, concerning the prospects for radically different computer architectures based upon parallelism more thoroughgoing and greater in degree than that evident in mainstream supercomputing. Here the laboratories have been and are a shaping force, but in an unintended fashion. Though first Los Alamos and then Livermore consciously sought to be in the forefront of novel architectural developments, the weight of the laboratories' presence was a factor tipping the scale toward evolutionary, incremental developments of computer architecture that would preserve the value of existing bodies of code and algorithms verified in nuclear testing.

This did not and will not decide the future of massive parallelism. The laboratories are not as important now as they were when the STAR-100 was, of necessity, selected rather than SOLOMON—an event that may well have been crucial in the array processor's exclusion from mainstream supercomputing. Yet the issue does have an interesting bearing on the role of the laboratories.

The end of the Cold War has already led to a considerable reduction in the support for and the significance of the laboratories' central activity: designing nuclear weapons. This is obviously a threat to them, but it is an opportunity as well—a chance to find a different, durable sense of purpose. In particular, a decline in the salience of the weapons-design codes permits a more thoroughgoing exploration and exploitation of novel computer architectures. This could be an important component of a new role for Los Alamos and Livermore.

Acknowledgments

The research upon which this article was based was supported by the U.K. Economic and Social Research Council under the Programme in Information and Communication Technologies and the Science Policy Support Group Initiative on Science Policy Research on Defence Science and Technology. I also wish to thank my "hosts" at Los Alamos and Livermore, Bill Spack and George A. Michael, and the following other current or past laboratory staff members who spoke to me about the laboratories' role in the development of computing: at Los Alamos, Norman Morse, Roger Lazarus, Robert Frank, Ira Agins, Kenneth Apt, Jim Jackson, Thurman Talley, Tom Dowler, and Nicholas Metropolis (the last-named by letter rather than in person); at Livermore, Sidney Fernbach, Roger E. Anderson, V. William Masson, Norman Hardy, Cecil E. Leith, Jr., Jack Russ, William A. Lokke, Francis H. McMahon, Tad Kishi, Harry Nelson, and Lowell Wood. I am also grateful to many of the above, to Bruno Latour, and to the referees of the *Annals of the History of Computing* for helpful comments on the original draft.

6

The Charismatic Engineer
(with Boelie Elzen)

The twentieth century's engineers have been anonymous figures. Few have captured public imagination like their nineteenth-century predecessors, their lives chronicled by admirers like Samuel Smiles.[1] Among the select few twentieth-century engineers whose names have become household words is Seymour Cray. "Cray" and "supercomputer" have become close to synonyms, and this verbal link is a barrier to other producers. When a film or a television program wishes to convey an image of "computer power," the most popular way of doing it is a picture of a Cray Research supercomputer, with its distinctive "love seat" design (figure 1)

Seymour Cray is a paradox. The prominence of his name makes him the most public of computer designers.[2] He is, simultaneously, the most private. Apart from very rare, strategic occasions, he (or, rather, a secretary acting on his behalf) steadfastly refuses interviews.[3] He restricts his very occasional "public" appearances to carefully selected audiences, usually made up largely of technical specialists from current or potential customers. These events are sometimes more like political rallies than scientific meetings, with Cray being greeted, like a party leader, by a standing ovation. Videotapes of these appearances circulate in the supercomputer community; they are the closest most members of that community, let alone a wider public, can get to the man.[4]

Cray's privacy is not that of an overwhelmingly shy or socially incompetent person. He is no archetypal computer nerd. The videotapes reveal a poised and witty man, a compelling public speaker, articulate within the deliberately low-key idiom of his native U.S. Midwest. A 1988 tape, for example, shows a fit, handsome Cray looking younger than his 63 years.

Around the privacy, the anecdotes proliferate. Cray has become a legend, a myth, a symbol. Tales (many no doubt apocryphal) of his doings and sayings are told and retold. Display boards of these sayings accompany the

Figure 1
Seymour Cray and the CRAY-1 Computer. Courtesy Cray Research, Inc.

exhibition devoted to Cray at Boston's Computer Museum. Rigorously rationed as they are, Cray's pronouncements take on exceptional significance. Again, the strategy of privacy has the consequence of public prominence, even fascination.

The Cray legend resonates with themes that are powerful in the American imagination: the lure of high technology, the individual against the organization, the country against the city. Both in itself and in the applications with which it is associated, the supercomputer is the epitome of the highest of high technology. For many years the United States unequivocally led the world in supercomputing. Because of the supercomputer's importance in the breaking of codes and in the designing of nuclear weapons, this lead has seemed an important foundation of American power.

Three times in his career (most recently in 1989) Cray has left the corporation for which he worked to strike out anew on his own. The first two times, at least, his venture was blessed by great success. Yet money has not taken to him to the corrupting city, nor does anyone imagine it is love of money that has driven Cray's work. He has eschewed the trappings of corporate success, preferring for most of his working life the quiet, rural surroundings of his home town of Chippewa Falls, Wisconsin.

When his startup company, Cray Research, went public in 1976, it had "no sales, no earnings, a $2.4 million deficit, and further losses looming." Yet the 600,000 shares of common stock it offered the securities market were snapped up "almost overnight," generating $10 million in capital.[5] As the years have gone by, Wall Street has come to apply more conventional criteria to Cray Research. Yet the appeal to the imagination persists.

Fourteen years on, in 1990, *Business Week* could still carry a front cover that captures the very essence of the legend of Cray. A color portrait depicting him as a rugged American individualist in an open-necked check shirt (certainly not a business suit), with hair scarcely touched with gray and with clear blue eyes looking resolutely into the future, is in the foreground. Behind this are an idyllic rural scene, with a small road winding through hills, and a computer-generated surface above which hovers a galactic spiral. Above is a simple, bold title: "The Genius."[6]

Charisma and Routinization

In Seymour Cray, then, we have an instance of a phenomenon little touched upon in social studies of technology—charisma (little touched upon, perhaps, because to some "charismatic engineer" embodies a contradiction). In the words of Max Weber, charisma is an "*extraordinary quality of a person,*" whether that person be prophet, warlord, or whatever.[7] For a sociologist, of course, charismatic authority inheres, not in the individual, but in the beliefs of others about that individual: charisma is the product of social relationships. This chapter will, therefore, inquire not into Cray's psyche (that is beyond both our competence and our data) but into the relationship between Cray and other actors. Agnostic on the question of whether Cray's unique style has psychological roots, we shall analyze it as a sociotechnical strategy, a way of constructing simultaneously both distinctive artifacts and distinctive social relations.

In seeing Cray as a "heterogeneous engineer" we are, of course, drawing on a theme that is important in the recent history and sociology of technology, notably in the work of Tom Hughes, Michel Callon, Bruno Latour, and John Law.[8] However, we shall also follow the central theme of Weber's discussion of charisma. Charismatic authority is an inherently transitory phenomenon. A network of social relationships can only temporarily express itself as the extraordinary characteristics of one person, because of human mortality if nothing else. If it is to develop and survive, other, more explicitly social, forms of expression must be found. As Weber wrote: "Just as revelation and the sword were the two extraordinary powers, so were they the two typical innovators. In typical fashion, however, both succumbed to routinization as soon as their work was done. . . . [R]ules in some form always come to govern. . . . The ruler's disciples, apostles, and followers became priests, feudal vassals and, above all, officials. . . ."[9] This "dialectic of charisma" is one of several patterns we detect in the history of supercomputing.[10]

Origins

The Chippewa River flows south through the woods of northwestern Wisconsin, eventually joining the Mississippi. On its banks grew the small town of Chippewa Falls. Seymour Cray was born there on September 28, 1925, the son of an engineer.[11] After military service as a radio operator and cryptographer, and a brief period at the University of Wisconsin at Madison, he studied electrical engineering and applied mathematics at the University of Minnesota, receiving a bachelor's degree in 1950 and a master's in 1951.

In 1950 he was recruited by Engineering Research Associates of St. Paul, Minnesota.[12] With its origins in wartime code breaking, ERA was one of the pioneers of digital computing in the United States, though the secrecy of cryptanalysis (its continuing primary market) meant that the firm's work was much less well known than, for example, that of J. Presper Eckert and John W. Mauchly in Philadelphia. In May 1952, however, ERA was sold to Remington Rand, which already owned Eckert-Mauchly, and in June 1955 Remington Rand merged with the Sperry Corporation to form Sperry Rand.

Little detail is known of Cray's work for ERA and Sperry Rand, though the young Cray quickly won considerable responsibility, notably for Sperry Rand's Naval Tactical Data System (NTDS) computer. He was thus already a figure of some importance to his first startup company, the Control Data Corporation (CDC), formed when Cray and eight others, most famously William C. Norris, left Sperry Rand in 1957. Cray was the chief designer of Control Data's first computer, the CDC 1604, announced in October 1959. Built with transistors rather than the previously pervasive vacuum tubes, the highly successful 1604 moved Control Data into profit and launched it on a path that was to enable it briefly to challenge IBM's dominance of the computer industry, a dominance that was already hardening by 1957.

Roots of the Cray Strategy

The NDTS, and especially the 1604, were considerable achievements, and secured Cray's growing reputation as a computer designer. Yet neither was the stuff of legend, nor—beyond the beginnings of anecdotes concerning his preference for simple designs and intolerance of those he considered fools[13]—is there much evidence of a distinctive Cray style in their development.

The origins of both legend and style, the earliest clear manifestation of what was to become Cray's distinctive sociotechnical strategy, can first be traced unequivocally in discussions within Control Data on what to do to follow the company's success with the 1604. The obvious step was to build directly on that success, offering an improved machine, but one compatible with the 1604 (so users of the latter could run their programs unaltered). While the 1604 had been oriented to the demands of "scientific" users, such as defense contractors and universities, there was a growing sense within Control Data of the need to orient at least equally to business data processing, where arithmetic speed was of less concern than the capacity to manipulate large data sets. Compatibility and business orientation were not necessarily at odds. By adding new instructions, specially tailored for commercial usage, to the instruction set of the 1604, Control Data could cater to business without sacrificing compatibility with the previous machine. This emerging strategy was perfectly sensible. It was indeed similar to, if less ambitious than, that to be announced in 1964 by IBM, with its famous System/360. This was a series of compatible machines, some oriented to the business and some to the scientific market, but all sharing the same basic architecture, and with an instruction set rich enough to serve both markets.

Cray, however, disagreed with all elements of the strategy—compatibility with the existing machine, orientation to the commercial as well as scientific market, a complex instruction set. His alternative strategy prioritized speed: in particular, speed at the "floating-point" arithmetic operations that were the dominant concern of defense and scientific users. In that prioritization, Cray did not wish to be constrained by choices made in the development of the 1604. Compatibility was to be sacrificed to speed. As one of his famous maxims has it, he likes to start the design of a new-generation machine with "a clean sheet of paper." He had no interest in business data processing, and abhorred the complexity that arose from trying to cater simultaneously to both scientific and business users.

The 1604 was making a lot of money for Control Data, and so it seemed possible to pursue both strategies simultaneously. One group of designers went on to develop a series of complex-instruction-set computers compatible with the 1604 (the Control Data 3600 series), with a primary orientation to the commercial market. A second group, led by Cray, set out to develop a machine that would prioritize speed.

Cray's status as the chief designer of the corporation's first and most successful computer, and the threat (possibly explicit) that he would leave,[14] enabled him to negotiate in 1961–62 a remarkable arrangement with Control Data chairman Norris. He was allowed to move, with the small team working on the 6600, a hundred miles away from Control Data's headquarters in Minneapolis-St. Paul, to a newly built laboratory on a plot of country land, owned by Cray personally and close to his house, in woods overlooking the Chippewa River. Cray thus won a remarkable degree of autonomy from corporate control. Even Norris had to seek Cray's permission to come to the Chippewa Falls laboratory, and Cray visited Control Data headquarters only every few months.

The technical and social aspects of Cray's strategy were tightly related. Chippewa-style isolation would not have been in harmony with successfully building a series of compatible, general purpose, computers. That required finding out the needs of different kinds of users, balancing one technical characteristic against another, giving attention to software as well as hardware, keeping different projects connected together, and harnessing all the different parts of a growing corporation to a common but diffuse set of tasks: "committee design."[15] By moving to Chippewa Falls, Cray created a geographical and social barrier between his team and all this negotiation and compromise. (Another reported motive for the move was Cray's fear of nuclear war: "I wanted to get out of the big city because I might get my head blown off."[16])

The instruction set of the computer designed at Chippewa Falls, the Control Data 6600, is emblematic of Cray's sociotechnical strategy. It contained only 64 instructions, at a time when 100 or more was common. When the attempt is being made to satisfy a variety of different user concerns, the easiest means of harmonization is to satisfy vested interests by adding instructions. The IBM Stretch computer, designed in the late 1950s, is an extreme example. An intensely ambitious project, intended to combine extreme speed with an attempt to straddle the scientific, cryptographic, and business markets, Stretch had no fewer than 735 instructions.[17] A simple instruction set for the 6600 permitted most of its instructions to have their own hardware support, tailor-made for speed.[18]

Striking though its overall design is, the 6600 by no means emerged, Athena-like, from the brain of Seymour Cray. The instruction-set simplicity of the 6600 became architectural complexity. For example, a sophisticated "scoreboard" unit had to be designed to keep independent hardware working harmoniously. Even Cray could not master all the details, so even within his small team a division of labor was needed. James Thornton took responsibility for much of the detailed design.

That kind of division of labor was not troublesome to Cray. In the isolation of Chippewa Falls, the team was a coherent one. Cray refused to be diverted by the few visitors allowed to come to the laboratory, and even family demands had to fit into their allotted place. Cray's children "remember that on long auto trips he demanded total silence, apparently to think through technical problems," and his wife, Verene, "worked hard to foster a sense of togetherness around Cray's obsessive work schedule." Family dinners were sacrosanct, but Cray would soon leave to return to his laboratory to work late into the night. "His eldest child, Susan Cray Borman, recalls leaving him questions about her algebra homework on his desk in the evening, knowing she would find answers waiting for her in the morning. 'It was like the elves had come,' she says."[19]

The combination of Cray's intense personal involvement and the laboratory's isolation lent coherence to the project: "A team spirit developed and carried over into sporting and recreational events in the community."[20] Developing the 6600, however, involved far more than the sociotechnical work of leading the team at Chippewa Falls. Cray did not attempt to develop the basic components for the machine; developing an innovative configuration or "architecture" for them was work enough. This placed his team on the horns of a dilemma. A conservative choice of components would reduce risks but might not give the speed that was necessary. The other Control Data employees working on the 3600 project were no slouches, despite the range of needs they were seeking to satisfy. To be justifiable within Control Data, much less to find a place in the market, the 6600 had to be a lot faster than the 3600. Components at the state of the art, or just beyond it, would give the edge in speed, but would place the fate of Cray's project in the hands of their developers, over whom he had no control.

Cray's preferred approach was conservative—"keep a decade behind" is one of his sayings on display at the Computer Museum—and his team began by trying to wring a 15- to 20-fold speed increase over the 1604 without a radical change in components. They found this impossible to achieve. Fortunately, a new silicon transistor, manufactured by Fairchild Semiconductor, appeared on the market in time to salvage the project, and design was begun again with that as its basis, though the speed goal of the delayed project had to be increased relative to the 3600 to make up for the lost time.

The problematic relationship between computer designer and component supplier is a theme that was to recur in the history of supercomputing. So is another issue that came to the fore in the development of the 6600. Like almost all other computers, the 6600's operations were synchronized by pulses in control circuitry; the intervals between those pulses was its "clock speed." The target clock speed for the 6600 was 100 nanoseconds (one ten-millionth of a second). In such a tiny interval of time, the finite speed of electrical signals became a constraint. If the wires were too long, a signal would not arrive at its destination within one cycle of the clock. So the circuitry of the 6600 had to be packaged very densely. The new silicon transistor gave a tenfold density improvement, but dense packaging meant intense heat. Cray grasped the centrality of what others might have considered a menial aspect of computer design and superintended the design of a special cooling system, with freon refrigerant circulating though pipes in the machine's structure to remove the heat.

To produce an artifact of the 6600's daunting complexity was no easy task. The wider computer industry had already started applying its own products in increasingly automated design and production systems. Cray took a step in the opposite direction. The most sophisticated computer of its day was, in effect, handcrafted. Cray was even reluctant to turn it over to the Control Data production facilities in the Minneapolis suburb of Arden Hills, and the first few 6600s were built in Chippewa Falls. Finally, the transition was successfully made, but it was no simple matter of handing over blueprints. The production process, and integrating the large network of suppliers whose parts went into the 6600, required the most careful attention, but not from Cray himself. His habit has been to delegate the task to others in his team, but he has been fortunate in the people to whom he has delegated it. Les Davis, Cray's chief engineer for almost three decades, is regarded by many in the supercomputer world as the man who made Cray's ideas work.

This connection to the outside world was not the only way the boundary around the Chippewa Falls lab had to be made permeable. If users in general were kept at arm's length, a few select people passed relatively freely between Chippewa Falls and sites where the 6600 might be used. The crucial such site was the nuclear weapons laboratory at Livermore, California. Livermore's director of computing, Sidney Fernbach, had easier access to Cray's laboratory than Cray's boss Norris had, and close liaison developed between the two sites. Cray, normally considered a "technical dictator," was prepared to listen to Fernbach's advice about how to shape the 6600 to ensure it met Livermore's unique needs for computer power.[21]

For all his apparent isolation, Cray was building potential users into what we, following Callon, Latour, and Law, might call his "network." His was not the slick marketing of the brochure and slide presentation, but the quiet link-making of a man who is reported to have said at the time of the 6600's development that he knew all his potential customers by their first names.[22] Its very lack of slickness—no conventional sales presentation appears ever to have been given at the Chippewa Falls laboratory—made it all the more convincing. One participant remembers an Army colonel "asking what would be the performance of the model

7600 compared to the 6600. Seymour replied that he would be happy if it just ran! Somehow the quiet low-key discussions were terribly impressive. The image of a man who knew *exactly* what he was doing came across clearly to the visitors, as they told me afterwards."[23]

The link-making paid off. Despite its rapid early growth, predictable financial stability seemed to evade Control Data, and there was always the possibility that the risky 6600 project might be sacrificed for the sake of the more mainstream 3600. Livermore's commitment to Cray and his machine was vital in preventing this.[24]

The 6600 repaid Fernbach's and Livermore's trust. It provided a quantum leap in the computing power available to Livermore and to its competitor, Los Alamos. It even gave the United States a temporary lever in its attempt to control France's nuclear weapons policy: in 1966 the U.S. government blocked the export of a Control Data 6600 destined for the French bomb program (though the requisite calculations were performed surreptitiously on an apparently nonmilitary 6600).[25] Though the term was not yet in widespread use, the 6600 was indeed a supercomputer, enjoying a significant advantage in arithmetic speed over all other machines of its day, worldwide.

Even Control Data's managers and shareholders, who had to proceed much more on faith than had Fernbach,[26] were repaid. The largest sale achieved by any previous supercomputer, IBM's Stretch, was eight. Before the decade was out, orders for the 6600 exceeded 100, at around $8 million a machine. In an acerbic memo to his staff, IBM's chairman, Thomas J. Watson, Jr., asked why Cray's team of "only 34— including the night janitor" had outperformed the computing industry's mightiest corporation.[27]

"Big Blue," as the rest of the industry called IBM, struggled to develop, out of the basic multi-purpose System/360 architecture, "top-end" machines to compete with Cray's. Although IBM controlled vastly more resources and had at its call considerable talent (including Gene Amdahl, a computer designer of great skill who was to become almost as famous as Cray), it succeeded only partially. Ultimately, IBM was not prepared to sacrifice compatibility for speed; nor, perhaps, were the "social" aspects of Cray's strategy replicable within the organization's corporate culture. The 1967 IBM 360/91 surpassed the 6600; however, Cray had his 7600 ready by 1969, and not until 1971 did IBM, with the 360/195, catch up to that. The competition, however, was by no means simply about speed. Control Data claimed that IBM was using unfair means to defend its market share, such as allegedly telling potential customers about attractive future machines far in advance of their readiness. The dispute led to the computer industry's most famous lawsuit, with Control Data charging that IBM's marketing of the most powerful System/360 machines violated the antitrust laws of the United States.

Yet all was not entirely well with Cray's strategy, and the competition with IBM was only part of the problem. The 7600 was built according to very much the same priorities as the 6600. But whereas the 6600 offered a speed advantage of as much as 20 over previous-generation computers, the 7600 was only four times faster than the 6600. Was it worth spending a further $10 million, and significant amounts of time modifying programs, to obtain that degree of speedup? Some 6600 users, notably the nuclear weapons laboratories, answered in the affirmative. But many said "no." Sales of the 7600, while still healthy, were only half those of the 6600. In particular, universities, a large sector of the 6600's market, declined to upgrade.

Initially, Cray seemed unperturbed. He and his team began designing the 8600, a bigger departure from the 7600 than the 7600 had been from the 6600. Despairing of achieving great enough speed increases by means of new components and incremental changes to architecture, Cray moved to embrace the much-discussed but as yet little-practiced principle of parallelism. The 8600 was to have four central processing units working simultaneously; all previous Cray machines (and nearly all previous computers of whatever kind) had had but one. Again, an idea that seems simple in principle turned out complex in practice. Ensuring adequate communication between the processors, and keeping them from contending for access to the computer's memory, were formidable problems.[28]

While Seymour Cray and his team were working on the 8600, another team within Control Data, led initially by Cray's former deputy James Thornton, was working on a rival machine: the STAR-100. Central to the STAR-100 was an idea at least as novel as multiple central processors: vector processing. In a vector processor, one instruction can be used to perform a certain operation, not just on one or two pieces of data (as in a conventional "scalar" computer), but on large, ordered arrays ("strings" or vectors). An example would be an instruction to add two strings each of 100 numbers to give one string of 100 numbers as a result. If the data could be organized in this way (and many of the problems that interested the weapons designers at Livermore appeared, at least at first sight, to have this kind of regularity), considerable gains in speed could be achieved without the complexity of multiple central processors.

With the backing of Fernbach and Livermore, the design of the STAR-100 began within Control Data—but at the Arden Hills site, not at Chippewa Falls. The strategy was not Cray's. Speed was indeed a priority, but users of different kinds had to be catered to. The cryptographers of the National Security Agency persuaded the STAR's developers to add hardware support for what those developers referred to as "spook instructions": data manipulations of particular interest to cryptoanalysis. Control Data management (who had much greater access to the STAR than to the Cray machines) saw the STAR as the centerpiece of an integrated, compatible set of computers analogous to, but more advanced than, the IBM System/360. The result was a large instruction set of over 200 instructions. For a long period, too, leadership of the project was ambiguous and communication and coordination poor. The machine became an "engineers' paradise" in which everybody could have novel ideas incorporated. Only determined action by project manager Neil Lincoln (Thornton had left Control Data to set up his own firm, Network Systems Corporation) finally achieved delivery of the STAR to Livermore in 1974, four years late.[29]

The STAR was—indeed still is—a controversial machine. Its adherents point out that it surpassed its impressive goal of 100 million results per second and note that the vector processing pioneered on it dominated at least the next two decades of supercomputing. Its detractors point out that it approached its top speed only on special programs that allowed extensive use of its vector capabilities. In scalar mode, in which one instruction produces one result, the machine was slower than the CDC 7600 of five years earlier. The judgment of the market at the time was with the detractors, and only three STARs were sold.

The very existence of the STAR project in the late 1960s and the early 1970s was, however, a further factor adding to the internal troubles of the 8600 project. Both factors were compounded by a change in direction at the top of Control Data. Diagnosing "a great change" taking place in the supercomputer market, William Norris, president and chairman of the board, said that Control Data's high-speed scientific computers had developed to a point where customers now needed little more in the way of increased speed and power. Instead, said Norris, supercomputer users were demanding service and software to help them get more effective use of the speed they already had. "In other words," he concluded, "the emphasis today shifts to applying very large computers as opposed to development of more power."[30] Although Control Data would continue to build and market large-scale scientific computers, investment in research and development would be curtailed.

Control Data's new corporate plan allowed for supercomputer development, but not at the pace Cray wanted—a new, significantly faster machine every five years. Control Data, which had gone from an innovative startup company to a large, diversified, customer-oriented corporation with a very important financial offshoot, the Commercial Credit Company, had no place for charisma. The increasingly tenuous ties between Seymour Cray and Control Data were severed in 1972. "Since building large computers is my hobby, I decided that with this shift in emphasis, it was time for me to make a change," Cray said. He and four of his colleagues left to start a new company.[31]

Cray Research

It is testimony to the respect in which Cray was held that the divorce was surprisingly amicable. Control Data's Commercial Credit Company even invested $500,000 in the new Cray Research, Inc., adding to $500,000 of Cray's own money and a total of $1,500,000 from 14 other investors. The computer business had been sufficiently profitable that Cray had several personal friends within it who were able to put in as much as $250,000 each. Cray was both president and chief executive officer. At last he could give pure expression to his sociotechnical strategy, without even the residual encumbrances of a large corporation.

The strategy was breathtaking simple: Cray Research would build and sell one machine at a time, and each machine would be a supercomputer. There would be no diversified product range; no attempt to make money (as Control Data very successfully did) primarily from the sale of peripherals, such as disk drives and printers; no dilution of the commitment to built the fastest possible machine. By delimiting the goal, and keeping to a single development team of perhaps 20 people under the undistracted command of Seymour Cray, costs could be kept small. A selling price sufficiently above cost would be set to cover R&D expenditures.[32]

That the customer base for the world's fastest computer was small—Cray estimated it at 50—did not disturb him. Indeed, it was an advantage that he already knew who his potential customers were: those purchasers of the 6600 and 7600, including the nuclear weapons laboratories, whose demand for speed was still unsatisfied and was perhaps insatiable. They would be prepared to pay the now-traditional supercomputer price of around $8 million to $10 million per machine. If the high-performance computer industry's traditional margin of a selling price three times the manufacturing cost could be achieved, the proceeds of a single sale would recoup the entire initial capital investment.

Cray could not afford to take any risks with component technology, and certainly could not afford to develop it within Cray Research. He chose a very simple but fast integrated circuit. It was, however, by no means fast enough on its own to give anything like the increase in speed needed to establish his new machine, which was (in a very public move from an apparently private man) to be called the Cray-1.

It was in the choice of the Cray-1's architecture that Seymour Cray displayed a flexibility often absent in single-minded, dominant technical entrepreneurs.[33] He abandoned the multiple-processor approach that had failed on the CDC 8600, and adopted the vector processing pioneered by its rival, the STAR-100.[34] However, Cray had the advantage over the STAR's designers that he had the failings of a real machine (or at least an advanced development project) to learn from. Like others, he concluded that the STAR had two interrelated flaws.

First, the STAR's scalar performance was far slower than its vector performance; thus, if even a small part of a program could not be made suitable for vector processing, the overall speed of running that program would be drastically reduced. Cray therefore decided to place great emphasis on giving the Cray-1 the fastest possible scalar processor. Second, the full vector speed of the STAR was achieved only if data could be packaged into regular vectors of considerable size. This was partly attributable to the fact that the STAR processed the vectors directly from memory and then sent the results back to memory, in effect using a "pipeline" that was very fast when full but which took a relatively long time to fill. Cray decided instead to introduce a small, intermediate storage level ("vector registers"), built from very fast but extremely expensive memory chips.

Other differences between the Cray 1 and the STAR were predictable consequences of their very different circumstances of development. Cray did not worry about compatibility with any other machine, whether designed by him or anyone else. One again, he had his "clean sheet of paper." The Cray-1's instruction set was more complex than the 6600's (here Cray's old links to cryptography may have come into play), but the great elaboration of the STAR was avoided. The old issues of physical size and cooling were once again central. The STAR's "memory-to-memory" pipeline and its relatively slow scalar unit permitted a physically large machine. Cray's fast scalar unit and vector register design did not.

The goal was a clock speed of 12.5 nanoseconds, well below the 40 nanoseconds of the STAR. In the former time interval, even light in free space travels less than 4 meters, and an electric signal in a wire is slower. This influenced several technical decisions. Along with the continuing fear of placing himself in the hands of others whom he could not control, it persuaded even Cray to override, as far as memory design was concerned, his motto about keeping a decade behind. In previous machines he had always used magnetic core memories. Under the impact of competition from the new semiconductor memories, however, the manufacturers of core memories were concentrating on the cheap, low-performance end of the market. Cray therefore decided to opt for slower, but physically smaller and reliably available, semiconductor memory chips.[35]

Shrinking the machine also intensified the familiar problem of heat. Cray's team developed a new cooling scheme for the Cray-1. Its integrated circuits were mounted on boards back to back on copper plates built onto vertical columns of "cold bars"—aluminum blocks containing stainless steel tubes through which freon coolant flowed. The complete machine consisted of twelve columns arranged in a 270° arc, thus giving the machine its now famously elegant C-shaped horizontal cross section.[36]

The Cray-1 was as much of a tour de force as the 6600. All but a few adherents of the STAR accepted that the Cray-1 was, by a large margin, the world's fastest machine when it appeared in 1976. A triumph of Cray's general sociotechnical strategy and of his insight into the weak points of previous designs rather than of specific invention (the only parts of the original design to be patented were the cooling system and the vector registers), it was nevertheless a technical triumph. This time IBM did not even try to compete directly: corporate pride was outweighed by the memory of past failures and by a sense that supercomputing was merely a market niche of limited size rather than the flagship of all computing. Control Data tried, with a reengineered and improved version of the STAR known as the Cyber 205. It was a strong technical rival to the Cray-1—faster on long vectors, though still not as fast on short ones—but it was too late.

Control Data's managers hesitated on whether the supercomputing market was really worth the trouble and risk they had found it to involve. An interim machine, the Cyber 203, appeared in 1979, and the Cyber 205 in 1981—five years after the Cray-1. By 1985 about thirty Cyber 205s had been sold—not bad by the standards of the 1960s, but, as we shall see, not good enough by the new standards of the 1980s, standards set by Cray Research. First Control Data and then its supercomputer spin-off, ETA Systems, were constantly chasing Cray Research from behind—an ultimately fruitless effort that culminated in the closure of ETA by its parent in April 1989, with total losses said to be $490 million.[37]

The Transformation of the Cray Strategy

What made the chase ultimately fruitless was not any entrenched speed advantage of the Cray-1 and the later Cray Research machines. ETA's successor to the Cyber 205, the ETA[10], was indeed faster in raw speed than its Cray competitors. Japanese industry's entry into supercomputing during the 1980s has, likewise, led to machines faster on some measures than Cray Research's, and the American suppliers of "massively parallel" computers regularly quote performance figures far higher than those of Cray Research's modestly parallel vector processors.

Rather, Seymour Cray's sociotechnical strategy was quietly transformed. In pursuit of success and stability, charisma was routinized. The resulting "network" (again in the sense of Callon, Latour, and Law) no longer expressed itself in the figure of Cray himself—indeed, ultimately it had no place for him—but it has, to date, proved remarkably durable.

The transformation began with the second Cray-1 to be sold. The first sale was the classic Cray linkage of raw speed with the needs and resources of a nuclear weapons laboratory (Los Alamos this time, since Fernbach had committed Livermore to a second STAR-100), together with the classic Cray confidence. Los Alamos had not budgeted for a new supercomputer before 1977, and by the mid 1970s the weapons laboratories' computer acquisition process had become much more bureaucratized. Cray Research, without any sales four years after it had been set up, had to establish itself, and few customers other than Los Alamos would have the interest and the resources to accept a new machine that was almost devoid of software and was not compatible with its existing computers. Cray gambled and offered Los Alamos Cray-1 Serial 1 on loan. "If the machine hadn't performed, Cray Research wouldn't have continued as a company."[38]

The customer identified for the next Cray-1, Serial 3, was not a nuclear weapons laboratory but the National Center for Atmospheric Research. (Construction of Serial 2 was halted when Serial 1 displayed a high level of memory errors when installed at Los Alamos, perhaps because of the high incidence of cosmic rays high in the mountains. Error detection and correction facilities were added to later machines.)

Although attracted by the Cray-1's speed, the Boulder meteorological bureau refused to buy the machine unless Cray Research supplied the systems software as well.

Cray's strategy of building "the fastest computer in the world" and letting the users worry about software was put to the test. By 1976, the management of Cray Research was no longer solely in Seymour Cray's hands. Cray could raise money from friends, but he was not the man to negotiate details with Wall Street. Another midwesterner, also an electrical engineer but holding a Harvard MBA degree, had been hired in 1975 as chief financial officer: the 34-year-old John Rollwagen, who organized the successful 1976 public flotation of Cray Research. Rollwagen and others concluded that a more accommodating approach to users had to be taken, and committed the company to supplying the National Center for Atmospheric Research with an operating system and a Fortran compiler as well as hardware. A major software-development effort (located, significantly, in Minneapolis, not Chippewa Falls) was initiated to parallel Cray's hardware development. In July 1977 Cray 1 Serial 3 was shipped to Boulder, and by the end of the year Serials 4, 5, and 6 had also been sold. The company made its first net profit ($2 million), and John Rollwagen became president and chief executive officer.[39]

Though sales of the Cray-1 were just beginning, thoughts within Cray Research already began to turn to what to do next. There was discussion in the company as to whether to move into the "low end" by making smaller machines (what in the 1980s would come to be called minisupercomputers), derived from the Cray-1, that would be cheaper and therefore perhaps command a larger market. Seymour Cray disagreed, explicitly turning his back on the lure of growth. In the 1978 Annual Report he wrote:

I would rather use the corporate resources to explore and develop newer and more unique computing equipment. Such a course will keep the Company in a position of providing advanced equipment to a small customer base in an area where no other manufacturer offers competitive products. This course also tends to limit the rate of growth of the company by moving out of market areas when competitive equipments begin to impact our sales. I think it is in the long-term best interests of the stockholders to limit growth in this manner and maintain a good profit margin on a smaller sales base.[40]

As always, Cray put his technological effort where his mouth was, starting work in 1978 on the Cray-2, with the goal of a sixfold increase in speed over the Cray-1—nearing the tantalizing target of the gigaflop, a thousand million floating-point arithmetic operations per second.

No one defied the founder by beginning work on a minisupercomputer. Nevertheless, Cray Research's development effort was not restricted to the Cray-2, and extensive efforts were made to make the existing Cray-1 attractive for a broader range of customers. The Cray-1S, announced in 1979, kept to the founder's remit by offering improved performance: an input-output subsystem to remove bottlenecks that had slowed the original Cray-1, and a larger memory. In 1982 it was followed by the Cray-1M, offering the same performance as the 1S but, through the use of a different, cheaper component technology, at a significantly lower price—$4 to $7 million, rather than the $8.5 to $13.3 million of the machines in the 1S series. The Cray-1M was no "mini," but it could be seen as a step in that direction.

Tailoring the hardware to expand the market was only one aspect of the new strategy that began to develop at Cray Research. Systematically, different categories of users (not just weapons laboratories and weather bureaus, important though those remained) were cultivated. Their needs were explored and, where necessary, existing technology was altered or new technology developed to meet them. The oil industry was first, with four years of cultivation between first contacts and the first sale, in 1980. Originally, the oil industry used supercomputers for reservoir engineering in order to extract as much oil from a well as possible. Soon, however, these machines were also used for the processing of seismic data to locate possible oil deposits. Enormous amounts of data had to be processed, in quantities exceeding even the weapons laboratories or weather bureaus. A specially developed additional memory, the Solid State Device (SSD), helped, but a crucial (and symbolically significant) step was Cray Research's investment in the development of a link between IBM magnetic-tape equipment and a Cray supercomputer, together with software to handle tapes that had suffered physically from the rigors of oil exploration.[41]

The aerospace, automobile, and chemicals industries were further targets for successive waves of focused cultivation and "network building." Though physical devices sometimes had to be developed to cement links, software development was far more important. It was no longer sufficient, as it had been when Cray-1 Serial 3 was sold, to supply an operating system and a Fortran compiler. Compilers for other programming languages were developed, and particular attention was devoted to "vectorizing" compilers, which would enable the users of Cray supercomputers to take advantage of their machines' vector speed without having to invest large amounts of time in rewriting programs (as the Livermore users of the STAR-100 had to do). Cray Research even took upon itself to convert for use on its machines specific computer packages that were important in areas being targeted for sales, such as the Pamcrash automobile crash simulation program. By 1985, 200 major applications packages had been converted to run on Cray machines.[42]

The result of all this effort was that during the 1980s Cray Research's expenditure on software development came to equal that on hardware development. Seymour Cray should not be thought of as having opposed this. Ever flexible about the means to achieve the goal of speed, if not about the goal itself, he could even be seen as having led this effort, abandoning for the Cray-2 Cray Research's own Cray Operating System and moving to Unix, which was rapidly becoming a standard operating system and which was much better at handling time-sharing and interactive computing than Cray Research's original system. But his readiness to make the shift also indicated a widening chasm. It was in part sensible because the Cray-2 differed so radically in architecture from the Cray-1 that converting the Cray Operating System was scarcely any easier than shifting to Unix. Once again, Seymour Cray had started with a clean sheet of paper in his search for speed. However, with the growing attention to the priorities of users disinclined to invest in rewriting programs, and with Cray Research's growing investment in software, the clean sheet of paper was beginning to seem a problem, not an advantage, to others in the company.[43]

Here, furthermore, the unpredictability of trying to implement a strategy in an only partially tractable world, rather than rational choice between strategies, played a decisive role, and added to the centrifugal forces already pushing Seymour Cray away from the center of the company he had founded. To gain his desired speed increase with the Cray-2, Seymour Cray was pursuing an ambitious design combining new, faster chips (developed by Motorola and Fairchild in cooperation with Cray Research), multiple central processing, and a processor architecture significantly different from that of the Cray-1. Despite their respect for Cray, others in the firm questioned whether he could succeed in all these innovations simultaneously, and difficulties and delays in the Cray-2 project reinforced their fears.

Les Davis, Cray's chief engineer, began the process from which an alternative to the Cray-2 emerged by suggesting the use of the new, faster chips in a multiprocessor design in which the basic processor was essentially the same as in the Cray-1. Such a machine would enjoy greater software compatibility with the Cray-1, and even if no product emerged it would allow the increasingly important software developers to experiment with parallel processing in advance of the Cray-2. On Davis's instigation a second design team was formed, headed by a young Taiwan-born engineer, Steve Chen, to pursue a low-level, cheap effort along these lines.

The effort surpassed by far the hopes of its instigator. John Rollwagen, concerned that Cray Research's future was being staked too exclusively on the Cray-2, decided to sell the Davis-Chen machine as a product. The Cray X-MP, announced in 1982, offered up to five times the performance of the Cray-1S, nearly as much as the Cray-2's as yet unmet goal, but with the advantage of substantial software compatibility with the Cray-1. The X-MP fitted its niche beautifully, in both a physical sense (the use of more advanced chips meant that multiple processors could be fitted into a cabinet very similar to that of the Cray-1) and a commercial sense, given the importance of software to Cray Research and its users. It became the Western world's most successful supercomputer, with almost 160 sales of different versions of the X-MP by the end of 1989.[44]

Seymour Cray managed to snatch partial success for the Cray-2 from the jaws of what was rapidly beginning to look like a path to oblivion: an advanced design route to a new computer offering little, if any, speed increase over the X-MP, which was fast becoming an established product. He managed to differentiate the Cray-2 from the X-MP by using slow but relatively cheap chips to offer a massive memory of up to 256 megawords, two orders of magnitude more than existing machines at the time. He compensated for the slowness of this massive memory by attaching small, fast memories to each of the Cray-2's four processors. This tactic worked only partially: very careful management of memory resources by users of the Cray-2 is needed to prevent the slow main memory from becoming a bottleneck. But a sufficient number of users wanted a massive memory (by November 1989 24 Cray-2 machines had been sold[45]) for there to be reasons other than sentiment for Cray Research to market its founder's design.

The most immediate reason for the Cray-2's dangerous delay (it came on the market only in 1985, three years after the X-MP) had been problems with cooling, and different approaches taken to these are further interesting indicators of the specificity of Seymour Cray's preferred sociotechnical style. The Cray-2's components were packaged even more closely than in previous machines, and Cray, despite repeated efforts, could not make his existing approach to cooling, based on circulating freon refrigerant, work well enough: "You don't know you are going down the wrong road on a design until you have invested six months or a year in it. I had about three of those false starts on the Cray-2. The cooling mechanism in those designs didn't work."[46]

These failures led Cray Research to initiate yet another alternative to what Seymour Cray was trying to do: a research subsidiary called Cray Laboratories in Boulder. Its job was to do what Seymour Cray appeared unable to do: design a Cray-2 that worked. While Cray continued to search for a way to cool his computer design, using existing chips with modest numbers of circuits on the chip, Cray Labs pursued the alternative path (more widely followed in the computer industry at large) of packing more computing power onto each chip, using very-large-scale integration.

"It was going to be either one design or the other," Cray later recalled. There was no commonality in the two approaches, and only his own fitted his style of work. "I can't take the high-technology [very-large-scale integration] approach because it requires a division of the work into specific areas, and I can't handle that as an individual," Cray said. "I was grasping at a low-technology approach to the Cray-2."[47]

In a "last desperate attempt," Cray tried a completely new cooling approach in which the whole machine would be immersed in a cooling liquid, which, by its forced circulation, would remove the heat produced. When he proposed the scheme nobody took it seriously. "Everyone on the project laughed, in fact they rolled in the aisles. Because everybody knew the boards would swell up and it just wouldn't work." Liquid immersion cooling had been tried before, and a variety of known coolants had been investigated, all of which had soon damaged the printed circuit boards. But Seymour Cray took a different approach: he did not select liquids primarily because of their cooling properties; he chose them on the basis of their known inertness. One of the liquids he tried was a substance that was used in artificial blood. This finally worked, allowing Cray to build a machine that was very densely packed but which could still be cooled.[48]

The Split

The great success of the X-MP and the partial success of the Cray-2 contributed to Cray Research's continuing growth. Despite its specific market, the company's sales began to nudge it into the ranks of the dozen or so leading computer suppliers in the world, albeit still far behind the giant

IBM. That success, however, masked the deepening of the divide between its founder and the bulk of the company he had founded. In 1989, four years after the Cray-2 went on the market, Seymour Cray left Cray Research.

During the latter part of the 1980s Seymour Cray's strategy and the strategy dominant at Cray Research continued to diverge. The Cray Research strategy was to build on the success of the X-MP and the growing user base that made that success possible, seeking systematically to develop new fields of application and strengthening relations with existing customers. The purchaser of an X-MP (or the improved but compatible Y-MP) was now buying not raw speed but access to extensive software resources and services. Cray Research helped customers new to supercomputing to plan their installations and provided on-site support for the life of the installation. The firm guaranteed that, should a Cray Research supercomputer fail, anywhere in the world, a Cray engineer would be there in 2 hours.

Far from keeping users at arm's length, Cray Research sought to bind them ever more tightly to the company. A Cray User Group was set up, and it held yearly meetings for all users of Cray supercomputers. These meetings were attended by Cray Research representatives, and the company sought to respond quickly to problems or desires that emerged. Cray Research also paid increasing attention to links with other suppliers. The announcement in 1990 of a "network supercomputing" strategy for Cray Research made this explicit: the supercomputer was now to be seen not as an artifact standing on its own but as a central part of complex network, many parts of which would be supplied by companies other than Cray Research.[49]

If this strategy begins to sound a little like the statements by William Norris that were instrumental in Seymour Cray's leaving Control Data, one crucial difference should be emphasized: Cray Research was (and still is) very concerned with speed. The Cray Y-MP, which came onto the market in 1988, was about two to three times faster than the X-MP, though comparison is difficult because both machines have been made available in many different processor numbers and memory sizes to suit different customers. The successor to the Y-MP, the C-90, is a further expression of this technological approach: it is compatible with its predecessors, and broadly similar in design; however, it is significantly faster, embodying improved component technologies, more central processors (sixteen, rather than the eight of the top-of-the-range Y-MP), and a larger memory.

Simultaneously, however, the heretical step "down" toward the minisupercomputer has been made. In 1990 Cray Research announced an air-cooled version of the Y-MP series. Slower but significantly cheaper (starting at $2.2 million) than the original Y-MP, it was still more expensive than most minisupercomputers. The company also bought Supertek Computers, a Santa Clara minisupercomputer company that specializes in machines that are compatible with Cray Research's but still cheaper. The acquisition was intended to speed up Cray Research's entry into the minisupercomputer market.[50]

Seymour Cray's commitment to speed, on the other hand, remained much more naked. His goal for the Cray-3 was a twelvefold increase in speed over the Cray-2, to 16×10^9 floating-point arithmetic operations per second. The machine was designed to be compatible with the Cray-2, but not with the X-MP, the Y-MP, or the C-90.

In one crucial respect, however, the Cray-3 was a departure from Cray's previous strategy. Up to a fourfold improvement in speed over the Cray-2 was expected to come from using sixteen processors rather than four, but that left a factor of at least 3 to come from faster components. As we have seen, Cray's preference in all previous machines was to remain well within the state of the art in component technology—to avoid both risk and also a complex division of labor. For the Cray-3, however, he concluded that the existing state of the art would not sustain the increase in component speed he sought, and so he took the "high-technology approach" he had eschewed for the Cray-2.

Nor, furthermore, is the form taken by this approach the silicon VLSI path, which, though still relatively new in supercomputing, was commonplace in the wider computer industry. Instead, Seymour Cray became the first computer designer to commit himself to using processor chips made out of gallium arsenide rather than silicon. Gallium arsenide had long been discussed as a faster substitute for silicon (Cray himself had investigated but rejected it for the Cray-2), but there were known to be daunting difficulties in the number of circuits that could be implemented on a gallium arsenide chip, in manufacturing the chips, and in their reliability. Around the computer industry it was joked that "gallium arsenide is the technology of the future, always has been, always will be."[51]

With the Cray-3, Seymour Cray gambled that the future was about to arrive, and that he could manage a more complex process of technological development, involving not just his own design team but also groups outside the company developing gallium arsenide components for the project. Reports suggest that he has been successful. The Cray-3 project, however, met with serious delays. The problems arose, it seems, not from the gallium arsenide but from the continuing miniaturization needed to sustain ever-increasing speed. The Cray-3 was planned to operate with a 2-nanosecond clock period, handling one scalar instruction every clock period. For this to be possible, given the finite speed of electrical signals, the maximum allowable wire length is 16 inches. So the Cray-3 modules are unprecedentedly small for a supercomputer.

Seymour Cray's design called for cramming 1024 chips into a package measuring just 4 inches by 4 inches by $\frac{1}{4}$ inch. A 16-processor Cray-3 would use 208 of those packages. Assembling them was so complex and delicate a task that it was believed to be beyond the capacity of human assemblers. Special robots had to be developed to do it; this again involved collaboration outside Cray's team. Cray worked with the Hughes Aircraft Company to develop robotic assembly equipment to attach 52 gold wires to each gallium arsenide chip. The wires, which served as both electrical connectors and fasteners, were far thinner than human hairs, and only $\frac{1}{4}$ inch long. They had to be perfectly straight, because after 52 of them were attached to each chip—each with a microscopic dot of solder—the chip was turned over and pressed directly into a 1-inch-square printed circuit board. Each board held nine chips and was composed of eight layers, each with its own circuitry. To cool the chips, which sat flush against the circuit board, thousands of holes were drilled to permit the coolant to contact each chip directly. As in the case of the Cray-2, the density was to be achieved by means of three-dimensional packaging. All in all, in each module 12,000 vertical jumpers had to be soldered, which posed a tremendous manufacturing challenge.

Sustaining both the Cray-3 and C-90 projects, together with the growing range of other development activities (especially in software) that Cray Research was becoming committed to, began to place strains on the company. During the first quarter of 1989, research and development expenses were about 35 percent higher than in the first quarter of the preceding year, even after the cancellation of another ambitious project, the MP project led by Steve Chen. At the same time, the supercomputer market's annual growth had slowed from about 50 percent in the early 1980s to about 10 percent in 1988.[52] The share price of Cray Research, long the darling of Wall Street, was beginning to slump. Either the Cray-3 or the C-90 had to go.

Which was it to be? With the delays in the Cray-3, it and the C-90 seemed likely to appear at roughly the same time, in 1990 or 1991, and were going to be similar in speed and memory sizes. The main difference was in compatibility. The Cray-3 would be compatible with the Cray-2, of which about 25 units had been sold. The C-90 would be compatible with the X-MP and the Y-MP, with an installed base of about 200. The logic was clear. The C-90 had to be preserved and the Cray-3 cancelled.

Yet it could not be as simple as that. John Rollwagen knew that "not choosing Seymour's machine would have torn the company apart." After "six months mulling over alternatives," Rollwagen proposed to Cray yet another "amicable divorce."[53] A new company would be incorporated to undertake the further development of the Cray-3. The new company would at first be a wholly owned subsidiary of Cray Research, with Cray Research transferring to it $50 million worth of facilities and up to $100 million in operating funds over two years. For Cray Research, with revenues of $750 million a year, that was a lot of money. However, Cray Research's shareholders were rewarded by receiving shares in the new company on a tax-free basis, Cray Research was able to concentrate its efforts and resources on the one project, and failure of the Cray-3 would not endanger Cray Research's existence.

While making it clear the split was not his idea, Seymour Cray agreed to it: "I don't mind this role. I kind of like starting over."[54] Once again he was working for a small startup company, this time called the Cray Computer Corporation. This time, however, it was not in Chippewa Falls. Before the split, the Cray-3 team had moved to Colorado Springs, in the foothills of the Rocky Mountains. Nor was the world faced by him the same as in 1957, or in 1972. In 1972, in particular, Cray Research's commitment to building the world's fastest computer was unique. It had no direct competitor, the less-than-wholehearted effort at Control Data aside. At the start of the 1990s, Cray Research, the Japanese supercomputer companies, Steve Chen (who left Cray Research when his MP project was cancelled, to form a new firm, backed by IBM) were all pursuing speed, as were a number of other companies that had developed or were developing machines with "massively parallel" architectures rather than the modestly parallel ones characteristic of mainstream supercomputing.

In this newly competitive marketplace, Cray Research still enjoyed a dominant position, not through speed alone, but rather through the diverse, entrenched links it had built with users. Seymour Cray, on the other hand, was taking on this new world with essentially two resources: the strategy of speed and the power of the name.

Charisma and Routinization

The charisma of Seymour Cray is real and continuing. When a local newspaper reported that the Cray-3 development team was moving to Colorado

Springs, 3500 people wrote asking for jobs before any formal advertisement appeared. The temptation of a purely psychological interpretation of that charisma—"The Genius"—is strong. Yet it must be resisted.

The charisma of Cray was the product of a network of relationships that stretched far beyond the man and his brain. Most obviously, the charisma was in good part the result of his computers, several of which were accepted unequivocally as the most powerful in the world of their time. The engineer's charisma, if it is not to be evanescent, must embody itself in the machine, just as the warlord's must embody itself in an army and the prophet's in a religious movement. And not just any machine, or any army, or any movement, but one that succeeds (or, at the very least, fails gloriously and tragically).

Building a machine and making it successful cannot be done by one person alone. Others must play their parts, from those who labored with Cray through the long Chippewa Falls evenings, to those (mostly women) who for months on end painstakingly connected up the Cray-1's complex wiring, to users like Fernbach who made Cray's first supercomputer work, to Cray's daughter who was satisfied with algebra done by elves. They, and the wider world, may be happy to attribute authorship of the machine and of its success to the charismatic engineer, but without them the machine and its success would not exist.

Hence the dialectic of charisma. If a network is to grow and survive (more machines, more sales, a growing firm; an empire; a church), its links must multiply, expand, and solidify. Not only are more actors involved, but also many more specialist functions, often far removed from the skills of the leader. However entrenched the image of the charismatic leader's authorship of everything, strains in the opposite direction develop. The specialist functions demand due resources and due recognition. Social organization, once implicit (Seymour Cray's personal links to key users), becomes explicit (the official Cray User Group).

Max Weber's term for all this—routinization—is too stark in the context we are discussing. The information technology industry changes too fast for bureaucracy (in the Weberian sense of a formal hierarchical division of labor, with clear-cut rules, roles, and responsibilities, and with a clear separation between the demands and duties of the role and the personal characteristics of its temporary incumbent) to stabilize successfully, as Tom Burns showed in 1961.[55] Explicit organization—albeit "organic" in Burns's sense, rather than bureaucratic—is nonetheless necessary, beyond a certain size. And explicit organization brings with it its own exigencies and its own priorities. Risks, for example, may still have to be taken; however, in the words of John Rollwagen, they must be taken "more carefully."[56]

The successful charismatic leader thus eventually faces a hard choice: to play a revered but different role in the new network, or to cut loose from it and begin afresh. Seymour Cray, yet again, has embarked upon the second path. Will he succeed as he has done before? The odds are stacked against him, but, precisely because of that, if he does succeed, his remarkable charismatic authority will grow yet further.

Addendum

In March 1995 the Cray Computer Corporation, having failed to find a customer for the Cray-3 or a firm order for its successor, the Cray-4, filed for bankruptcy protection. The end of the Cold War (and the consequent decline of much of the traditional supercomputer market), design and manufacturing problems with the Cray-3, the entrenchment of Cray Research, and increasing competition from massively parallel machines appear to have combined to render Cray Computer's task too hard.

7

The Fangs of the VIPER

Computer systems are increasingly taking on roles in which their failure could have catastrophic results, for example in medical care and in the control systems of aircraft and nuclear power stations. How can such systems be known to be safe? Certainly, they can be tested. For a system of any complexity, however, the number of possible combinations of external inputs and internal states is too large for even the most highly automated testing to be comprehensive. Nor does it necessarily help to install systems in triplicate, as is often done with crucial electromechanical systems. This is a good insurance against physical failure, but not against a hardware or software design flaw common to all three systems. Computer scientists have, therefore, been seeking ways to prove mathematically that the design of a computer system is correct. In 1986 the U.K. Cabinet Office's Advisory Council for Applied Research and Development suggested that such mathematical proof should eventually become mandatory for any system whose failure could result in more than ten deaths.[1]

A major step in this direction came in the late 1980s when a team of researchers employed by the U.K. Ministry of Defence developed a novel microprocessor called VIPER (verifiable integrated processor for enhanced reliability). Though VIPER had several other features designed to make it safe (such as stopping if it encountered an error state), what was crucial about it was the claimed existence of a mathematical proof of the correctness of its design—something no other commercially available microprocessor could boast.

The claim of proof became controversial, however. There was sharp disagreement over whether the chain of reasoning connecting VIPER's design to its specification could legitimately be called a proof. Lawyers as well as researchers became involved. Charter Technologies Ltd., a small English firm that licensed aspects of VIPER technology from the

Ministry of Defence, took legal action against the ministry in the High
Court, alleging among other things that the claim of proof was a mis-
representation. The ministry vigorously contested Charter's allegations,
and Charter went into liquidation before the case could come to court.
If it had, the court would have been asked to rule on what constitutes
mathematical proof, at least in the context of computer systems.
Lawyers and judges would have had to grapple with esoteric issues previously in the province of mathematicians and logicians.

The home of VIPER was the Ministry of Defence's Royal Signals and
Radar Establishment (RSRE), Britain's leading research and development laboratory in radar, semiconductor physics, and several fields of
information technology. VIPER was developed by a team of three at
RSRE, led by John Cullyer, a man concerned by the potential for com-
puter-induced catastrophe.[2] In the VIPER project, Cullyer and his colleagues, John Kershaw and Clive Pygott, aimed to design a
microprocessor whose detailed, logic-gate-level design could be proved
to be a correct implementation of a top-level specification of its intended behavior. With the methods available in the 1980s, a direct proof of
correspondence was out of the question, even for the relatively simple
VIPER chip. Instead, the team sought to construct the proof in the form
of a chain of mathematical reasoning connecting four levels of decreas-
ing abstraction.

Most abstract is the top-level specification (level A), which lays down
the changes that should result from each of the limited set of instructions provided for use by VIPER's programmers. Level B, the major-state
machine, is still an abstract description but contains more details on the
steps gone through in executing an instruction. Level C, the block
model, is more concrete and consists of a diagram (of a kind familiar to
designers of integrated circuits) of the major components of VIPER,
together with a specification of the intended behavior of each component. In level D the description is sufficiently detailed that it can be used
to control the automated equipment employed to construct the "masks"
needed to fabricate VIPER chips.

At the end of 1986, the RSRE team began its authoritative account:
"VIPER is a 32-bit microprocessor invented at RSRE for use in highly
safety-critical military and civil systems. To satisfy certification authori-
ties of the correctness of the processor's implementation, formal mathematical methods have been used both to specify the overall behavior
of the processor and to prove that gate-level realizations conform to this
top-level specification."[3] The paper made clear that part of the chain of

proof was still being constructed. Simultaneously, however, VIPER began to be sold, both literally and metaphorically. The two processes were to collide disastrously.

During 1987 and 1988, VIPER moved rapidly toward the market. Two leading U.K. electronics firms, Ferranti Electronics Ltd. and Marconi Electronic Devices Ltd., undertook to make prototype chips, using different processes as an insurance against the introduction of errors in their physical production. Several firms took up the equally essential work of making it possible for users to write programs for VIPER and to check their correctness. Most central was Charter Technologies, which in 1988 signed a contract with the Ministry of Defence to sell and support the VIPER software tool set developed by RSRE. Through Defence Technology Enterprises (a firm set up to help implement the U.K. government's goal of having more of the research done at defense establishments turned into commercial products), Charter also purchased the license to develop and sell the VIPER prototyping system. Charter marketed VIPER actively.[4] The technical press was unrestrained in its welcome, claiming that VIPER had been mathematically proved to be free of design faults.[5]

Commercialization and glowing media accounts of VIPER had a double-edged effect. In 1986, RSRE had awarded a contract to the University of Cambridge to investigate the possibility of mechanically checking the higher-level VIPER correctness proofs. Cullyer had used a system called LCF-LSM, developed at Cambridge, to write VIPER's specification and to construct an outline "paper-and-pencil" proof that the major-state machine (level B) was a correct implementation of it. By January 1987, Avra Cohn, at Cambridge, had successfully used HOL,[6] LCF-LSM's successor, to formalize and automate Cullyer's outline proof linking levels A and B. The pencil-and-paper proof took Cullyer three weeks. The mechanized proof, in which every step of logical inference is made explicit, took six person-months to set up and consisted of more than a million inferences.[7]

Cohn then embarked on the more ambitious task of proving that the block model (level C) was a correct implementation of the top-level specification. She constructed a partial proof of more than 7 million inference steps, but by May 1988 she had concluded that to complete the proof would mean a great deal more work for what she called a "dwindling research interest" as well as requiring further development of the HOL system. Cohn had also become worried about the way VIPER was being described in marketing material and in the media.

The very language of proof and verification could, she felt, convey a false sense of security. To say that the design of a device is "correct" does not imply that the device will do what its designers intended; it means only that it is a correct implementation of a formal specification. Even the most detailed description of a device is still an abstraction from the physical object. Gaps unbridgeable by formal logic and mathematical proof must always potentially remain between a formal specification and the designers' intentions, and between the logic-gate-level description and the actual chip (figure 1).[8]

VIPER's developers were perfectly aware of these unavoidable limitations of formal proof. The gap between gate-level descriptions and physical chips was the very reason they had sought independent physical implementations of VIPER, implementations that could (in the 1A version) be run in parallel, checking one another. However, a highly critical assessment from the verification specialists of Computational Logic

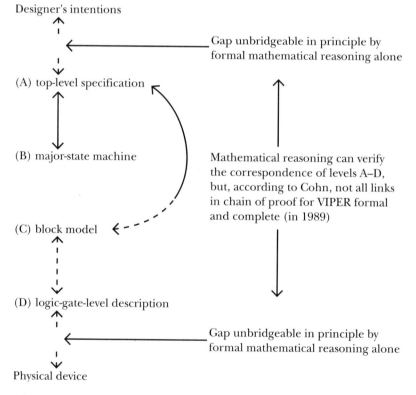

Figure 1
Cohn's arguments on the limits of formal verification.

Inc., of Austin, Texas, seems to have been something of a shock to them. The American firm had been commissioned by NASA to assess the VIPER proof. The report's authors, Bishop Brock and Warren Hunt, were prepared to grant the status of formal proof only to Cohn's HOL proof linking levels A and B. All the other links in the chain were, in their view, either incomplete or informal. For example, the RSRE team had initially established the correspondence of levels C and D by simulation and case-by-case checking, using a method they called "intelligent exhaustion."[9] The simulation, however, required LCF-LSM specifications to be translated into the RSRE-developed ELLA language, and there was, Brock and Hunt said, no formal proof of the correctness of the translation. Nor, they claimed, was there a proof that all possible cases had been considered. VIPER, they concluded, had been "intensively simulated and informally checked" but not "formally verified."[10]

News of the impending report reached the VIPER team in September 1989, when Charter's managing director, Digby Dyke, met Professor Cullyer (who had by then left the Royal Signals and Radar Establishment) at a seminar. When the RSRE team received a copy of the draft report, it appeared that Cullyer conceded Brock and Hunt's criticisms. The version sent by fax from Austin had been shown to Cullyer and contained his handwritten annotations and his signature. The word "agreed" appeared repeatedly in the margin.

Cullyer's response was generous, especially given that Computational Logic could be seen as a rival, since the firm was developing a formally proven microprocessor of its own. The RSRE team was happy to acknowledge that "more remains to be done, both to build up confidence in the existing VIPER design and to develop new techniques of design and verification which avoid the limitations of present methods."[11]

By the autumn of 1989, however, VIPER was no longer just a research project. It was a commercial product, and one that was not meeting with great success, quite apart from any problems of proof. Potential users were reluctant to abandon tried and trusted microprocessors (even with their known bugs) for a novel chip and new software, and VIPER was too simple and slow for many applications. Government policy prohibited the Ministry of Defence from mandating the use of VIPER, and by the time of the lawsuit only one defense project had adopted it. The only civil adoption had been for a system to control signals on automated railroad crossings in Australia.

The lack of commercial success and the criticism of the claim of proof broke the previously close ties between RSRE and Charter

Technologies. Charter began by seeking informal redress for the losses it believed it had suffered but then took its grievances to members of parliament, the media, and (in January 1991) the law courts.

In the ensuing publicity, the central issue was often submerged. No "bug" had been found in the VIPER chips. Indeed, their design had been subjected to an unprecedented degree of scrutiny, checking, simulation, and mathematical analysis—work that has continued since the litigation.[12] At issue in the litigation, however, was whether the results of this process—as it stood immediately before the lawsuit began— amounted to a formal, mathematical proof. As we have seen, to Brock and Hunt they did not. Cohn, similarly, wrote in 1989: ". . . no formal proofs of Viper (to the author's knowledge) have thus far been obtained at or near the gate level."[13] Others, however, would counter that most of mathematics has not been proved in a fully formal sense, and would claim that it is an unduly restrictive, even distorted, notion of "proof." (See chapter 8 of the present volume.)

Because of Charter's bankruptcy, the High Court was not, in the end, called on to take sides in this dispute over the nature of mathematical proof. Yet the underlying issues did not disappear along with Charter. The Ministry of Defence, in April 1991, issued a standard for "The Procurement of Safety Critical Software in Defence Equipment."[14] Although many other measures are proposed, and although there is provision for exceptions, the standard clearly points to the desirability of fully formal proof in the most crucial applications.

As the Ministry of Defence and other major users of safety-critical computer systems move in this wholly praiseworthy direction, it is important that the lessons of the past be learned rather than suppressed. The VIPER episode reminds us that "proof" is both a seductive word and a dangerous one. We need a better understanding of what might be called the "sociology of proof": of what kinds of argument count, for whom, under what circumstances, as proofs. Without such an understanding, the move of "proof" from the world of mathematicians and logicians into that of safety-critical computer systems will surely end up, as VIPER nearly did, in the courts.

8

Negotiating Arithmetic, Constructing Proof

Computer systems have been a subject of considerable interest to social scientists since the 1960s. Their diffusion, their likely effects on organizations, on employment levels and on society at large, the evolution of the computer industry—these and other topics have received considerable attention. Computer systems as *mathematical* entities have, however, remained almost entirely the province of technical specialists. Here I seek to redress that balance by arguing that computer systems offer interesting and counterintuitive case studies in the sociology of mathematics.

Two different aspects of computer systems as mathematical entities will be discussed. The first is the computer (and the advanced digital calculator) as an arithmetical tool. Intuition might suggest that arithmetic done by calculator or computer would be wholly unproblematic. Arithmetic, after all, is the very paradigm of secure, consensual knowledge, and surely the calculator or computer simply removes the tedium and error-proneness of arithmetic performed by human beings! Not so. Not only is considerable skill, normally taken entirely for granted, needed in order reliably to perform arithmetic even on simple calculators[1]; in addition, there has been significant dispute as to the nature of the arithmetic to be implemented, at least when the numbers to be worked with are not integers. Different computer arithmetics have been proposed, and the nearest approximation to a consensual computer arithmetic, the Institute of Electrical and Electronics Engineers' standard for floating-point arithmetic, had to be negotiated, rather than deduced from existing human arithmetic.

The second aspect of computer systems that will be discussed here is mathematical proof of the correctness of a program or a hardware design. As computer systems are used more and more in situations where national security and human life depend on them, there have been increasing demands for such mathematical proofs in place of, or in addition to, empirical testing. This is of interest from the viewpoint of the sociology of knowledge because moves "proof" from the world of mathematicians and logicians to that of engineers, corporations, and lawyers.

Although mathematical proof is being sought precisely because of the certainty it is ordinarily held to grant, constructing proofs of the correctness of computer system again turns out to be no simple "application" of mathematics. It involves negotiating what proof consists in. In 1987, Peláez, Fleck, and I predicted that the demand for proofs of the correctness of computer systems would inevitably lead to a court ruling on the nature of mathematical proof.[2] This chapter develops the preceding chapter's discussion of the controversy that led to this prediction's having already come close to confirmation; it also explores more general issues of the "sociology of proof" in the context of computer systems.

Negotiating Arithmetic

Human arithmetic is consensual in advanced industrial societies. Typically, agreement can be found on the correct outcome of any calculation. For example, to my knowledge there have been no scientific disputes in which the parties have disagreed at the level of the arithmetic. Furthermore, there are certain "ideal" laws that we all agree must hold, independent of any particular calculation. For example, we agree that addition and multiplication of real numbers should be both commutative and associative—i.e., that

$a + b = b + a,$

$(a + b) + c = a + (b + c),$

$a \times b = b \times a,$

and

$(a \times b) \times c = a \times (b \times c)$

whatever the values of a, b, and c.

The consensual status of arithmetic has indeed been taken as indicating a self-evident limit on the scope of the sociology of knowledge.[3] It might seem to make implementing arithmetic on a calculator or a computer straightforward. Yet that has not been the case.

The difficulties are most striking in the form of arithmetic used in the kind of calculation typically found in scientific work: floating-point arithmetic. This is the computer analogue of the well-known "scientific notation" for expressing numbers. In the latter, a number is expressed in three parts: a positive or negative sign (the former usually implicit); a set of decimal digits (the significand or mantissa), including a decimal point; and a further set of digits (the exponent), which is a power of 10. Thus 1,245,000 could be expressed in "scientific notation" as $+1.245 \times 10^6$, and -0.0006734 as -6.734×10^{-4}. The advantage of scientific notation is that allowing the decimal point to "float" in this way leads to a more economical and easily manipulable format than the standard representation, where the decimal point is "fixed" in its position.

Computer floating-point arithmetic carries over this flexibility, and is quite similar, except in the following respects:

Decimal (base 10) representation is now unusual; hexadecimal (base 16) and binary (base 2) representation are more common. Since the episode I am about to discuss concerns binary arithmetic, let us concentrate on that. Every digit is either 1 or 0: the decimal 3, for example, is expressed as 11, the decimal 4 as 100, and so on. The exponent is a power of 2, and the equivalent of the decimal point is known as the *binary point*. The sign, similarly, is expressed as a binary digit, typically with a 0 for positive numbers and a 1 for negative numbers.[4]

A firm decision has to be taken as to the number of binary digits (*bits*) in the significand and in the exponent. In the arithmetic to be discussed, the basic format[5] has one binary digit to express the sign, eight to express the exponent,[6] and 23 to express the significand, adding up to a total of 32 (figure 1). The total number of bits (sometimes called "word length") is of considerable importance, since most modern computers seek to process all the bits making up a number in parallel rather than one after another. The more bits to be processed simultaneously, the greater the complexity of the hardware needed.

Any floating-point system, unless constrained, allows multiple representations of the same number. For example, -0.0006734 could be expressed by -67.34×10^{-5} as well as by -6.734×10^{-4} or -0.6734×10^{-3}. Computer floating-point systems, however, typically employ a unique "normal" representation of each

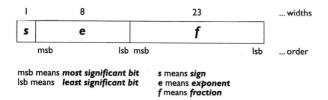

msb means **most significant bit**   s means **sign**
lsb means **least significant bit**   e means **exponent**
                                f means **fraction**

Figure 1
Number representation in IEEE floating-point arithmetic. Source: *IEEE Standard for Binary Floating-Point Arithmetic,* American National Standards Institute/Institute for Electrical and Electronics Engineers Standard 754-186 (Institute for Electrical and Electronics Engineers, August 12, 1985), p. 9.

nonzero number. In that representation, there is always one nonzero bit to the left of the binary point. Since that bit must be a 1, it need not be stored explicitly, and only the parts of the significand to the right of the binary point (known for obvious reasons as the *fraction*) are explicit.

There are several decisions, then, to be taken in implementing computer floating-point arithmetic. What base shall be used? What word length, what size of significand, and what size of exponent? Shall a sign bit of 1 represent negative numbers, or positive numbers? How shall zero be represented? What methods of rounding shall be used? What should be done if the result of a calculation exceeds the largest absolute value expressible in the chosen system (i.e., if it "overflows"), or if it falls below the smallest (i.e., if it "underflows")? What should be done if a user attempts an arithmetically meaningless operation, such as dividing zero by zero?

Different computer manufacturers (and then, as it became possible to implement floating-point arithmetic on electronic calculators, different calculator manufacturers) answered these questions differently. This was patently a source of some practical difficulty, since it made it difficult to use a numerical program written for one computer on another, even when a standard programming language such as Fortran was used. But did it matter more profoundly than that? Were the results of the different decisions really different arithmetics, or were they simply different but essentially equivalent ways of implementing the one true arithmetic?

The answer depended on how one reacted to what might be called anomalous calculations. Under some circumstances, different machines yield substantially different results for the same calculation. In other cases, machine arithmetic violates consensual laws of human arithmetic. Small discrepancies between the results of the same calculations performed on different machines are common, and specialists in the field can produce cases of results differing considerably. One specialist cites a compound-interest problem producing four different answers when done on calculators of four different types: \$331,667.00, \$293,539.16, \$334,858.18, and \$331,559.38. He identifies machines on which $a/1$ is not equal to a (as, in human arithmetic, it always should be) and $e\pi - \pi e$ is not zero.[7]

Different reactions to anomalous calculations can be categorized according to the schema developed by Imre Lakatos in his celebrated analysis of the evolution of Euler's theorem concerning the relationship between the number of faces (F), edges (E), and vertices (V) of a polyhedron. Lakatos showed that attempts to prove the relationship $V - E + F = 2$ were plagued by counterexamples: figures which could be claimed to be polyhedra but which did not obey the law.[8] One response to these "anomalous figures" was what Lakatos calls "primitive exception barring": simple indifference to their existence. That response characterizes well what seems to have been the majority response to anomalous computer or calculator calculations. Most users have probably been either unaware of the possibility of anomalous calculations or unconcerned about them in the same sort of sense that we continue happily to cross bridges even though we are aware that some bridges have collapsed. For many computer designers, too, anomalous calculations seem to have been well down the list of matters needing attention, if they were on it at all.

A more sophisticated "exception barring" strategy was to cite the vast bulk of calculations that were performed perfectly satisfactorily, and to argue that anomalous calculations were instances of problems that were not "well posed." A well-posed problem was one in which a slight change of data caused only a slight change of result; the solution employed an algorithm that was in this sense "numerically stable." The newly developed technique of "backward error analysis" was used, in justification of this response, to discriminate between well-posed problems and those that were not well posed. Computers and calculators worked reliably on well-posed problems. If "pathological" calculations and "degenerate" problems were avoided (use of these terms might be taken as indicating that exception barring was sliding into what Lakatos calls monsterbarring), no difficulties would arise.[9]

A small number of computer scientists, however, positively sought "monsters."[10] A leader among them was Professor W. Kahan, who holds a joint appointment in the Mathematics Department and the Department of Electrical Engineering and Computer Science at the University of California at Berkeley. Kahan's approach is an example of what Lakatos calls the "dialectical" strategy, in that "anomalies and irregularities are welcomed and seen as the justification for new approaches, new concepts, and new methods."[11] Kahan has been a reformer, not content with the current state of computer and calculator floating-point arithmetic and constantly seeking to devise, and build support for, ways of improving it. He has quite deliberately sought to discover and publicize anomalies that can be used to show that differences between computer arithmetics are serious and consequential.

What first gave Kahan the opportunity to reform arithmetic in the direction he desired was competition between two leading manufacturers of sophisticated calculators, Texas Instruments and Hewlett-Packard. TI questioned the accuracy of HP's calculators; HP responded by claiming that calculation on its competitor's machines manifested more anomalies. A Hewlett-Packard executive, Dennis Harms, saw an advantage in attempting to strengthen his company's position in this respect, and employed Kahan as a consultant on the design of the arithmetic of the corporation's new-generation calculators, thus enabling Kahan to get his ideas incorporated into them.[12]

Kahan's next opening came around 1977 when the leading microprocessor firm, Intel, started to develop a silicon chip specifically to perform floating-point arithmetic. Existing microcomputers implemented floating-point arithmetic in their software rather than in their hardware, while the hardware floating-point units in large computers were multichip. The Intel i8087, as the chip was eventually christened, was intended as a "floating-point coprocessor," working alongside the main processing chip in a microcomputer to improve its arithmetic performance.

John Palmer, the engineer leading the design of the i8087, had attended lectures by Kahan as an undergraduate, and turned to him for advice.[13] Palmer rejected the idea of adopting "IBM arithmetic," despite its widespread use; Kahan believed this arithmetic to be inferior. The arithmetic of the leading minicomputer maker, the Digital Equipment Corporation, was also rejected. Palmer was, however, not simply seeking "product differentiation." He was worried that if the wrong decisions were made it would be impossible to share some programs between "different boxes all bearing the Intel logo," and he wanted to avoid for floating-point arithmetic on microprocessors "the chaotic situation that now exists in the mainframe and minicomputer environments."[14]

Intel and other Silicon Valley chip manufacturers supported the establishment of a committee to consider standards for floating-point arithmetic. The initiative for the committee had come from an independent consultant, Robert Stewart, who was active in the Computer Society of the Institute of Electrical and Electronics Engineers (IEEE), under whose aegis the committee was established. Stewart recruited to the committee Kahan and representatives of Intel, other chip manufacturers, and minicomputer makers. Richard Delp was appointed by Stewart as the first chair of the working group.[15] Intel—which was hard at work on other projects—agreed to delay finalizing the arithmetic of the i8087 while the committee deliberated, even though Intel clearly hoped that the final standard would be similar to what it had already developed.

Negotiating arithmetic proved to be a lengthy process. The committee started work in September 1977, and IEEE Standard 754, Binary Floating-Point Arithmetic, was adopted only in 1985.[16] The general nature of the vested interests involved is clear. Unless the committee took the easy route of writing a general standard that would "grandfather" all widely used existing arithmetics (an option that was considered but rejected), or unless it opted for an arithmetic radically different from any in existence, whatever it decided would be bound to advantage those companies whose existing practice was closest to the standard and disadvantage those whose practice differed widely from it. The latter would be forced into an unpleasant choice. If they retained their existing arithmetic, their market could diminish as a result of users' preferring machines implementing the standard. If they changed, considerable investment of time and money would have to be written off, and there might be troublesome incompatibilities between their new machines and their old ones.

Ultimately the choice came down to two arithmetics closely aligned with major corporate interests. One was essentially the arithmetic employed by the Digital Equipment Corporation (DEC), the leading manufacturer of minicomputers. DEC's VAX machines were very widely used in scientific computing, and their arithmetic was admitted even by its critics to be "sound" and "respectable."[17] The other was an arithmetic proposed by Kahan, his graduate student Jerome Coonen, and Harold Stone, Manager of Advanced Architectures at IBM's Yorktown Heights Laboratory. Not surprisingly, in view of the collaboration between Kahan and Intel's Palmer, that proposal was very similar to what Intel was already well on the way to implementing.[18]

The Kahan-Coonen-Stone scheme has several interesting features, such as the handling of zero. In their basic format they opted for what is called a "normalized zero." Zero is expressed only by a zero significand and zero exponent (0×2^0). The combination of zero significand and nonzero exponent (0×2^1, 0×2^2, etc.) is not permitted. But they permitted the sign bit to take both values, and allowed its value to be meaningful. In other words, unlike consensual human arithmetic, which contains only one zero, their arithmetic contains both a positive and a negative zero, with, for example, the rule that the square root of -0 is -0.[19]

This and other features of their arithmetic were, however, relatively uncontroversial. The battleground between their proposal and the main alternative arithmetic was underflow. Unlike the arithmetic of real numbers, where there is no number "next to zero" and indefinitely small numbers are possible, computer arithmetics contain a lower bound, albeit tiny, on the size of number that can be represented. For example, 2^{-126}, or roughly 10^{-38}, is the smallest number possible in normal representation in the Kahan-Coonen-Stone scheme's basic format. Like the majority of existing computer arithmetics, DEC's arithmetic simply represented all numbers as precisely as possible until the number next to zero was reached. Should a calculation yield a result smaller than that very small number, the result was stored as zero. "Flush-to-zero underflow" is what this scheme was generally called.

Kahan and his colleagues advocated the different principle of "gradual underflow."[20] They introduced a special set of "denormalized numbers" smaller in size than the normal-format number next-to-zero. As was noted above, in normal floating-point format the digit to the left of the binary point is always 1. In a denormalized number it is 0. Denormalized numbers are created by right-shifting the significand so that the exponent always remains within the expressible range. In a system where the smallest normal number is 2^{-126}, therefore, 2^{-127} could be given denormalized expression as $\frac{1}{2}$ (0.1 in binary) $\times 2^{-126}$; 2^{-128} as $\frac{1}{4}$ (0.01 in binary) $\times 2^{-126}$, and so on. Of course, this meant that accuracy would usually be lost, as one or more significant digits would have to be discarded in right-shifting the significand. But this, to its proponents, seemed an acceptable price to pay for a more gradual approach to zero. Their argument against what many took to be the "obvious" DEC procedure was that, using the latter, as one approached zero the differences between successive numbers diminished until one reached the number-next-to-zero, whose distance from zero would be much greater than its distance from the next larger number. In gradual underflow the differences between successive numbers diminished monotonically all the way down to zero (see figure 2).

Gradual underflow became the focus of attacks on Kahan, Coonen, and Stone's proposed arithmetic:

The interconnectedness of the proposed standard's basic features complicated attempts to oppose it. Early challenges within the subcommittee were not easily focused on single aspects of the proposed number system and its encoding, since so many of the design choices were interconnected. These challenges ultimately addressed the proposal as a whole and, quite naturally, tended to drift to its points of least resistance. Thus it was possible for gradual underflow—one of the system's less compelling features—to become its most contentious.[21]

Flush to Zero

Gradual Underflow

Figure 2
Small numbers in flush to zero and gradual underflow. Based on Jerome T. Coonen, "Underflow and denormalized numbers," *Computers*, March 1981, p. 77.

There was no wholly compelling way in which one scheme could be proved superior to the other. Proponents of the Kahan-Coonen-Stone scheme could point to anomalies caused, they argued, by flush-to-zero—anomalies that would be corrected by gradual underflow:

Consider the simple computation $(Y - X) + X$ where $Y - X$ underflows. The gradual underflow always returns Y exactly, flush to zero returns X. . . . We could look at this as another isolated example, but I prefer to look at it as the preservation of the associative law of addition to within rounding error. That is, under gradual underflow we always have $(Y - X) + X = Y + (-X + X)$ to within rounding error. This is compelling, in my opinion.[22]

The defenders of the more traditional DEC scheme could, however, also point to potential problems with gradual underflow:

Multiplication of denormalized numbers by numbers greater than 1 (or division by numbers less than 1) can generate significant inaccuracies. If such a product (or quotient) is in the ordinary range of numbers, it cannot be represented in denormalized form, because of the hidden bit used in KCS [Kahan-Coonen-Stone arithmetic]. However, the denormalized operand has fewer (perhaps many fewer) than the prescribed number of bits for its level of precision. Thus the product (or quotient) could in the worst case contain only one valid bit. KCS specifies two modes to deal with this problem. "Warning mode" is mandatory: the invalid flag is set, and a symbol NaN (Not a Number) is stored for the result. . . . The other mode, "normalize," is optional: if present, and selected, the possibly very inaccurate product is stored as an ordinary number, no flag is set, and, of course, further tracking of the effect of the original underflow is impossible.[23]

As of this time, an illegal operation exception is raised when a denormalized number is multiplied by a value 2 or greater. But on closer inspection, there are denormalized numbers which lie close to the normalized range which exhibit more erratic behavior. The denormalized number $\frac{3}{4} \times 2^{-126}$, for example, will generate an invalid operation exception when multiplied by 5, but not when multiplied by 6. When multiplied by 8 an exception will again be generated. . . . This effect is caused because the exception for the multiplication occurs when attempting to store an unnormalized number into a basic format. When multiplying by $8 = 1 \times 2^3$, the result is $\frac{3}{4} \times 2^{-123}$, which is unnormalized. But multiplication by $6 = \frac{3}{2} \times 2^2$ gives $\frac{9}{8} \times 2^{-124}$, which is normalized.[24]

These objections could be dismissed in their turn:

Like any new tool, it is possible to misuse this facility and to have a malfunction. . . . I do not believe that the facility introduces malfunctions into processes that previously worked [with flush-to-zero].[25]

The proneness of the two arithmetics to generating errors and anomalous calculations was not the only issue to be considered. There was, for example, little doubt that gradual underflow was more complicated to implement than flush-to-zero; it would therefore have costs in money and (perhaps) in the time taken by arithmetic operations such as multiplication. It might make the proposed standard harder to police, since, given its complication, manufacturers might choose to implement it in software rather than (demonstrably present or absent) hardware.[26] Finally, DEC's scheme simply had the enormous advantage of essentially being that already employed in the world's most popular scientific minicomputers.

Thus, nothing abstract guaranteed that the Kahan-Coonen-Stone scheme would win: in Professor Kahan's words "it was not a foregone conclusion."[27] In its favor were the composition of the working group, the facts of geography, and its status as the group's original working document. Mary Payne of the Massachusetts-based DEC commented:

The active (and voting) membership of the Working Group is largely from minicomputer and semiconductor manufacturers, Academia, and purveyors of portable software. There is virtually no representation from Mainframe manufacturers and "ordinary users"—people writing their own programs for their own (or their employers') use. Most of the active membership is from the San Francisco Bay area, and all but one of the meetings have been in this area.[28]

Others, however, point to the influence of Kahan's persuasiveness and forceful personality. Kahan himself regards as important two demonstrations of the technological feasibility of gradual underflow (Intel implemented it in the microcode software of the prototype i8087, and one of

Kahan's students, George Taylor, designed a processor board for DEC's own VAX that was acknowledged as successfully performing gradual underflow), together with the qualified support for gradual underflow given by a well-known error analyst, Professor G. W. Stewart III of the University of Maryland, who had actually been hired to investigate the topic by DEC.

In a spring 1980 ballot of the committee, the Kahan-Coonen-Stone scheme received the necessary two-thirds majority support for adoption. The scheme then took several years to pass through higher committees of the Institute of Electrical and Electronics Engineers, but it was finally approved by the IEEE Standards Board in March 1985, and by the American National Standards Institute as ANSI/IEEE Standard 754 in July 1985.

It is not a universal standard. Most supercomputers (such as Crays), mainframes (such as IBM machines), and minicomputers (such as DEC VAXs) are not claimed to comply with it. However, no competing collective standard has been agreed. Thus, a virtuous cycle exists: as the IEEE 754 Standard becomes more popular, the problems involved in moving numerical programs from one machine to another diminish, and more and more software is thus written with the 754 Standard in mind, increasing its popularity. The proponents of new technologies adopt the 754 Standard so that users do not have to rewrite programs to move to these new technologies.[29]

What has happened is thus a version of a form of "closure" typical within technology. In the words of Brian Arthur:

Very often, technologies show increasing returns to adoption—the more they are adopted the more they are improved. . . . When two or more increasing-returns technologies compete for adopters, insignificant "chance" events may give one of the technologies an initial adoption advantage. Then more experience is gained with the technology and so it improves; it is then further adopted, and in turn it further improves. Thus the technology that by "chance" gets off to a good start may eventually "corner the market" of potential adopters, with the other technologies gradually being shut out.[30]

There are those who deny that what has been institutionalized is the best possible computer arithmetic,[31] and who would indeed attribute the standard's adoption to "chance events" rather than to the intrinsic merits its proponents would claim. That dispute, however, is now in a sense irrelevant: the very process of the institutionalization of the standard gives it practical advantages that makes being overturned by a competitor unlikely.

Constructing Proof

As was noted in chapter 7, it has been argued influentially that the complexity of computer systems limits the extent to which empirical testing can demonstrate the correctness of computer software or the design of computer hardware. Because program testing cannot normally be exhaustive, it "can be a very effective way of showing the presence of bugs, but it is hopelessly inadequate for showing their absence."[32] "Mathematicizing" computer scientists have felt that there is only one sure route to programs or hardware designs that are demonstrably correct implementations of their specifications: deductive, mathematical proof. "The only effective way to raise the confidence level of a program significantly is to give a convincing proof of its correctness."[33]

By the 1980s, these originally academic arguments were beginning to be taken up by those responsible for computer systems critical either to national security or to human safety. First to act on them was the U.S. Department of Defense, which in 1983 laid down its Trusted Computer System Evaluation Criteria, known from the color of the cover of the document containing them as the "Orange Book." The Orange Book set out a hierarchy of criteria to be applied to computer systems containing information critical to national security. To attain the highest evaluation—Class A1 ("Verified Design") systems—requires mathematical proof that the design of a system conforms to a formal model of what constitutes "security."[34]

In Europe the demand for mathematical proof has been heard more strongly for computer systems critical to human safety than for those critical to national security (although European criteria loosely analogous to the Orange Book have been issued). In 1986 the U.K. Cabinet Office's Advisory Council for Applied Research and Development called for mathematical proof in the case of systems whose failure could lead to more than ten deaths, and in 1991 Interim Defence Standard 00-55 demanded formal mathematical proof that the programs most crucial to safety are correct implementations of their specifications.[35]

In such documents, with the exception of the most recent (Defence Standard 00-55, discussed below), the notion of "proof" has typically been used as if its meaning were unproblematic. In 1987, Peláez, Fleck, and I speculated that this unproblematic usage would not survive the entry of proof into the commercial and regulatory domains. We predicted that it might not be long before a "court of law has to decide what constitutes a mathematical proof procedure."[36] This prediction was based on the sociology-of-knowledge considerations outlined in chapter 1 and on the considerable variation, revealed by the history of mathematics, in the forms of argument that have been taken as constituting proofs. For example, Judith Grabiner has shown how arguments that satisfied eighteenth-century mathematicians were rejected as not constituting proofs by their nineteenth-century successors, such as Cauchy.[37] Our prediction rested on the assumption that attempts to prove the correctness of the design of computer systems would bring to light similar disagreements about the nature of proof.

By 1991, the prediction of litigation was borne out in the dispute, discussed in chapter 7, over whether the chain of reasoning—as it then stood[38]—connecting the design of the VIPER microprocessor to its specification could legitimately be called a "proof." Only the bankruptcy of the litigant, Charter Technologies Ltd., kept the case from coming to court.[39]

The dispute over VIPER should not be viewed as entirely *sui generis*. What was (at least potentially) at issue was not merely the status of one specific chain of mathematical reasoning, but also what mathematical "proof" should be taken as meaning—a matter that clearly goes beyond the particularities of this episode. This will be the focus of the remainder of the present chapter.

One meaning of "proof" is summarized by Robert Boyer and J. Strother Moore, leading proponents of the use of computer systems to prove mathematical theorems (and colleagues of two of the critics of the VIPER proof, Bishop Brock and Warren Hunt), as follows: "A formal mathematical proof is a finite sequence of formulas, each element of which is either an axiom or the result of applying one of a fixed set of mechanical rules to previous formulas in the sequence."[40] The application of this criterion to VIPER was never publicly challenged before or during the litigation. The Ministry's defense against the litigant's claims is a confidential document. The one published response (known to this author) by a member of the VIPER team to criticism of the claim of proof did not attempt a rebuttal.[41] In any case, the defendant in the lawsuit was the Ministry rather than the individual members of the team, so the line of argument adopted might not have been theirs.

Nevertheless, an attack on the formal notion of proof was indeed the basis of the defense of VIPER mounted, after the litigation halted, by Martyn Thomas, head of the software house Praxis:

We must beware of having the term "proof" restricted to one, extremely formal, approach to verification. If proof can only mean axiomatic verification with theorem provers, most of mathematics is unproven and unprovable. The "social" processes of proof are good enough for engineers in other disciplines, good enough for mathematicians, and good enough for me. . . . If we reserve the word "proof" for the activities of the followers of Hilbert, we waste a useful word, and we are in danger of overselling the results of their activities.[42]

David Hilbert (1862–1943) was a formalist mathematician whose definition of "proof" was in most respects similar to that given above by Boyer and Moore.[43] The formalist tradition spearheaded by Hilbert sought to break the connection between mathematical symbols and their physical or mental referents. Symbols, the formalist holds, are just marks upon paper, devoid of intrinsic meaning.[44] Proofs are constructed by manipulating these symbols according to the rules of transformation of formal logic—rules that take a precise, "mechanical" form.[45]

Despite formalism's considerable influence within mathematics, not all mathematical proofs take this form. Most in fact are shorter, more "high-level," and more "informal." Part of the reason for this is the sheer tedium of producing formal proofs, and their length; this is also a large part of the attraction of automatic or semi-automatic proof-generating systems, such as the HOL system used in the VIPER proof or the automated theorem prover developed by Boyer and Moore.

The relatively informal nature of much mathematical proof was a resource for the defense of the claim of proof for VIPER, as the above quotation from Thomas shows. It was also the basis for a widely debated general attack on formal verification of programs, a 1979 paper by Richard De Millo of the Georgia Institute of Technology and Richard Lipton and Alan Perlis of Yale University's Department of Computer Science.[46] Proofs of theorems in mathematics and formal verifications of computer programs were radically different entities, they argued:

A proof is not a beautiful abstract object with an independent existence. No mathematician grasps a proof, sits back, and sighs happily at the knowledge that he can now be certain of the truth of his theorem. He runs out into the hall and looks for someone to listen to it. He bursts into a colleague's office and commandeers the blackboard. He throws aside his scheduled topic and regales a seminar with his new idea. He drags his graduate students away from their dissertations to listen. He gets onto the phone and tells his colleagues in Texas and Toronto. . . .

After enough internalization, enough transformation, enough generalization, enough use, and enough connection, the mathematical community eventually decides that the central concepts in the original theorem, now perhaps greatly changed, have an ultimate stability. If the various proofs feel right and the results are examined from enough angles, then the truth of the theorem is eventually considered to be established. The theorem is thought to be true in the classical sense—that is, in the sense that it could be demonstrated by formal deductive logic, although for almost all theorems no such deduction ever took place or ever will. . . .

Mathematical proofs increase our confidence in the truth of mathematical statements only after they have been subjected to the social mechanisms of the mathematical community. These same mechanisms doom the so-called proofs of software, the long formal verifications that correspond, not to the working mathematical proof, but to the imaginary logical structure that the mathematician conjures up to describe his feeling of belief. Verifications are not messages; a person who ran out into the hall to communicate his latest verification would rapidly find himself a social pariah. Verifications cannot readily be read; a reader can flay himself through one of the shorter ones by dint of heroic effort, but that's not reading. Being unreadable and—literally—unspeakable, verifications cannot be internalized, transformed, generalized, used, connected to other disciplines, and eventually incorporated into a community consciousness. They cannot acquire credibility gradually, as a mathematical theorem does; one either believes them blindly, as a pure act of faith, or not at all.[47]

The De Millo–Lipton–Perlis paper provoked sharp criticism from defenders of the evolving practice of program verification. One wrote: "I am one of those 'classicists' who believe that a theorem either can or cannot be derived from a set of axioms. I don't believe that the correctness of a theorem is to be decided by a general election."[48] Edsger Dijkstra, one of the leaders of the movement to mathematicize computer science, described the De Millo–Lipton–Perlis paper as a "political pamphlet from the middle ages." Interestingly, though, Dijkstra's defense was of short, elegant, human (rather than machine) proofs of programs. He accepted that "communication between mathematicians is an essential ingredient of our mathematical culture" and conceded that "long formal proofs are unconvincing."[49] Elsewhere, Dijkstra had written: "To the idea that proofs are so boring that we cannot rely upon them unless they are checked mechanically I have nearly philosophical objections, for I consider mathematical proofs as a reflection of my understanding and 'understanding' is something we cannot delegate, either to another person or to a machine."[50]

At least three positions thus contended in the debate sparked by De Millo, Lipton, and Perlis: the formal, mechanized verification of programs and hardware designs; the denial that verification confers certainty akin to that conferred by proof in mathematics; and the advocacy of human rather than machine proof. No wholly definitive closure of the debate within computer science was reached, and the validity of the analogy between proofs in mathematics and formal verification of computer systems remains controversial.[51]

Within mathematics, too, the status of computer-supported proofs has been the subject of controversy. The controversy crystallized most clearly around the 1976 computer-based proof by Kenneth Appel and Wolfgang

Haken of the four-color conjecture.[52] The developers of this proof summarized at least some of the objections and their defense as follows:

Most mathematicians who were educated prior to the development of fast computers tend not to think of the computer as a routine tool to be used in conjunction with other older and more theoretical tools in advancing mathematical knowledge. Thus they intuitively feel that if an argument contains parts that are not verifiable by hand calculation it is on rather insecure ground. There is a tendency to feel that the verification of computer results by independent computer programs is not as certain to be correct as independent hand checking of the proof of theorems proved in the standard way.

This point of view is reasonable for those theorems whose proofs are of moderate length and highly theoretical. When proofs are long and highly computational, it may be argued that even when hand checking is possible, the probability of human error is considerably higher than that of machine error.[53]

Although the general issue of the status of computer-generated formal proofs remains a matter of dispute, there are signs that at the level of the setting of standards for safety-critical and security-critical computer systems the dispute is being won in practice by the proponents of formal verification. The demand for verification in the Orange Book represented a victory for this position, albeit a controversial one, since there has been criticism both of the model of "security" underlying the Orange Book and of the procedures for certification according to Orange Book criteria.[54] Nor did the Orange Book directly address the question of the nature of proof. Most recently, however, Def Stan 00-55, representing official policy of the U.K. Ministry of Defence, has done so, explicitly tackling the issue of the relative status of different forms of mathematical argument. It differentiates between "Formal Proof" and "Rigorous Argument":

A Formal Proof is a strictly well-formed sequence of logical formulae such that each one is entailed from formulae appearing earlier in the sequence or as instances of axioms of the logical theory. . . .

A Rigorous Argument is at the level of a mathematical argument in the scientific literature that will be subjected to peer review. . . . [55]

According to the Ministry, formal proof is to be preferred to rigorous argument:

Creation of [formal] proofs will . . . consume a considerable amount of the time of skilled staff. The Standard therefore also envisages a lower level of design assurance; this level is known as a Rigorous Argument. A Rigorous Argument is not a Formal Proof and is no substitute for it. . . .[56]

It remains uncertain to what degree software-industry practices will be influenced by Def Stan 00-55 and by similar standards for other sectors that may follow—a procedure for granting exceptions to 00-55's stringent demands is embodied in the document. Formal proofs of "real-world" programs or hardware designs are still relatively rare. If they do indeed become more common, I would predict that a further level of dispute and litigation will emerge. This will concern, not the overall status of computer-generated formal proofs (though that issue will surely be returned to), but an issue that has not hitherto sparked controversy: the internal structure of formal proofs. Even if all are agreed that proofs should consist of the manipulation of formulas according to "mechanical" rules of logic, it does not follow that all will agree on what these rules should be. The histories of mathematical proof and formal logic reveal the scope for significant disagreement.

The best-known dispute concerns the law of the excluded middle (which asserts that either a proposition or its negation must be true) and thus the acceptability of proving that a mathematical object exists by showing that its nonexistence would imply a contradiction. Formalists, such as Hilbert, did not regard such proofs as problematic; "constructivists" and "intuitionists," notably L. E. J. Brouwer, refused to employ them, at least for infinite sets.[57]

Other examples are what are sometimes called the Lewis principles, named after the logician Clarence Irving Lewis.[58] These principles are that a contradiction implies any proposition and that a tautology is implied by any proposition. They follow from intuitively appealing axiomatizations of formal logic, yet they have seemed to some to be dubious. Is it sensible, for example, to infer, as the first Lewis principle permits, that "The moon is made from green cheese" follows from "John is a man and John is not a man"? In the words of one text: "Different people react in different ways to the Lewis principles. For some they are welcome guests, whilst for others they are strange and suspect. For some, it is no more objectionable in logic to say that a [contradiction] implies all formulae than it is in arithmetic to say that x^0 always equals 1. . . . For others, however, the Lewis principles are quite unacceptable because the antecedent formula may have 'nothing to do with' the consequent formula."[59] Critics have to face the problem that any logical system which gives up the Lewis principles appears to have to give up at least one, more basic, "intuitively obvious," logical axiom.

These controversial rules of logic are to be found in systems upon which formal proof of programs and hardware depends. The law of the excluded middle is widely used in automated theorem proof (for example, in the HOL system used for the VIPER formal proof). The first Lewis principle—that a contradiction implies any proposition—is to be found in nearly all automated reasoning systems (e.g., among the basic inference rules of the influential Vienna Development Method).[60]

To date, these rules have not provoked within computer science the kind of controversy that has surrounded them in metamathematics and formal logic. There has been some intellectual skirmishing between the proponents of "classical" theorem provers, which employ the law of the excluded middle, and "constructivist" ones, which do not.[61] That skirmishing has not, to date, taken the form of entrenched philosophical dispute, and, to this author's knowledge, no computer-system proof has been objected to because of its reliance on excluded middle or the Lewis principles. Pragmatic considerations—getting systems to "work," choosing logics appropriate to particular contexts—have outweighed wider philosophical issues.

Can we assume, however, that a situation of pragmatism and peaceful coexistence between different logical systems will continue? My feeling is that we cannot; that this situation is a product of the experimental, academic phase of the development of proof of computer system correctness. As formal proofs become of greater commercial and regulatory significance, powerful interests will develop in the defense of, or in criticism of, particular proofs. Sometimes, at least, these interests will conflict. In such a situation, the validity of rules of formal logic will inevitably be drawn into the fray, and into the law courts.

Conclusion

There is an important difference between computer floating-point arithmetic and the proof of computer systems. In the former there was a stable, consensual human arithmetic against which computer arithmetic could be judged. Human arithmetic was, however, insufficient to *determine* the best form of computer arithmetic. It was indeed a matter of judgment which was best, and contested judgment at that. Human arithmetic provided a resource, drawn on differently by different participants, rather than a set of rules that could simply be applied in computer arithmetic. There is even tentative evidence that social interests, notably the different interests of the Intel and Digital corporations, influenced the judgments made. Similarly, the outcome—"closure" in favor of the Kahan-Coonen-Stone arithmetic scheme—may have been influenced by contingent factors such as the proximity of the meetings of the relevant committee to Silicon Valley, home to Intel and other semiconductor firms, and to Kahan's Berkeley base.

In the case of the proof of computer systems, pre-existing practices of proof, within mathematics, have been less compelling. The reputation of mathematics for precision and certainty has been an important rhetorical resource for those who sought to move from an empirical to a deductive approach to computer-system correctness. However, critics have argued that proof of computer-system correctness and proof of a mathematical theorem are different in kind.

One dispute over the mathematical proof of a computer system has already reached the stage of litigation: the controversy concerning the VIPER microprocessor. The prediction of this chapter is that the VIPER case will not be unique. Nor will it be sufficient to reach consensus on the general form to be taken by proofs—for example, to demand that they take the form of sequences of symbol manipulations performed according to the transformation rules of a logical system. If the position adopted in this chapter is correct, that will simply drive dispute "downward" from the status of general types of argument to the validity of particular steps in those arguments. Specifically, dispute is to be expected over the logical systems that underpin formal proofs.

Formal proof of computer-system correctness is, therefore, an interesting test case for the sociology of knowledge, for this prediction is contrary to our ordinary intuitions about mathematical certainty. It concerns not informal or semiformal mathematics of the sort that has to date provided most of the empirical material for the sociology of mathematics, but mathematical deduction of the most formal kind: precisely the kind of reasoning that, we might imagine, must simply compel consent. As computer-system proof grows in significance and moves into the commercial and regulatory worlds, we will have a chance to see whether our ordinary intuitions about mathematics, or the conclusions of the sociology of mathematical knowledge, are correct.

Acknowledgments

The research on floating-point arithmetic and on VIPER was supported by the U.K. Economic and Social Research Council (ERSC) under the Programme on Information and Communication Technologies (PICT), grants A35250006 and WA35250006. Current work on the topic of the second part of the paper is being supported by a further grant from the ESRC on "Studies in the Sociology of Proof" (R000234031).

9

Computer-Related Accidental Death

Just how safe, or how dangerous, are the computer systems on which lives depend? How many lives have been lost through failures of such systems? What are the causes of such accidents? Although there is a large literature on computer-system safety, it contains little in the way of systematic, empirical answers to these questions. Published discussions tend to highlight a handful of dangerous failures but fail to place these in the context of any wider record. There is, it is true, widespread awareness of the potential dangers of computer systems, and considerable research work and substantial sums of money are being devoted to technical means for making computer systems safer. This effort to find a solution is entirely necessary and desirable. Its chances of success might, however, be enhanced by detailed investigation of the problem.

My aim in this chapter is to indicate what might be involved in an empirical investigation of fatal accidents involving computer systems. The chapter's contribution to our knowledge of these accidents is at best modest. The fact that it is based on patently incomplete data sources renders its quantitative conclusions dubious. There are, moreover, both conceptual and empirical difficulties with its central category of "computer-related accidental deaths." Nevertheless, I hope that, precisely by showing how little systematic information is available, I can spark further work on this topic. One of the chapter's conclusions—that there is a pressing need for public agencies to begin systematic, cross-sectoral data collection in this area—indeed seems to follow irresistibly from the very inadequacies of the existing record. Other conclusions— such as that computer-related fatalities have, to date, seldom been caused by technical failure alone—seem reasonably robust, despite the deficiencies in the data drawn on here.

Defining "Computer-Related Accidental Death"

What is meant by "computer-related accidental death"? Each of the four words in this phrase requires some justification or elaboration, beginning with the last.

"Death"

There are three reasons for focusing on accidents involving death, rather than simply on computer-related injury. First, the latter would be too broad a category for sensible analysis. It would, for example, be necessary to include the large numbers of cases of ill health resulting from the use of computer terminals, of which cases of upper limb disease (or "repetitive strain injury") are perhaps the most prominent. Second, the only available source of international, cross-sectoral data (described below) is indirectly dependent on press reports. Deaths are, to put it crudely, more newsworthy than nonfatal injuries, and so there is a far better chance of obtaining reasonable coverage of deaths than of injuries. Third, accidental deaths often trigger formal inquiries, which then provide useful information that is absent in many cases of nonfatal injury.

To allow a reasonable period for reports of such deaths to enter the public domain, I have set the cutoff point of this analysis at the end of December 1992. As far as possible, I have attempted to encompass all earlier cases of computer-related accidental death, worldwide.

"Accidental"

Some computer systems are meant to kill people. Since my interest is in unintended and erroneous behavior in computer systems, it would not be appropriate to include in the analysis deaths caused by military computer systems when these function as intended.

A more difficult issue is deaths of civilian bystanders caused by computer-controlled offensive military systems whose primary targets are opposing military forces. Such deaths have clearly been substantial in number, from the Vietnam War, in which computerized military systems first found major use, to the Gulf War and its aftermath. In one sense, these are accidental deaths: the designers and operators of such systems would, ideally, prefer them not to take place. On the other hand, a certain level of "collateral" civilian death is typically an anticipated and tacitly accepted feature of some kinds of military operations. Furthermore, it is extremely difficult to obtain reliable data on such incidents. I have, therefore, reluctantly decided to exclude such deaths from my analysis.

I have, however, sought to include in the data set deaths related to military operations where those deaths result from system failures that are in some more clear-cut sense accidental in nature (rather than "by-products" of normal system operation). Thus, the analysis includes deaths resulting from computer-related failures of defensive military systems and from computer-related accidental crashes of military aircraft. It also includes the 1983 shooting down of a Korean airliner by Soviet air defenses (where the accidental element is the navigational error that led the plane to stray into Soviet air space) and the 1988 downing of an Iranian airliner by the U.S.S. *Vincennes* (where the accidental element is the misidentification of the plane as an attacking military aircraft).

"Computer"

I have deliberately taken a broad view of what constitutes a computer, including in my definition any programmable electronic device or system, and not only those incorporating a full general-purpose digital computer. An industrial robot (so long as it is both electronic and programmable), a numerically controlled machine tool, and a programmable cardiac pacemaker all fall under my definition of systems that incorporate a computer. Nevertheless, some problems remain. For example, the first-generation industrial robots installed in the 1960s typically had pneumatic and electromechanical, rather than electronic, control systems.[1] Strictly speaking, these would fall outside my definition; however, in reports of cases of robot-related death it is often unclear whether this kind of robot or a more sophisticated electronic device was involved.

"Related"

The above definitional problems are negligible in comparison with the problem of saying when a given accidental death is computer-*related*. The mere presence of a computer (even one playing a safety-critical role) in a system that suffers an accident is not sufficient for any reasonable categorization of a death as computer-related. Rather, the presence of the computer must be causally important to the accident.

On the other hand, it would be too narrow to class an accident as computer-related only when a computer-system problem was its sole cause. Major accidents often, and perhaps usually, have multiple causes.[2] It would, in my opinion, also be too narrow to include only cases of "technical" failure of a computer system. I have included cases where no technical failure is evident but there has been a breakdown or error in human interaction with the system. Of course, such accidents can be, and often are, attributed simply to "human error." Yet system design often contributes to human error—for example, where the user interface of a computer system increases the probability of certain kinds of mistake, or where the safe functioning of a system requires its human operators to perform perfectly on tasks that are known to be error-prone.[3] Also included in my definition of "computer-related" are accidents where false confidence in computer systems, or specific misunderstandings of them, seems to have been a dominant factor in leading operators to adopt or persist in courses of action that they otherwise would have avoided or abandoned.

These considerations mean, however, that there is inevitably a degree of judgment involved in the categorization of such cases as computer-related. Just when does the role of a computer system in the sequence of events leading to an accidental death become important enough to justify calling the death "computer-related"? While seeking to exclude cases where the computer system's role was minor, I have also tried to avoid being overly stringent, on the ground that it is easier for a critical reader to exclude a case as insufficiently computer-related than to scrutinize for possible inclusion all the possible "marginal" cases.

This kind of (obviously contestable) judgment is not the only difficulty involved in deciding whether any given death is computer-related. The widely publicized failure in late 1992 of the new computerized dispatch system at the London Ambulance Service indicates another problem. There is no doubt that considerable suffering and some degree of physical harm to patients resulted from this failure. Patients also unquestionably died in London on the crucial days of October 26 and 27 and November 4. Yet there are matters of delicate medical judgment involved in assessing whether the lives of those who died might have been saved had ambulances reached them earlier. The coroners involved seem to have taken the view that they would not have been saved. Therefore, the London Ambulance Service case has to be excluded from my list of computer-related deaths. (However, were that case to be included, the findings of the inquiry into this incident, which highlight the interaction of technical and organizational failings, would reinforce, rather than undermine, the qualitative conclusions below,[4] and the number of deaths involved would not alter the quantitative totals greatly.) Similarly (to take a case that is included in the data set), many cancer patients died after receiving underdoses in computerized radiotherapy at the North Staffordshire Royal Infirmary between 1982 and 1991, but there are clearly difficult clinical judgments to be made as to which of those deaths are attributable to the underdosing. No figure more precise than "tens . . . rather than hundreds" has been given.[5]

Furthermore, there is often sharp disagreement over the causes of an accident. On the outcome of such disagreement may hinge issues of civil liability and even criminal culpability. Unless a researcher has the resources to mount an investigation, the best he or she can do is turn to the most authoritative available source: an official inquiry or, in some cases, an independent report. In practice, however, it is often wise to be skeptical of even these sources. For example, Martyn Thomas, a leading commentator on computer-system safety, suggests that "the probability of the pilot being blamed for [an air] crash is more than twice as high if the pilot died in the crash."[6] In a substantial number of cases, furthermore, I have been able to find neither the report of an official inquiry nor that of a thorough independent investigation.

In these latter cases, I have erred on the side of inclusion, at least so long as there seemed to me to be a not wholly implausible case for their computer-relatedness. Unlike many official inquiries, research such as this does not seek to allocate blame, and I have felt it better to include cases that *may* be computer-related than to exclude them because of lack of information. Critical readers may, however, wish to excise from the totals those cases where I have described the data as "poor" or "very poor," as well as drawing on the bibliographic materials cited here to form their own opinion of the degree of computer-relatedness of the better-documented cases.

A more particular problem concerns what this data set suggests are the two most important "technical" causes of computer-related accidental death: electromagnetic interference and software error. A broken part will often survive even a catastrophic accident, such as an air crash, sufficiently well for investigators to be able to determine its causal role in the sequence of events. Typically, neither electromagnetic interference nor software error leaves physical traces of this kind. Their role can often be inferred only from experiments seeking to reproduce the conditions leading to an accident. Though this can on occasion be done convincingly, it is sometimes far from easy, and the suspicion therefore remains that these causes are underreported.

Method

My primary source of cases was the remarkable compilation of reports of computer-related accidents and other failures that has, as a result of the efforts of computer scientist Peter Neumann, accumulated over the years in the pages of the Association for Computing Machinery's newsletter *Software Engineering Notes*, established in 1976. To begin with, these reports were a sporadic feature of Neumann's "Letter from the Editor." In the early 1980s, however, the volume of such reports grew sharply, and in August 1985 an on-line electronic news group, called RISKS Forum, was set up, moderated by Neumann, with many contributors. This forum (accessible via Internet) has become the basis of a section on "Risks to the Public" in each issue of *Software Engineering Notes*. Although the resultant record has deficiencies from the point of view of systematic analysis, this material forms a unique and valuable data source. There is no doubt that its very existence has been a spur to a great deal of the research work relevant to computer safety. Inspection of existing articles dealing with the topic makes clear how important *Software Engineering Notes* and the RISKS forum have been in publicizing accidents involving computers.[7]

The method I used to gather cases was very simple. I examined each issue of *Software Engineering Notes* carefully for cases of apparent computer-related accidental death. The cases thus collected were cross-checked against the helpful indexes regularly produced by Peter Neumann in case one should be missed in the sheer volume of material. Wherever possible, I then sought the report of an official inquiry into, or an independent investigation of, the incident described. At the very least, an attempt was made to check the original published source whenever this was quoted.

Apart from the general issues raised in the previous section, there are clearly two potential problems in this use of *Software Engineering Notes*: the overreporting and the underreporting there of computer-related accidental deaths. Overreporting is more common than might be imagined. Computer professionals have shown commendable zeal in searching for and publicizing cases of computer-system failure. (There is, indeed, an interesting puzzle for the sociology of the professions in the contrast between this attitude and what seems to be the typically less zealous attitude of other professionals, such as physicians or lawyers, in uncovering and publicizing errors by their colleagues.) Reasonably often, incidents reported in *Software Engineering Notes* that appear to be computer-related accidental deaths subsequently turn out not to have been computer-related. The newsletter has often published corrections, and in other cases my own research suggested that the role of computers was small or negligible. Such cases are excluded.

In other cases, no reliable source of information could be found on which to base such a judgment. As noted above, most of these are included in the data set, with warnings as to the poverty of information on them. A handful of cases that appeared *prima facie* merely apocryphal were, however, excluded; the number of deaths at issue is small, so the effect on the overall pattern of the data of either including or excluding them is not great.

Unfortunately, underreporting is a far more intractable problem than overreporting. *Software Engineering Notes* makes no pretense to be comprehensive in its coverage. Neumann, for example, is careful to title his indexes "Illustrative Risks to the Public." The cases reported in the RISKS forum and *Software Engineering Notes* are typically culled from press coverage: only a minority come from the reporter's personal experience (and these are almost always the less serious incidents, not those involving death). Furthermore, there is an enormous preponderance of English-language newspapers and journals among the sources quoted. At best, therefore, *Software Engineering Notes* appears to cover only those computer-related fatal accidents that find their way into the English-language press.

In the absence of any comparable alternative source, however, there is no straightforward way of investigating the extent of underreporting in *Software Engineering Notes*. The impression I formed while conducting the research was that coverage of "catastrophic" accidents such as crashes of large passenger aircraft is good. These will always be reported in the press, extensive inquiries will typically ensue, and the subscribers to RISKS seem carefully to scrutinize reports of such accidents and inquiries for any suggestion of computer involvement.

It seemed likely, however, that coverage of less catastrophic accidents, such as accidents involving robots or other forms of automated industrial equipment, would be poorer. These will typically involve only a single death; they take place on the premises of an employer who may have no wish to see them widely publicized; and they may be regarded by the media as too "routine" to be worth extensive coverage. Accordingly, I investigated these separately through contacts in the firms producing robots and in the Health and Safety Executive, the organization responsible for enforcing industrial safety regulations in the United Kingdom.

It turns out that the coverage of fatal accidents involving robots by *Software Engineering Notes* is reasonable: indeed, there seems to have been a degree of overreporting. This good coverage probably arises because robot accidents have been regarded by the media as newsworthy. On the other hand, even the small amount of systematic data I have found on fatal industrial accidents involving more general types of computer-controlled machinery makes it clear that this kind of accident is greatly underreported in *Software Engineering Notes*. I would indeed hypothesize that this is the most important systematic gap in the data recorded below.

The Data

Overall Total

There are around 1100 computer-related accidental deaths in the overall data set generated by the above methods: to be precise, 1075 plus the "tens" of the North Staffordshire radiation therapy incident (see table 1). The data's limitations, discussed above, mean that these figures are far from definitive. Despite extensive literature searches, information on a substantial number of the incidents remains poor. Those inclined to attribute accidents to human error alone would probably deny that many of the "human-computer interaction" cases are properly to be described as computer-related. It might be argued that some of the deaths (for example, those resulting from failure to intercept a Scud missile and from the Soviet downing of the Korean airliner) should not be classed as accidental. There are, furthermore, a variety of particular problems in the diagnosis of other incidents (some of these problems are discussed below) which might lead a critic to exclude them too. Only a small minority of incidents—perhaps only the Therac-25 radiation therapy incidents—seem entirely immune from one or other of these exclusionary strategies, although to force the total much below 100 would require what seem to me to be bizarre definitions, such as a refusal to accept the North Staffordshire deaths as computer-related.

In other words, more stringent criteria of what is to count as a computer-related accidental death could reduce the overall total to well below 1100. On the other hand, the fact that the mechanisms by which a death reaches *Software Engineering Notes* are far from comprehensive means that there is almost certainly a substantial degree of underreporting in this data set. In particular, there must have been more fatal industrial accidents involving computer-controlled industrial equipment than the 22 cases recorded here. Systematic data were available to me only for Britain and France, and for limited periods of time. Comprehensive coverage of other advanced industrial nations would increase the overall total considerably. Furthermore, the relatively small number of cases from outside the English-speaking world (particularly from the former Soviet bloc) is suspicious. Reliance on computers is more pervasive in Western industrial nations than in the former Soviet bloc and Third World, but probably not to the extent the geographic distribution of the accidents recorded here might suggest.

Any attempt to correct for this underreporting is obviously problematic. It seems to me unlikely, however, that any plausible correction could boost the total by much more than about a further 1000. For that to happen would require that one or more catastrophic computer-related accidents, involving at least several hundred deaths, has been misclassified by me or has gone unrecorded. The latter is certainly possible, but, given the number and diligence of Neumann's correspondents, unlikely.

Therefore, the findings of this analysis on the total number of computer-related accidental deaths, worldwide, to the end of 1992, can be expressed, in conventional format, as 1100 ± 1000. The relatively large error band appropriately conveys the twin problems inherent in this exercise: more stringent definition would reduce the total considerably, while correction for underreporting could plausibly just about double it.

Aside from the total number of deaths, the other most salient aspect of this data set is the causes of the incidents it contains. I have divided the accidents into three rough categories, according to the apparent nature of their dominant computer-related cause: physical failure of a computer system or physical disturbance of its correct functioning; software error; or problems in human-computer interaction. Although inadequate data prohibit description of every individual incident, some discussion of the type of accident to be found in each category may be of interest.

Physical Causes: 48 Deaths
Apart from one case of capacitor failure and one dubious case in which a safety-critical computer system may have failed because of fire, all deaths involving physical causes have been due to electromagnetic interference (i.e., a programmable system's being reprogrammed or having its normal operation impeded by stray radio signals or other electromagnetic emissions). Two deaths have been attributed to accidental reprogramming of cardiac pacemakers. Several military accidents have been alleged to have been caused by electromagnetic interference, although (perhaps because of the particular difficulty of diagnosing electromagnetic interference retrospectively) these cases are almost all controversial. In only one of them has electromagnetic

Table 1
Cases of possible computer-related accidental death (to end of 1992).

Date(s)	No. of deaths	Location	Nature of incident	Probable main cause(s)	Main ref(s).	Data quality
				Physical causes		
?	1	US	Accidental reprogramming of cardiac pacemaker	Interference from therapeutic microwaves	Dennett (1979)	Poor
?	1	US	Accidental reprogramming of cardiac pacemaker	Interference from antitheft device	SEN[a] 10(2), p. 6 SEN 11(1), p. 9	Poor
1982	20	South Atlantic	Sinking of *Sheffield* after failure to intercept Argentinean Exocet missile	Interference from satellite radio transmission	*Daily Mirror* 5/15/86; Hansard 6/9/86	Fair
1982	1	US	Car accident	Fire may have caused failure of antilock braking system	*San Francisco Chronicle* 2/5/86	Very poor
1986	2	Libya	Crash of US F-111 during attack on Tripoli	Possible electromagnetic interference	SEN 14(2), p. 22	Very poor
1982–1987	22	?	Crashes of US military helicopters	Possible electromagnetic interference, denied by makers and by US Army	*AW&ST*[b] 11/16/87, 27–28	Poor, controversial
1988	1	UK	Operator killed by computer-controlled boring machine	Machine restarted unexpectedly due to faulty capacitor	Edwards (n.d.)	Good

Software error

Year	No.	Country	Event	Details	References	Rating
1986	2	US	Overdoses from radiation therapy machine	Error in relationship between data-entry routine and treatment-monitoring task	Leveson and Turner (1992)	Very good
1991	28	Saudi Arabia	Failure to intercept Iraqi Scud missile	Omitted call to time-conversion subroutine; delayed arrival of corrected software	GAO[c] (1992), Skeel (1992)	Good

Human-computer interaction problems

Medical

Year	No.	Country	Event	Details	References	Rating
1982–1991	"in the tens"	UK	Underdosing by radiation therapy machine	Correction factor for reduced source-skin distance in isocentric therapy applied twice (already present in software).	West Midlands Regional Health Authority (1992), North Staffordshire Health Authority (1993)	Good

Military

Year	No.	Country	Event	Details	References	Rating
1987	37	Persian Gulf	Failure to intercept attack on *Stark* by Iraqi Exocet missile	Alleged lack of combat readiness; possible defective friend/foe identification or switching off of audible warning	Sharp (1987), Committee on Armed Services (1987), Adam (1987), Vlahos (1988)	Fair
1988	290	Persian Gulf	Shooting down of Iran Air airliner by *Vincennes*	Stress; need for rapid decision; weapon system–human interface not optimal for situation	Fogarty (1988)	Good

Table 1 (continued)

Date(s)	No. of deaths	Location	Nature of incident	Probable main cause(s)	Main ref(s).	Data quality
				Air		
1979	257	Antarctica	Crash of airliner on sightseeing trip	Communication failure re resetting of navigation system; continuation of flight in dangerous visual conditions	Mahon (1981)	Fair, but aspects controversial
1983	269	USSR	Shooting down of Korean Air Lines airliner after navigational error	Autopilot connected to compass rather than inertial navigation system	AW&ST 6/21/93, p. 17	Fair
1988	4	UK	Collision of two RAF Tornado aircraft	Use of identical navigational cassettes by different aircraft	Sunday Times 3/11/90	Fair
1989	12	Brazil	Crash of airliner after running out of fuel	Incorrect input to navigation system?	SEN 15(1), p. 18	Poor, controversial
1992	87	France	Crash of airliner into mountain during night approach	Vertical speed mode may have been selected instead of flight-path angle; limited cross-checking between crewmembers; possible distraction; no ground-proximity warning system	Sparaco (1994)	Fair

Robot-related

Date	No.	Country	Description	Details	Source	Data quality
1978–1987	10	Japan	Workers struck during repair, maintenance, installation, or adjustment of robots	Workers entered envelope of powered-up robots; in some cases, deficiencies in training and absence of fences	Nagamachi (1988)	Fair
1984	1	US	Heart failure after being pinned by robot	Worker entered envelope of powered-up robot	Sanderson et al. (1986)	Fair

Involving other automated plant

Date	No.	Country	Description	Details	Source	Data quality
1979	1	US	Worker struck by automated vehicle in computerized storage facility	Absence of audible warning; inadequate training; production pressure	Fuller (1984)	Good
1983–1988	13	France	Accidents to operators, installers, repairers in automated plant	Insufficient individual details given in source	Vautrim and Dei-Svaldi (1989)	Good, but too aggregated for current purpose
1988	1	UK	Maintenance electrician killed by unexpected movement of automatic hoist	Maintenance electrician disconnected proximity switch, which sent signal to controller; machine not isolated	Edwards (n.d.)	Good
1989	1	UK	Setter/operator killed by palletizer	Machine cycled when boxes interrupting photoelectric beam removed; transfer table not isolated	Edwards (n.d.)	Good

Table 1 (continued)

Date(s)	No. of deaths	Location	Nature of incident	Probable main cause(s)	Main ref(s).	Data quality
1991	1	UK	Maintenance fitter killed by hold-down arm of feed unit to log saw	Fitter's body interrupted beam of process sensor; machine not isolated	Edwards (n.d.)	Good
1991	1	UK	Maintenance fitter killed in automated brick plant	Fitter inside guarding enclosure observing cause of misalignment of bricks	Edwards (n.d.)	Good
?	3	Netherlands	Explosion at chemical plant	Typing error caused addition of wrong chemical to reactor	SEN 18(2), p. 7	Fair
				Insufficient data		
1986	1	US	Overdose of pain-relieving drugs	Error in medical expert system (?)	Forester and Morrison (1990)	Very poor
1989	1	US	Failure of school-crossing pedestrian signals	Breakdown in radio communications link to computer (?)	Emery (1989)	Poor
1990	1	US	Collision of automated guided vehicle and crane	Unclear	SEN 16(1), p. 10	Very poor
1990	1?	US	Delayed dispatch of ambulance	Logging program not installed (?) (unclear whether death was due to delay)	SEN 16(1), p. 10	Poor
c 1983	1	West Germany	Woman killed daughter after erroneous medical diagnosis	"Computer error"	SEN 10(3), p. 8	Very poor

c 1984	1	China	Electrocution	Unclear	*SEN* 10(1), p. 8	Very poor
c 1989	1	USSR	Electrocution	Unclear	*SEN* 14(5), p. 7	Very poor
?	2?	?	Sudden unintended acceleration of car	Unclear	*SEN* 12 (1), pp. 8–9 *Business Week* 5/29/89, p. 19	Poor; controversial

Sources not listed in notes to chapter: J. T. Dennett, "When toasters sing and brakes fail," *Science 80* 1 (November–December 1979), p. 84; Report of the Independent Inquiry into the Conduct of Isocentric Radiotherapy at the North Staffordshire Royal Infirmary between 1982 and 1991 (West Midlands Regional Health Authority, 1992); Report on the Staff Investigation into the Iraqi Attack on the USS *Stark* (House of Representatives, Committee on Armed Services, 1987); J. A. Adam, "USS *Stark*: What really happened?" *IEEE Spectrum*, September 1987, pp. 26–29; P. T. Mahon, Report of the Royal Commission to Inquire into the Crash on Mount Erebus, Antarctica (Hasselberg 1981); M. Nagamachi, "Ten fatal accidents due to robots in Japan," in *Ergonomics of Hybrid Automated Systems I*, ed. W. Karwoski et al. (Elsevier, 1988); J. G. Fuller, "Death by robot," *Omni*, March 1984, pp. 45–46 and 97–102; T. Forester and P. Morrison, "Computer unreliability and social vulnerability," *Futures*, June 1990, pp. 462–474; E. Emery, "Child's death spurs safety inquiries," *Colorado Springs Gazette Telegraph*, January 11, 1989.

a. Association for Computing Machinery's *Software Engineering Notes*

b. *Aviation Week and Space Technology*

c. General Accounting Office

interference been stated officially to be the cause: the failure of H.M.S. *Sheffield*'s defensive systems to intercept an attacking Argentinean Exocet missile during the Falklands War. At the time of the attack, the *Sheffield* was in urgent radio communication, by satellite, with another vessel in the British task force. Interference from this transmission prevented the *Sheffield* from picking up warning signals on its electronic support measures equipment until it was too late to intercept the Exocet attack. Published reports leave unclear what precise aspect of the equipment was interfered with (although the distinction is difficult for a modern system of this kind, it clearly could have been the radar rather than the information-processing aspect), but there seems to me to be sufficient indication here of possible "computer-relatedness" to merit the inclusion of this case in the data set.

Software Error: 30 deaths

Much of the discussion of the risks of safety-critical computing has focused on software error, and the data set contains two incidents which are clearly of this kind. Two deaths resulted from overdoses from a computer-controlled radiation therapy machine known as the Therac-25. (A third patient also died from complications related to a Therac-25 overdose, but he was already suffering from a terminal form of cancer. The autopsy on a fourth overdosed patient revealed her cause of death to have been cancer rather than radiation overexposure.)

The Therac-25 has two therapeutic modes: the electron mode (used for treating tumor sites on or near the surface of the body) and the x-ray mode (used for treating deeper tumor sites). The latter involves placing in the path of the electron beam a tungsten target (to produce the x-rays) and a "beam flattener" (to ensure a uniform treatment field). Because the beam flattener greatly reduces the intensity of the beam, x-ray therapy requires about 100 times more electron-beam current than electron-mode therapy. If the stronger current were used without the target and the beam flattener in place, the patient would receive a massive overdose. Because of a software error,[8] there was a particular form of data entry on the Therac-25 that caused precisely this to happen, because it shifted the mode from x-ray to electron while leaving the intensity at the current required for x-ray therapy. The data that appeared on the system's display did not reveal that this had taken place, and the fatal error was diagnosed only with some difficulty. Investigation also revealed another dangerous software error, although this seems not to have been implicated in the two deaths included in the data set.[9]

A software error also caused the failure of the Patriot air defense system at Dhahran during the 1991 Gulf War, which led to the deaths of 28 American troops in an Iraqi Scud missile attack. When tracking a target, sophisticated modern radar systems, such as that used for the Patriot, process not the entire reflected radar beam but only a portion of it known as the "range gate." An algorithm embedded in the system software shifts the range gate according to the velocity of the object being tracked and the time and location of its last detection. An error in the implementation of the range-gate algorithm was the cause of the failure to attempt to intercept the attacking Scud.[10]

The Patriot's internal clock keeps time as an integer number of tenths of seconds. That number is stored as a binary integer in the registers of the Patriot's computer, each of which can store 24 binary digits (bits). For use in the range-gate algorithm, this integer number of tenths of a second is converted into a 48-bit floating-point[11] number of seconds—a conversion that requires multiplication of the integer by the 24-bit binary representation of one tenth. The binary representation of $\frac{1}{10}$ is nonterminating, and so a tiny rounding error arises when it is truncated to 24 bits. That error, if uncorrected, causes the resultant floating-point representations of time to be reduced by 0.0001% from their true values.[12]

The Patriot was originally designed to intercept relatively slow targets, such as aircraft. Among the modifications made to give it the capacity to intercept much faster ballistic missiles was a software upgrade that increased the accuracy of the conversion of clock time to a binary floating-point number. At one place in the upgraded software a necessary call to the subroutine was accidentally omitted, causing a discrepancy between the floating-point representations of time used in different places in the range-gate algorithm. The result was an error that was insignificant if the system was used for only a small amount of time but which steadily increased until the system was "rebooted" (which resets time to zero).

The problem was detected before the Dhahran incident. A message was send to Patriot users warning them that "very long run times could cause a shift in the range gate, resulting in the target being offset."[13] A software modification correcting the error was dispatched to users more than a week before the incident. However, the matter was reportedly treated as not one of extreme urgency because Army officials "presumed that the users [of Patriot] would not continuously run the batteries for such extended periods of time that the Patriot would fail to track targets."[14] (Rebooting takes only 60–90 seconds.) Unfortunately, on the night of February 25, 1991, Alpha Battery at Dhahran had been in uninterrupted operation for more than 100 hours, long enough for the error to cause loss of tracking of a target moving as fast as a Scud. As a result, no defensive missiles were launched against the fatal Scud attack.[15] The modified software arrived one day too late.

Human-Computer Interaction Problems: 988 Plus "Tens" of Deaths

Accidents caused by failures in the interaction between human beings and a computer system are typically "messier" in research terms than those caused by clear-cut technical errors or faults. Precisely because such accidents were caused by failures in human-computer interaction, fixing the blame can be contentious. System designers may see the failure as being due to "human error" on the part of the operators. Operators sometimes make allegations of defective technical functioning of the system—often allegations for which no decisive evidence can be found, but which cannot be ruled out *a priori*.

These blame-seeking disputes cloud over what is typically the key point. Many safety-critical systems involving computers rely for their safe functioning upon the correctness of the behavior of both their technical and their human components. Just as failure of technical components is typically regarded as a predicable contingency (and guarded against by duplication or triplication of key parts, for example), so human failure should be expected and, as far as possible, allowed for.

Medical   For the sake of convenience, I have divided the problems of human-computer interaction into five broad categories: medical, military, air, robot-related, and those involving other automated industrial equipment. The medical case is the most clear-cut of the incidents. Systematic underdosing in isocentric radiotherapy for cancer took place at the North Staffordshire Royal Infirmary between 1982 and 1991. Isocentric therapy is a form of treatment in which the system's focal distance is set at the center of a tumor and the machine is rotated so that the tumor is "hit" from several different angles. In calculating the required intensity of radiation for isocentric therapy, it is necessary to allow for the fact that the distance between the source of the beam and the skin of the patient will be less than the 100 cm standard in forms of radiotherapy where each beam is directed not at the tumor but at a point in the skin overlying it. If not, the patient will be overdosed. Before computerization, this correction was always calculated and entered manually. This practice continued at the North Staffordshire hospital after a computerized treatment plan for isocentric radiotherapy was introduced in 1982, because it was not realized that the correction was already being made by the system software. The error was not detected until a new computer planning system was installed in 1991. The result was the underdosing by various amounts of approximately 1000 patients. Subsequent investigation[16] suggests that 492 patients may have been adversely affected by underdosing, of whom 401 had died by the middle of 1993. However, radiation therapy for cancer has a far from total success rate even when conducted perfectly, and so many of these patients would have died in any case. As noted above, the clinical verdict was that the deaths resulting from the error were likely to be "in the tens rather the hundreds."[17]

Military   The two military cases are much less clear-cut in their causes, and their interpretation has been controversial. While patrolling the Persian Gulf in 1987, during the Iran-Iraq war, the U.S. frigate *Stark* was struck by two Exocet missiles fired by an Iraqi aircraft. Like the *Sheffield*, the *Stark* was equipped with computerized systems designed to detect and intercept such an attack. The subsequent U.S. Navy investigation focused mainly on the *Stark*'s alleged lack of combat-readiness[18] ; it should be noted, however, that the United States was at war with neither party to the conflict, and indeed was widely seen as a *de facto* supporter of Iraq. More particularly, it remains puzzling that, although the *Stark*'s electronic warfare system detected the Iraqi Mirage fighter, its crew appear not to have received a warning from the system about the incoming missiles. Each of the main candidate explanations of this would lead to the classification of the incident as computer-related. One possibility is that the system may have detected the missiles but had been programmed to define the French-made Exocet as "friendly" rather than "hostile." (This suggestion was also made in attempts to explain why the *Sheffield* failed to intercept the Exocet attack on it, but was denied by the U.K. Ministry of Defence.) The *Stark*'s SLQ-32 electronic warfare system "had Exocet parameters in its software library, but this software might have been flawed or out of date, a problem the Navy has admitted."[19] Another possibility is that the system did produce a warning, but that this was not noticed by its operator. The operator had switched off its audible alarm feature because the system was issuing too many false alarms.

In the case of the Iranian airliner there is no evidence of any technical malfunction of the sophisticated Aegis computerized combat system aboard the *Vincennes*. Data tapes from the system are consistent with what in retrospect we know to have been the true course of events. It is clear that the crew of the *Vincennes* was operating under considerable stress. While fighting off several small, fast boats, the *Vincennes* had to turn abruptly at full speed to keep its weapons engaged on the targets (it had a fouled gun mount). Such turns cause a vessel such as the *Vincennes* to keel sharply. Furthermore, memories of the surprise airborne attack on the *Stark* were still fresh, and there was little time available in which to check the identification of the radar contact as a hostile Iranian military aircraft.

However, the human error that occurred may bear at least some relation to the computerization of the *Vincennes*. A key role in the misidentification of the Iranian airliner as a military threat was played by the perception of it as descending toward the *Vincennes*, when in fact it was (and was correctly being analyzed by the Aegis system as) rising away from it. Stress undoubtedly played a major role in this misperception. However, the U.S. Navy's report on the incident suggested that "it is important to note, that altitude cannot be displayed on the LSD [large screen display] in real time." After the investigation of the incident, the chairman of the Joint Chiefs of Staff recommended that "a means for displaying altitude information on a contact such as 'ascending' or 'descending' on the LSD should . . . be examined" and that "some additional human engineering be done on the display screens of AEGIS."[20] More generally, it is noteworthy that it was the highly computerized *Vincennes* that misidentified the radar contact, while its technologically more primitive sister ship, the *Sides*, correctly identified the Iranian aircraft as no threat.[21] A possible reason for this is discussed in the conclusion.

Air   The air incidents are also cases where there is typically no evidence of technical malfunction, but where problems seem to have arisen in human interaction with an automated system. The most recent of them has been the focus of intense scrutiny because it involved the first of the new generation of highly computerized "fly-by-wire" aircraft, the Airbus A320,[22] one of which crashed in mountainous terrain after an over-rapid nighttime descent in bad weather to Strasbourg-Entzheim Airport. That there had been a technical failure of the A320's Flight Control Unit computer system was not ruled out by the crash investigators but was judged a "low probability."[23] Instead, the investigators' central hypothesis is that the pilot and the co-pilot, both of whom died in the accident, may have intended to instruct the flight-control system to descend at the gentle angle of 3.3° but, by mistake, instructed it to descend at the extremely rapid rate of 3300 feet per minute. A letter

designation on the Flight Control Unit, and distinct symbols on the primary flight displays, indicate which mode has been selected, but the particular angle or speed chosen were both represented by two-digit numbers. (The interface has since been redesigned so that the vertical speed mode is now represented by a four-digit number.)

Analysis of the cockpit voice recorder suggests that "there was limited verbal communication, coordination and cross-checking between the two pilots,"[24] who had never previously flown together and whose attention may have been distracted from their speed of descent by a last-minute air-traffic-control instruction to change runways and terminal guidance systems. The carrier operating the particular aircraft in question had declined to install automated ground-proximity warning systems in its A320 fleet, at least in part because it believed such systems to give too many false alarms in the type of operation it conducted, so no warning of imminent impact was received by the crew.

The cases involving air navigation errors are, in a broad sense, similar to the case just discussed. Modern long-range civil air transports and nearly all modern military aircraft are equipped with automatic navigation systems, most commonly inertial systems (which are self-contained, not reliant on external radio signals). Inertial navigators are now extremely reliable technically—perhaps to such an extent that undue reliance is placed on their output, with other sources of navigational data not always checked, and flights sometimes continued under what might otherwise be seen as overly dangerous conditions.

Yet such automated systems do have vulnerabilities. Inertial navigation systems need to be fed data on initial latitude and longitude before takeoff. In civil airliners, inertial navigators are typically triplicated to allow the isolation of individual errors. However, some configurations contain an override that allows data to be entered simultaneously to all three systems instead of individually to each. Furthermore, if the inertial system is to "fly" the plane (via an autopilot), details of the requisite course must also be entered (typically in the form of the latitude and longitude of a set of way points, and often as a pre-prepared tape cassette) and the correct "connection" must be made between the inertial system and the autopilot.

The best known of the resulting incidents is the 1983 episode in which a Korean airliner strayed into Soviet air space and was shot down. The fact that the Korean plane was flying over Soviet territory attracted much speculation and led to some lurid conspiracy theories. Data tapes from the airliner released recently by Russia, however, seem to point to

a simple, undetected mistake: the autopilot was connected to the plane's compass rather than to its inertial navigation system. The aircraft therefore followed a constant magnetic heading throughout its flight rather than the intended flight plan.

Robot-Related   The robot-related deaths in the data set seem to manifest a common pattern—one also seen in nonfatal robot-related accidents, on which considerable amounts of data are available. The key risk posed by robotic systems, in contrast to more conventional industrial machinery, is that the movements of the latter are typically repetitive and predictable (the danger points being obvious), whereas robot motion is much less predictable.[25] A robot may suddenly start after a period of inactivity while internal processing is going on; the direction of movement of a robot "arm" may suddenly change; and all points in a robot's work envelope (the three-dimensional space which it can reach) are potentially hazardous. Deaths and other serious accidents involving robots thus nearly always involve the presence of a worker within the envelope of a powered-up robot. Often the worker is struck from behind and is pushed into another machine or against a fixed obstacle.

Workers are typically instructed not to enter the envelopes of powered-up robots, so it is tempting to ascribe all such accidents to "human error" alone. But to do this would be to miss several points. First, the human error involved is an entirely foreseeable one, and so one that should be anticipated in system design. However (this is my second point), in some early installations no barriers were present to inhibit workers from entering the envelope, and training was sometimes inadequate. Third, there is little reason to think that workers enter robot envelopes gratuitously. They may, for example, be cleaning or attending to some small snag in the robot installation. It may be that there are pressures in the situation, such as to maintain productivity, that encourage workers to do this without switching off the power supply. Fourth, some fatal accidents have occurred when a worker did indeed switch off power to the robot but it was switched back on either inadvertently by him or by another worker. Installation design could guard against this, at least to some extent.[26]

Other Automated Industrial Equipment   While robot-related accidents have attracted considerable interest, there has been much less attention to fatal accidents involving other kinds of automated industrial equipment, although the latter appear likely to be considerably more numerous. Again, a particularly dangerous situation (the situation, for example, in three of the five U.K. fatalities identified by Edwards[27]) arises when

workers enter or reach into computer-controlled machinery when it has stopped but is still powered up, so that it can be restarted by sensors, by faults in the control system, or by signals from other locations.[28]

As in the robot case, accidents of this type should not be disregarded as gratuitous and unpredictable "human error." The two systematic studies of this type of accident which I have been able to locate[29] both suggest that accidents with automated equipment typically involve system designs that make some necessary work activities—such as finding and rectifying faults, adjusting workpieces, and (especially) clearing blockages—dangerous. Sometimes the guarding is deficient or there are defects in isolation systems. Other dangers arise from having a process "stop" device that halts the machine but does not isolate it; the resultant accidents are far from unpredictable. More generally, accidents involving unsafe work systems typically point to organizational rather than individual failures. For example, the maintenance electrician killed in Britain in 1988 by unexpected movement of an automatic hoist was reportedly "expected to maintain a system which had been supplied without an interlocked enclosure, and without any form of operating or maintenance manual."[30]

Conclusions

How Safe Are Computers?

The data presented here are clearly insufficient for any quantitative measure of levels of risk associated with computer systems. For that to be possible, we would need to know not just numbers of accidental deaths but also levels of "exposure": total usage of computerized radiation therapy machines, total passenger miles or hours flown in fly-by-wire aircraft or in planes reliant upon inertial navigators, total hours of work spent in proximity to industrial robots or close to automated plant, and so on. I do not possess such data. Nor am I sure that the aggregate result of such an exercise would be meaningful: the risks involved in such different activities are scarcely commensurable. Furthermore, even the crudest quantitative assessment of the benefits and dangers of computerization would also require data on the risks of analogous activities conducted without the aid of computers. In limited spheres such as radiotherapy and (perhaps) civil aviation the comparison might be an interesting research exercise,[31] but often it is impossible. For example, effective defense against ballistic missiles without the aid of computers is hard to imagine; thus, there is no comparator case.

I can answer the question of the overall safety of computer systems in only the crudest sense: the prevalence of computer-related accidents as a cause of death. In that sense, a total of no more than about 2000 deaths so far, worldwide, is modest. For example, in 1992 alone, there were 4274 deaths in the United Kingdom in traffic accidents.[32] By comparison, computer-related accident has not, up until now, been a major cause of death.

Nevertheless, there are no grounds here for complacency. In the context of activities with a generally excellent safety record, such as scheduled air transport, even a small number of major accidents becomes most worrying. In addition, deaths are sometimes only the visible tip of what can be a much larger "iceberg" of serious injuries, minor injuries, and "near misses." This is, for example, clearly the case for accidents involving robots and other forms of automated industrial equipment. Edwards's data set contains 14 major injuries and 40 minor ones for each fatality.[33] These multipliers would most likely be smaller in other sectors, notably air travel,[34] but there have clearly been a substantial number of computer-related injuries to add to the total of fatalities. Furthermore, even a cursory reading of the "risks" reports in *Software Engineering Notes* leaves one convinced that the number of "near misses" is likely to be considerable.

In addition, we are dealing here with a relatively new problem, where the record of the past is unlikely to be a good guide to the future, since the incidence of computerization, its complexity, and its safety-criticality are increasing.[35] True, an unequivocal trend in time in the data set cannot be established: the numbers of deaths are dominated too much by the three incidents in 1979, 1983, and 1988 in each of which over 200 people were killed. It is, however, striking that there is no well-documented case of a computer-related accidental death before 1978. Of course, that may to some degree be an artifact of the reporting system: "risks" reports in *Software Engineering Notes* were only beginning then. But attention to the problem of computer safety goes back at least to the late 1960s,[36] and so it seems unlikely that large numbers of deaths before 1979 have gone unrecorded in the literature.

The Need for Systematic Data Collection

The attempt to conduct an exercise such as this quickly reveals the need for systematic collection of data on computer-related accidents. There are occasional pieces of excellent scientific detective work, such as Robert Skeel's uncovering of the precise role of rounding error in the

Dhahran incident[37] (a role not fully evident even in the otherwise useful report by the General Accounting Office).[38] There is one superb detailed case study: Leveson and Turner's investigation of the Therac-25 accidents. There are also "islands" of systematic data on particular sectors, such as Edwards's study of accidents involving computer-controlled industrial equipment in Britain. But the RISKS Forum and *Software Engineering Notes* remain the only cross-sectoral, international database. Remarkable and commendable efforts though they are, they are no substitute for properly resourced, official, systematic data collection.

A large part of the problem is the diversity of regulatory regimes which cover safety-critical computing. By and large, what has happened is that computer use is covered by the regulatory apparatus for its sector of application—apparatus which normally will predate the use of digital computers in that sector and which will naturally be influenced strongly by the history and specific features of the sector.

Yet there is a strong argument to be made that the introduction of digital computers, or of programmable electronic devices more generally, introduces relatively novel hazards which have common features across sectors. Software-controlled systems tend to be logically complex, so operators may find it difficult to generate adequate "mental models" of them. Their complexity also increases "the danger of their harboring potentially risky design faults," and "the largely discrete nature of their behavior . . . means that concepts such as 'stress,' 'failure region,' [and] 'safety factor,' which are basic to conventional risk management, have little meaning."[39] Digital systems are characterized by the "discontinuity of effects as a function of cause. There is an unusual amplification of the effects of small changes. Change of a single bit of information (whether in a program or data) can have devastating effects."[40] Installing identical programmable systems in duplicate or triplicate offers only limited protection, since errors in software or hardware design can be expected to produce "common-mode failures" that manifest themselves in each system simultaneously. Even installing different systems may be less of a protection against common-mode failures than might be imagined, because in some cases the different programs produced by separate programmers can still contain "equivalent logical errors."[41]

If this is correct (and some of these phenomena can be found among the cases presented here[42]), the risks associated with computer systems can be expected to have generic, technology-specific features as well as sector-specific, application-specific ones. It could thus be that a great deal of important information is being lost through the partial and pre-dominantly intrasectoral nature of current information gathering.

Nor is this simply a matter of the need for an empirical basis for research. There is evidence from other areas that the existence of independent data-gathering systems in itself makes systems safer, especially when data is collected on "incidents" as well as on actual "accidents," when the gathering of data on the former is on a no-fault and confidential basis (to reduce to a minimum the motivations to underreport), and when results are well publicized to relevant audiences. The incident-reporting system in civil air transport is a good example.[43] The British Computer Society has recently called for a system of registration of safety-related computer systems with mandatory fault reporting. Such a system would be an important contribution to improving the safety of such systems as well as a valuable basis for research.[44]

The Technical and the Human

Computer-related accidental deaths caused *solely* by technical design flaws are rare. The fatalities in the data set resulting from problems of human-computer interaction greatly outnumber those resulting from either physical causes or software errors. True, some of the "interaction" cases may mask software errors or hardware faults; on the other hand, one of the cases of software error and some of the cases of physical causes also have "interaction" aspects. The Dhahran deaths were not due entirely to the omitted call to the time-conversion subroutine; assumptions about how the system would be operated in practice and delays in the arrival of the corrected software were also crucial. Leveson and Turner argue that even in the Therac-25 deaths—whose cause was perhaps the closest in the well-documented cases in this data set to a "pure" technical error—software error "was only one contributing factor." They argue that organizational matters, such as what they regard as inadequacies in the procedure for reporting and acting upon incidents, were also important, as were beliefs about system safety.[45]

Indeed, multi-causality may be the rule rather than the exception. More computer-related accidental deaths seem to be caused by *interactions* of technical and cognitive/organizational factors than by technical factors alone; computer-related accidents may thus often best be understood as *system* accidents.[46] In the absence, in many cases, of the depth of understanding now available of the Therac-25 and Dhahran deaths, or of the systematic coverage of Edwards's study of industrial accidents, this hypothesis cannot be verified conclusively, but such data as are available make it plausible.

There is, however, another worrying category: accidents in which there is unimpaired technical operation of a computerized system, as far as we can tell, and yet disastrous human interaction with it. Contrasting the *Vincennes*'s erroneous identification of its radar contact and the *Sides*'s correct one, Gene Rochlin argues that computerization can result in a changed relationship of human beings to technology, and his argument has wider implications than just for the analysis of this particular incident.[47] In a traditional naval vessel or aircraft, human beings play a central role in processing the information flowing into the vehicle. By contrast, as computerization becomes more intensive, highly automated systems become increasingly primary. Ultimate human control—such as a human decision to activate the firing mode of an automated weapon system—is currently retained in most such systems.[48] But the human beings responsible for these systems may have lost the intangible cognitive benefits that flow from their having constantly to integrate and make sense of the data flowing in.

In such a situation, danger can come both from stress and from routine. Under stress, and pressed for time, the human beings in charge of automated military systems cannot be expected always to question whether the situation they face is one that "the elaborate control system in which they were embedded, and for which they were responsible"[49] was designed to meet. We should not be surprised if sometimes they act out "the scenario compatible with the threat the system was designed to combat."[50] Nor should we be surprised if, after hundreds or thousands of hours' personal experience of flawless functioning of automated flight equipment, pilots begin to trust that equipment too much and then fail to check other information available to them.

To make computer systems safer, we need to address not merely their technical aspects but also the cognitive and organizational aspects of their "real-world" operation. Psychologists and organizational analysts have to be involved in this effort, along with computer scientists. If this does not happen, then there is a risk that purely technical efforts to make computer systems safer may fail. Not only are such efforts addressing only part of the problem; they may conceivably even increase the risks through their effect on *beliefs* about computer systems. There is a danger of what several contributors to *Software Engineering Notes* have called the "*Titanic* effect": the safer a system is believed to be, the more catastrophic the accidents to which it is subject.

Self-Negating Prophecies

Although I have focused on the risks of computerization in this chapter, it is of course necessary to bear in mind the latter's very considerable benefits. The use of computer systems clearly offers considerable economic advantages. In some applications it may also be beneficial environmentally—for example, in reducing aircraft fuel consumption and resulting environmental damage. There are, furthermore, already examples of programmable electronic systems whose safety records, in extensive practical use, are impressive.[51] In many contexts computer use can actually enhance human safety—e.g., in automating the most dangerous parts of industrial processes or in warning of potentially dangerous situations. Wisely used, relatively simple forms of automation, such as ground-proximity warning systems on aircraft, can potentially save many lives: the most common cause of death in scheduled air travel is now "controlled flight into terrain" by technically unimpaired aircraft.[52]

There is thus every reason for optimism: with good research, careful regulation, and intelligent application, the computer's risk-benefit account can be kept positive. However, the relatively modest number so far of computer-related accidental deaths—particularly the small number caused by software error—is in one sense puzzling. While computer systems appear empirically to be reasonably safe, there are, as noted above, grounds for regarding them as inherently dangerous:

A few years ago, David Benson, Professor of Computer Science at Washington State University, issued a challenge by way of several electronic bulletin board systems. He asked for an example of a real-time system that functioned adequately when used for the first time by people other than its developers for a purpose other than testing. Only one candidate for this honor was proposed, but even that candidate was controversial. . . . As a rule software systems do not work well until they have been used, and have failed repeatedly, in real applications.[53]

The reason for this apparent paradox (an error-ridden technology that nevertheless has a reasonably good safety record in practice) is almost certainly conservatism in design: "restraint . . . in introducing [computers] into safety-critical control loops" and "defense-in-depth"—hardware interlocks, backup systems, and containment devices which reduce the impact of computer failure.[54] If this is correct, then we have an interesting case of a self-negating prophecy. I have already noted one side of this prophecy: the extent that operators and users believe the computer to be safe (completely reliable, utterly trustworthy in its output, and so on) may make it dangerous. Here is the prophecy's other side: until now, system designers have generally believed the computer to be dangerous, and therefore have fashioned systems so that it is in practice relatively safe. Those who work in this field, therefore, have a narrow path to tread. They must do the necessary research to make computer systems safer, and they also must ensure that the results of this research are well implemented, bearing in mind that much of the problem is not technical but cognitive and organizational. At the same time, they must do nothing to encourage complacency or overconfidence in regard to the safety of computer systems. To make computer systems safer while simultaneously keeping alive the belief that they are dangerous: that is the paradoxical challenge faced by the field of computer-system safety.

Acknowledgments

This work was supported financially by the Joint Committee of the U.K. Economic and Social Research Council (ESRC) and Science and Engineering Research Council (SERC) (grant J58619). While the work was underway, the Joint Committee's projects were transferred to the ESRC, where our work became grant R00029008. Further support came from the ESRC Programme on Information and Communication Technology (grant WA35250006), ESRC grant R000234031, and the U.K. Safety Critical Systems Research Programme (grant J58619).

I am grateful for bibliographic help, data, ideas and pointers received from Robin Bloomfield, Nick Curley, Bob Lloyd, Peter Mellor, Peter Nicolaisen, Gene Rochlin, Scott Sagan, and, especially, Moyra Forrest and Rosi Edwards. I also owe a broader debt to Peter Neumann and the many contributors to "Risks" reports in *Software Engineering Notes*. Without that body of data, this chapter could not have been written.

10

Tacit Knowledge and the Uninvention of Nuclear Weapons
(with Graham Spinardi)

Over the last three decades, an alternative account of scientific knowledge has gradually emerged to rival the traditional view. In the latter, scientific knowledge and science-based technology are universal, independent of context, impersonal, public, and cumulative; the practice of science is (or ought to be) a matter of following the rules of the scientific method. The alternative account emphasizes instead the local, situated, person-specific, private, and noncumulative aspects of scientific knowledge. Scientific practice is not the following of set rules; it consists of "particular courses of action with materials to hand"[1]—action that is fully understandable only in its local context; and materials that are inescapably heterogeneous, including human and nonhuman elements.[2] Universality and context independence, in this new view, are not to be taken as given but must be analyzed as precarious achievements—for example, as the result of the successful construction of wide-ranging networks linking human and nonhuman actors.[3]

This chapter focuses on a single thread in the extensive, tangled, and sometimes contradictory web of arguments that constitute this alternative account of science.[4] That thread is the contrast between explicit and tacit knowledge. Explicit knowledge is information or instructions that can be formulated in words or symbols and therefore can be stored, copied, and transferred by impersonal means, such as written documents or computer files. Tacit knowledge, on the other hand, is knowledge that has not been (and perhaps cannot be) formulated completely explicitly and therefore cannot effectively be stored or transferred entirely by impersonal means. Motor skills supply a set of paradigmatic examples of tacit knowledge in everyday life. Most of us, for example, know perfectly well how to ride a bicycle, yet would find it impossible to put into words how we do so. There are (to our knowledge) no textbooks of bicycle riding, and when we come to teach children to ride we do not give them long lists of written or verbal instructions; instead we attempt to show them what to do, and we encourage them in the inevitably slow and error-ridden process of learning for themselves.

That many human activities depend upon tacit knowledge is widely recognized. It is one reason why many occupations are learned by apprenticeship to a skilled practitioner. Tacit knowledge is also a major barrier to the encapsulation of human knowledge in artificially intelligent machines.[5] However, the focus on *method* in the traditional view of science downplayed the role of tacit knowledge, and the image of technology as "applied science" led to a similar deemphasis there.[6] Nevertheless, several authors have suggested that tacit knowledge is crucial to the successful pursuit of science and technology.[7]

H. M. Collins, above all, has shown the connections between an emphasis on tacit knowledge and other aspects of the alternative account of science. The dependence of successful scientific experiment upon tacit knowledge makes experiment a less solid bedrock of science than the traditional view assumes.[8] Because tacit knowledge is transmitted from person to person, rather than impersonally, there are greater barriers to the spread of competence than the traditional view might lead us to expect. If science rests upon specific, hard-to-acquire, tacit skills, then there is a sense in which scientific knowledge is always local knowledge. It is, for example, often small "core sets," rather than wider scientific communities, that resolve scientific controversies.[9]

Most important is how an emphasis on tacit knowledge indicates one way in which science and technology are not simply cumulative endeavors that result in permanent advances.[10] Barring social catastrophe, explicit knowledge, if widely diffused and stored, cannot be lost. Tacit knowledge, however, *can* be lost. Skills, if not practiced, decay. If there is no new generation of practitioners to whom tacit knowledge can be passed on from hand to hand, it may die out.

Of course, such a loss need not be permanent. Some modern archaeologists, for example, believe themselves to have recaptured the skills, long extinct in industrial societies, of Paleolithic flint knappers. The key point, however, is that the re-creation of tacit knowledge after its loss cannot simply be a matter of copying the original, because there is no sufficient set of explicit information or instructions to follow. The reacquisition of tacit knowledge after its extinction is, therefore, not necessarily any easier than its original acquisition, and may well be protracted and difficult. Furthermore, it is hard to know whether the original skill has been reacquired or a new, different skill created: there are, for example, clearly limits on the extent to which we can tell whether modern archaeologists knap in the same way as their ancestors.[11]

Such considerations may seem very distant from modern science and technology, especially in the area of nuclear weapons. The conventional wisdom about the latter is that knowledge of nuclear weapons cannot plausibly be lost—that nuclear weapons cannot be uninvented. In the words of a group of prominent U.S. defense and international relations scholars, "the discovery of nuclear weapons, like the discovery of fire itself, lies behind us on the trajectory of history: it cannot be undone. . . . The atomic fire cannot be extinguished."[12]

Implicitly, however, this conventional wisdom rests on the traditional view of science and technology as impersonal and cumulative. True, if explicit knowledge were sufficient for the design and production of nuclear weapons there would be little reason to doubt the conventional wisdom. Half a century of official and unofficial dissemination of information from the nuclear weapons laboratories, together with the normal publication processes in cognate branches of physics and engineering, mean that much of the relevant explicit knowledge is now irrevocably in the public domain.

Suppose, though, that the alternative view of science was true of nuclear weapons—in particular, that specific, local, tacit knowledge was crucial to their design and production. Then there would be a sense in which relevant knowledge could be unlearned and these weapons could be uninvented. If there were a sufficiently long hiatus in their design and production (say, two generations), that tacit knowledge might indeed vanish. Nuclear weapons could still be re-created, but not simply by copying from whatever artifacts, diagrams, and explicit instructions remained. In a sense, they would have to be reinvented.[13]

Our concern here is only with these possible consequences of a lengthy hiatus in the development of nuclear weapons; we do not discuss the desirability, durability, or likelihood of such a hiatus (none of which, of course, is self-evident). However, considerations of tacit knowledge are not relevant only to comprehensive nuclear disarmament. Although the majority of current nuclear weapons states show no inclination to disarm entirely, they may well in the near future turn current voluntary moratoria into a permanent ban on nuclear testing.

As we shall see, nuclear testing has been a crucial part of the "epistemic culture"[14] of nuclear weapons designers. Testing has made visible—to them and to others—the quality (or otherwise) of the non-explicit elements constituting their "judgment." In its absence, certification of the safety and reliability of the remaining arsenals, and the design of any new nuclear weapons, will have to rely much more heavily on explicit knowledge alone—in particular, on computer simulation. This is a prospect that many of the current generation of nuclear designers view with trepidation.

Furthermore, the balance of explicit and tacit knowledge in the design of nuclear weapons has clear implications for their proliferation. Hitherto, the most prominent barrier to proliferation has been control over fissile materials. There is alarming though not yet conclusive evidence that such control has broken down seriously in the former Soviet Union.[15] If it becomes possible for aspirant nuclear states or terrorist groups simply to buy fissile material in the requisite quantities, then clearly a great deal hangs on precisely what knowledge they need to turn that material into weapons.

Before we turn to such matters, however, we need to assess the evidence concerning the role of tacit knowledge in nuclear weapons design. Most of the chapter deals with this evidence. After this introduction, we begin with brief accounts of the main types of nuclear weapons and of the current extent of explicit public knowledge of their design. We then take a first cut at the question of whether knowledge of that sort is, on its own, sufficient for designing and constructing an atomic bomb. The evidence drawn on in this section is the history of the wartime effort by the Los Alamos laboratory to turn explicit knowledge of nuclear physics into working bombs.

We then move to a second form of evidence concerning the role of tacit knowledge in nuclear weapons design: designers' own accounts of the nature of the knowledge they deploy. This section is based on a series of semi-structured interviews we conducted with nearly fifty current or retired members of nuclear weapons laboratories, including nuclear weapons designers and computing experts specializing in support for the computer modeling of nuclear explosive phenomena. These interviews dealt only with unclassified matters; we sought no security clearance of any kind, and none was granted, and we neither asked for nor received information on the design features of particular weapons. However, we were able to discuss, in reasonable detail, the *process* of design and the knowledge used in that process.[16]

The third form of evidence about the role of tacit knowledge in designing nuclear weapons is less direct, and it concerns the spread of nuclear weapons. Despite efforts to prevent the movement of personnel between nuclear weapons programs, five states, in addition to the technology's American originators, have successfully conducted nuclear explosions, and three more are widely agreed to have—or, in the case of South Africa, to have had—the capacity to do so. *A priori*, this record of successful (and relatively impersonal) spread seems to imply that the role of local, tacit knowledge in nuclear weapons design is minimal. We draw on what is known of the histories of these programs to suggest that this is not so. Even the Soviet and British programs, both of which began by trying to reproduce an existing American design, have more of the characteristics of reinvention than of simple copying.

Our argument is that these three bodies of evidence, although not conclusive, strongly suggest that tacit knowledge has played a significant role in nuclear weapons design. The final section of the chapter goes on to consider whether the availability of "black box," "off the shelf" technologies eliminates this role. We contend that the history of the Iraqi nuclear weapons program suggests that it does not. We concede, however, that there are three reasons not to overstate the consequences of the role of tacit knowledge in nuclear weapons design: previous programs provide useful information on the "hardness"[17] of the task; relevant tacit knowledge can come not only from previous nuclear weapons programs but also from civilian nuclear power and non-nuclear military technologies; and we cannot rule out *a priori* the possibility of simpler routes to the construction of crude but workable weapons.

We conclude, therefore, that it is necessary to take a broader view of what it would be deliberately to uninvent nuclear weapons. However, even if deliberate uninvention does not take place, an accidental uninvention, in which much current tacit knowledge is lost, seems quite plausible, and its consequences, we suggest, may well be of considerable significance in the years to come. At the very least, we hope that this investigation of the role of tacit knowledge in nuclear weapons design demonstrates that the sociology of science and technology, sometimes condemned as apolitical and even amoral,[18] need possess neither of those characteristics.

The Science and Technology of Nuclear Weapons

Two physical processes are fundamental to nuclear weapons: fission and fusion. Fission is the splitting of an atomic nucleus by a neutron; fusion is the joining of two nuclei to form a single heavier one. "Atomic" bombs, such as the ones dropped on Hiroshima and Nagasaki, rely on fission. In such weapons, chemical explosives are used to turn a "subcritical" mass or masses of fissile material (in practice usually uranium 235 and/or plutonium 239) [19] into a "supercritical" mass, in which nuclear fission will become a self-sustaining, growing chain reaction.

One way of doing this is the *gun* method, in which the supercritical mass is created by shooting one subcritical piece of fissile material into another by means of propellant explosives. That was the basic design of the bomb dropped on Hiroshima on August 6, 1945. However, the first atomic bomb (exploded at the Trinity site, near Alamogordo, New Mexico, on July 16, 1945), the bomb that devastated Nagasaki, and most modern atomic bombs are of the *implosion* design (figure 1).

At the heart of an implosion weapon is a subcritical fissile core, typically of uranium 235 and/or plutonium 239. Around this core is a shell of chemical high explosives, built into a lens structure designed to focus its blast into a converging, inward-moving shock wave. Electrical systems detonate the chemical explosives as close to simultaneously as possible, and the resulting blast wave compresses the inner fissile core, the consequent increase in density making it supercritical. In the very short time before the core starts to expand again, an "initiator" (now normally external to the core, but in early designs inside it) produces a burst of neutrons to begin the fission chain reaction. The reaction is reinforced by an intermediate shell made of a material that reflects neutrons back inward, and this (or another) intermediate shell also acts as a "tamper," helping to hold the core together. If the bomb has been designed correctly, the fission reaction in the core is self-sustaining and

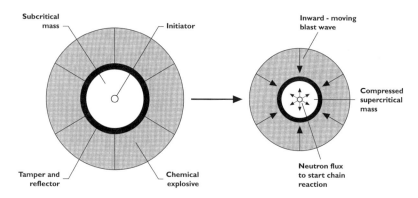

Figure 1
A highly schematic illustration (not to scale) of an atomic or fission bomb of implosion design.

growing in intensity, and it releases enormous amounts of energy as radiation, heat, and blast.

In a "thermonuclear" or "hydrogen" bomb, the destructive energy is provided by fusion as well as by the fission employed in an atomic bomb. The total release of energy, and thus the destructive power of a thermonuclear weapon, can be expected to be many times larger than that of a fission weapon; hence, it was originally referred to as the "Super." When the latter was first discussed in the 1940s, the design envisaged— the "classical Super"—relied for the initiation of fusion essentially upon the heating, by a fission explosion, of liquid deuterium (one of the isotopes of hydrogen). In early 1951, however, the mathematician Stanislaw Ulam and the physicist Edward Teller proposed a design in which the explosion of the fission "primary" compresses, as well as heats, a fusion "secondary." That design, or its independently developed equivalents, appears to be the basis of all modern hydrogen bombs.

Public Knowledge

At this general level, the design of a fission bomb is fully public knowledge, and little about the hydrogen bomb remains secret. A mixture of an idealistic desire for informed public debate and a pragmatic concern to avoid lurid speculation led the U.S. government (to the alarm of the more cautious British government) to release, in 1945, a reasonably detailed history of the effort to construct an atomic bomb: Henry D. Smyth, *Atomic Energy: A General Account of the Development of Methods of using Atomic Energy for Military Purposes under the Auspices of the United States Government.*[20] This history, commonly referred to as the Smyth Report, outlined the military significance of the process of nuclear fission, described the basic principle of the "gun" weapon, and described in general terms the various processes used to produce fissile materials. Implosion designs were not discussed in the Smyth Report. More recently, however, officially sanctioned publications have freely described implosion weapons at a level of detail roughly equivalent to that employed here,[21] and unofficial sources[22] have discussed their designs in far greater detail.

Even without such publications, much could be inferred from relatively elementary physics. As long ago as 1946 it was reported that a "Midwestern teacher of high-school physics" had used the information contained in the Smyth Report successfully to calculate the size of an atomic bomb.[23] Since then, there have been reports that "undergraduates at Princeton and MIT have drafted roughly feasible atomic weapon designs, drawing only from unclassified documents,"[24] as had scientists awaiting security clearance at the nuclear weapons laboratories.[25]

Although the precise workings of the Teller-Ulam configuration have never been disclosed officially, the basic role of fusion in hydrogen bombs was discussed openly from the 1950s on. In 1979 the radical U.S. magazine *The Progressive* sought to publish an article (Howard Morland, "The H-bomb secret"[26]) which contained conjectures about the nature of the Teller-Ulam configuration. Through the law courts, the U.S. Department of Energy tried, ultimately unsuccessfully, to prevent its publication. That effort backfired, since it drew much attention to and gave *de facto* official confirmation of some of Morland's inferences[27]; indeed, it made the gathering and disseminating of information on hydrogen bomb design something of a libertarian cause. A student working on behalf of the American Civil Liberties Union discovered, in the public-access stacks of the library at Los Alamos, a mistakenly declassified 1956 technical report on nuclear weapons development, UCRL-4725, which contained detailed information on hydrogen bomb design.[28] By the late 1980s, enough information had entered the public domain for hydrogen, as well as atomic, bomb design to be discussed in detail in an illustrated "coffee table" book.[29]

From Idea to Artifact

Would public knowledge of this kind be sufficient to build a nuclear weapon? Let us narrow the question to a fission bomb—as we have noted, all mainstream[30] hydrogen bomb designs rely upon a fission bomb to initiate fusion, so if a fission bomb cannot be built neither can a hydrogen bomb.

One way of approaching the question is historical. Let us first consider the state of relevant, explicit knowledge about nuclear physics as it stood at the time of the establishment of the Los Alamos laboratory, in 1943, and then examine what more the laboratory had to do to permit the explosion of the first atomic bombs in the summer of 1945.

In April 1943, the theoretical physicist Robert Serber gave a five-lecture "indoctrination course" to Los Alamos's first recruits in which he summarized the most salient aspects of the available knowledge relevant to the task before them.[31] Serber's lectures show that the "idea" of an atomic bomb, as described above, was essentially in place by early 1943. Indeed, the lectures, whose intended audience consisted primarily of physicists, were considerably more detailed and quantitative than our verbal description. They summarized relevant aspects of a recent but rapidly maturing body of knowledge, already "normal science" in the terminology of Thomas Kuhn.[32] Much "puzzle solving"[33] had still to be done; in particular, detailed investigations of the interactions between neutrons and the nuclei of uranium and plutonium were necessary. However, by the spring of 1943, while "there was still much work to be done in nuclear physics proper . . . enough was known to eliminate great uncertainties from this side of the picture."[34]

The physicists involved were confident enough of the status of their knowledge to feel reasonably sure of the likely destructive power of the weapon they hoped to build. George Kistiakowski, a professor of chemistry at Harvard, had argued that "a fission weapon would be only one-tenth as effective" as a chemical one, but the physicists produced calculations predicting that an atomic weapon could have a force at least a thousand times that of a chemical explosive.[35] Indeed, they were more perturbed by Edward Teller's 1942 speculation that the atomic bomb might be *too* powerful, extinguishing all life on earth by setting off runaway fusion of the nitrogen in the atmosphere. However, the "common sense"[36] of the elite physicists involved or consulted suggested that this was implausible. Detailed calculations based on well-established explicit knowledge of nuclear forces supported that common sense.[37]

To some physicists, indeed, it seemed that the relevant explicit knowledge was mature enough to make Los Alamos's remit essentially trivial. To produce usable quantities of plutonium and uranium 235 was clearly a major industrial task, but that was not the laboratory's job. Edward Teller recalls being warned by a friend, the theoretical physicist and future Nobel Laureate Eugene Wigner, not to join the new laboratory: ". . . the only difficulty, according to Wigner, was the production of the needed nuclear explosive material, that is, plutonium. Once we had enough of that, he asserted, it would be easy and obvious to put together an atomic bomb."[38]

Even those who set up the new laboratory seem initially to have underestimated greatly the task they were undertaking. In May 1942, the future director of Los Alamos, J. Robert Oppenheimer, wrote that the theoretical problems of designing an atomic bomb probably could be solved by "three experienced men and perhaps an equal number of younger ones."[39] When the experimental physicist John H. Manley drew up the first plans for the new laboratory in the fall of 1942, he provided accommodation for "six theoretical physicists with six assistants, twelve experimentalists with fourteen assistants, and five secretaries." Oppenheimer originally enlarged Manley's plans only marginally, allowing space for a little expansion, for a low-temperature laboratory for research on the "Super," and for a small engineering and machining facility.[40]

Less than three years later, however, the technical staff of the Los Alamos laboratory numbered around 3000.[41] One reason was the decision that it made more sense to purify plutonium at Los Alamos rather than beside the reactors at Hanford in Washington State.[42] More generally, though, what had seemed in advance to be simple practical matters turned out to be far less straightforward than anticipated. To begin with, it was assumed that, once the necessary fissile materials were available, fabricating a bomb would be straightforward, at least if the "obvious"[43] gun design were adopted (implosion was acknowledged to be more complicated): "We thought we could just go to the military and buy a gun that would blow a couple of pieces [of fissile material] together fast enough to make an explosion. But fast enough turned out to be really very fast. On top of that, the whole business had to be carried by a B-29 and dropped . . . and the Navy or Army just don't make guns for those purposes. All of this put very stringent size and shape and weight requirements on a gun. The upshot was that for the most part the gun was designed and tested at Los Alamos."[44] Even with help and advice from the Naval Gun Factory, the Naval Ordnance Plant, the Navy's senior gun designer, and the Bureau of Mines, the task was a demanding one. Furthermore, the Los Alamos team had to learn both how to refine the uranium 235 produced by the separation plant at Oak Ridge, Tennessee, and how to form it into the necessary shapes—tasks that led them into matters such the design of crucibles and vacuum furnaces.[45]

The "really big jolt,"[46] however, came in the first half of 1944, when it became apparent that reactor-produced plutonium differed in a crucial respect from the same element produced earlier, in tiny quantities, in laboratory cyclotrons.[47] Finding the properties of the latter type of plutonium had been demanding enough, and to help in the work Los Alamos hired an entomologist and other biologists skilled in handling small samples.[48] The new problem was that the reactors were producing not just plutonium 239, the dominant isotope in the cyclotron samples, but also significant quantities of plutonium 240. That had been anticipated, but what was unexpectedly found in the spring of 1944 was that the heavier isotope seemed to have a much higher rate of spontaneous neutron emission. The planned plutonium gun, nicknamed Thin Man, seemed likely to "fizzle"—to suffer a premature, partial chain reaction—and in July of 1944 it was abandoned. It was a painful crisis, and Oppenheimer had to be persuaded not to resign his directorship.[49]

The plutonium gun's problems did not affect the feasibility of a uranium gun, which had originally been given less priority but which was now moved to center stage. However, the physicists involved were reluctant to jettison plutonium entirely. The new element was, quite literally, their community's creation: unlike uranium, it does not exist in nature. As Manley later put it: "The choice was to junk the whole discovery of the chain reaction that produced plutonium, and all of the investment in time and effort of the Hanford plant, *unless* somebody could come up with a way of assembling the plutonium material into a weapon that would explode."[50]

In implosion, the idea of how to do that already existed. With a gun design, only a relatively low-powered propellant explosive could be used, for fear of simply blowing the device apart before the nuclear chain reaction had time to develop. Implosion, however, would permit the use of a high explosive, and the resultant sudden creation of a critical mass by compression reduced the risk of a fizzle. But implosion moved the Los Alamos scientists onto new terrain.

In part, the move was into areas of physics with which they were less familiar: implosion is a problem in hydrodynamics rather than just in nuclear physics. To begin with, the members of the Los Alamos team— perhaps the most talented group of physicists ever to be gathered together at a single site to achieve a single goal—seem to have felt that this should not be an insuperable barrier. However, "their work suffered from being too formal and mathematical."[51] Rescue came from the British delegation to Los Alamos, which included an immensely experienced hydrodynamicist, Geoffrey Taylor. "Most of the simple intuitive considerations which give true physical understanding" are reported to have come from discussions with Taylor.[52]

Of course, the Los Alamos team could not responsibly proceed on the basis of intuition alone. Frantic efforts were also made to achieve a mathematical and experimental understanding of implosion. The former was greatly assisted by a batch of IBM punched-card machines received by the laboratory in April 1944, but their results were not entirely trusted. For weeks a group of women (largely wives of the almost exclusively male Los Alamos scientists) ground their way through the massive quantities of arithmetic needed to flesh out a mathematical model of implosion, using hand-operated mechanical calculators. Different women were assigned different tasks—adding, multiplying, cubing, and so on—in a kind of reconfigurable arithmetical assembly line.[53]

The implosion experiments were demanding in a different way. By using an inert core instead of plutonium, implosion could be investigated without risking a nuclear explosion. However, new procedures and new instrumentation had to be developed in order to record what went on in implosion: x-ray "flashes," ultrafast cameras, placing a gamma-ray source at the center of the sphere and detecting the resultant rays after they passed through the shell and high explosive, and various other methods. Each of these, in turn, required other problems to be solved; for example, the gamma-ray source (radiolanthanum 140) had itself to be isolated from radioactive barium, and a "hot" laboratory in which test implosions could take place without contaminating large areas had to be built.[54]

The results of the experiments were less reassuring than those of the mathematical model. It was worrisome that the experimentally measured velocity of implosion appeared to be less than the model predicted. A hollow shell was more attractive than the solid sphere eventually employed, because a shell required less plutonium. However, jets of molten material seemed to squirt ahead of an imploding shell, upsetting symmetry and creating turbulence. (The possibility that they were optical illusions was considered.[55]) Detonation waves also seemed to reflect at the surface of the imploding shell, causing solid pieces of it to break off.

Furthermore, the metallurgy of plutonium turned out to be considerably more complicated than that of uranium. Learning how to mold it into whatever shape was eventually chosen was felt to require a separate research program (largely conducted at the Massachusetts Institute of Technology) on the design of suitable crucibles and materials for coating them. Much work also went into determining how to construct a three-dimensional lens structure of high explosives that would adequately focus the imploding blast. The basic design of a suitable structure was drawn up by the mathematical physicist John von Neumann. However, extensive research and development on the high explosives themselves was necessary, since no previous military or civilian application had called for the high precision needed for implosion. Learning how to mold high explosive into the required shapes without cracks or bubbles appearing was a major difficulty. Most basic of all, in order for implosion processes to stand a chance of being sufficiently symmetrical to achieve a full nuclear explosion, the explosive shell had to detonate virtually simultaneously at all points—this required much work on the electric detonators, on the development of firing circuits, and on the timing equipment.

The initiator also posed difficult problems. Again, the basic concept employed—a device that would create a sudden large neutron flux by mixing the elements beryllium and polonium together at the crucial moment—had been outlined in Robert Serber's lectures, but, as his later annotations dryly put it, actually designing and making the initiators for the gun and implosion weapons took "a great deal of effort."[56] Polonium was highly radioactive, decayed quickly, and, like plutonium, had to be made in nuclear reactors. Getting the design of the initiator right required extensive experiments on ways of achieving the sudden mixing—experiments analogous but not identical to those on implosion.

As a consequence of all these processes, the Los Alamos laboratory changed radically from its original intended form, which was not unlike a big university physics department. The constant flow of new recruits—especially to the ever-expanding Engineering Ordnance Division—had to be assigned to particular, narrowly delimited tasks. To a degree, the overall weapon still bore the marks of individuals. For example, the Trinity and Nagasaki design, "Fat Man," was also referred to as the "Christy gadget" after the original proponent of its solid core, Robert Christy.[57] Yet its design and that of the simpler uranium gun were products not of individuals but of a complex, differentiated organization.

Tacit Knowledge and the Design and Production of Nuclear Weapons

After reports of the horrors of Hiroshima and Nagasaki reached Los Alamos, the individuals involved had to face (often for the first time) the full human meaning of what they had done. Some simply left to resume distinguished academic careers. Oppenheimer reportedly wanted to give the mesa, with its beautiful vistas and dramatic canyon, "back to the Indians."[58]

Of course, Oppenheimer's wish was not granted. The Los Alamos laboratory continued, as did the design of further atomic (and soon hydrogen) weapons, and a similar laboratory was created in 1952 at Livermore, California. Let us, therefore, now move on in time to the late 1980s, and to the process of nuclear weapons design as institutionalized in these laboratories, focusing on common features rather than on differences in style.[59]

"Institutionalized" is indeed the appropriate word, and on the face of it some of the changes suggest that the role of tacit knowledge in the process should be minimal. By the 1980s, designing nuclear weapons had lost much of its flavor of virtuoso innovation and had become a more routine task—one, indeed, that some in the laboratories feel to have become bureaucratized, unchallenging, even "dull."[60]

Even more striking is the enormously expanded role of computers. As we have seen, during the Manhattan Project a "computer" was originally a woman, supplemented by a mechanical calculator or perhaps a punched-card machine. Digital computers, introduced in the late 1940s and the early 1950s, soon gave weapons designers computational capabilities unthinkable a decade earlier—capabilities that continued to grow exponentially in the decades to come. In turn, that permitted the development and use of vastly more detailed and sophisticated mathematical models. The computer programs (referred to by those involved as "codes") used in designing nuclear weapons are now very large and complex A modern American code will typically involve from 100,000 to 1,000,000 lines of program,[61] and many such codes are available to the designers.

Such codes have both a theoretical and an empirical basis. The theoretical basis is predominantly in well-established physics—"normal science," not regarded as a matter for debate and doubt. However, the code, and not merely the theory, is needed because the *implications*[62] of that well-established knowledge for nuclear weapons as particular, concrete artifacts are not always transparent. Even today, nuclear weapons designers feel that they do not have a full "first principles prediction capability"[63]: "you certainly can't do the calculations from first principles, basic physics principles. . . . That's a very frustrating thing."[64]

The most obvious form taken by this problem is computational complexity. It is one thing to have sound, quantitative knowledge of physical phenomena available, for example in well-established partial differential equations. It can be quite another matter to infer from those equations what exactly will happen in an attempted explosion of a particular nuclear weapon. Often, interactions between different physical processes, and nonlinearities in the underlying equations, take desired solutions far out of the reach of traditional physicists' methods of mathematical manipulation and paper-and-pencil calculation; hence the need for computer assistance.

The designers we spoke to, however, argued that even the most powerful computer—they have always enjoyed unique access to the world's fastest machines[65]—does not entirely bridge the gap between physical theory and concrete reality. One "can't even write down all the relevant equations, much less solve them," one designer told us, adding that even in the most modern codes "major pieces of physics" were still left out.[66] The codes "only explain 95 percent of physical phenomena at best; sometimes only 50 percent."[67]

All codes, they say, involve approximations. This is more the case for the "primary" (an atomic bomb or the fission component of a hydrogen bomb) than for the "secondary" (the fusion component of a hydrogen bomb): "The primary is less well understood than the secondary. Material physics is cleaner in the secondary: everything happens at high temperatures and pressures. The primary involves transitions from cold metal at low pressure and temperatures to high pressures and temperatures."[68]

The difficulty of predicting on the basis of explicit knowledge alone seems to be at its perceived peak with "boosting"—the injection of gaseous fusion materials into a fission weapon as it begins to detonate.[69] The neutrons generated by the fusion of these materials can considerably intensify the fission chain reaction. According to one U.S. weapons designer, "it is boosting that is mainly responsible for the remarkable 100-fold increase in the efficiency of fission weapons" since 1945.[70] If, however, the effects of boosting are insufficient, the small boosted primary in a modern thermonuclear bomb may simply fail to ignite the secondary, and the resultant explosion will be many times weaker than anticipated. Yet boosting is both hard to model numerically and hard to study in laboratory experiments, since the fusion reaction begins only when the fission explosion is underway. Because of the difficulty of accurate prediction, "the design of boosted fission devices is an empirical science."[71]

More generally, though, our interviewees saw *all* codes as needing an empirical as well as a theoretical basis, because they are approximations to reality rather than simply mirrors of it. Although non-nuclear experiments such as test implosions play an important role in providing this empirical basis, the ultimate check on the validity of the codes is nuclear explosive testing, which allows particular parameters whose values cannot be deduced from theory to be estimated empirically and which permits a code to be "normalized" (i.e., its predictions are checked against measurements made during testing, and the code is adjusted accordingly). Tests "almost never hit the predicted numbers exactly"; a prediction is reckoned to be "pretty good" if the actual yield (explosive energy released) is "within 25 percent of prediction."[72]

"No new code is used until it predicts the results of previous tests."[73] Although the modeling process is seen as having improved greatly over the years, even with modern designs and modern codes the measured yield sometimes falls significantly short of predicted values for reasons "we have not yet been able to explain."[74] On other occasions, codes "would give the right answer [i.e. correctly predict yield], but you didn't know why it gave you the right answer."[75]

The need for testing to develop and check codes does not, however, make testing an entirely unambiguous arbiter of their validity. The yield of a nuclear explosion is not a self-evident characteristic of that explosion; it has to be measured. Even in the mid 1980s, such measurements were seen as subject to uncertainties of as much as 5 percent[76]; another source suggested to us (in a private communication) that the uncertainty is actually greater than this. Furthermore, even an entirely successful prediction of the yield or of other "global" characteristics of a nuclear explosion does not conclusively demonstrate the correctness of a code or a model:

> . . . there are many aspects of the designs that we still don't understand well enough, and the reason for that is that most of the data we get is what you might call an integrated result, in that it's the sum of what happened over a period of time. You never know in detail what happened during that short time interval, and because of that there could be several different calculational models that actually explain what happened. And each one of those might look OK for a given set of circumstances but could be completely wrong for some other set of circumstances; and you don't know what those circumstances are, and so you're vulnerable.[77]

Between 10 percent and 30 percent of U.S. nuclear tests were not direct tests of a weapon design; they were "physics understanding tests," specifically designed to investigate theoretical or computational models of nuclear explosive phenomena.[78] But even these tests had their limitations. Nuclear explosions are both very fast and very destructive, and so they are hard to study empirically: they destroy sensors placed close to the blast almost immediately. Above all, "you . . . don't have the ability to put your instruments inside [the bomb] in the places where you really would like to get the detailed measurements. If you put your instruments in, then the device won't work."[79]

The Role of Judgment

With the implications of theory not entirely clear cut, with a continuing gap between model and reality, and with the results of experimentation and testing not always decisive, what remains is "judgment."[80] Judgment is the "feel" that experienced designers have for what will work and what won't, for which aspects of the codes can be trusted and which can't, and for the effects on a weapon's performance of a host of contingencies (e.g., the ambient temperature, the aging of the weapon, vagaries of production processes). These contingencies are so numerous, and the number of nuclear tests is so limited by their great expense and by increasing political sensitivity, that "nuclear warheads cannot be 'thoroughly' tested; the resources simply are not available. As a result, the functional capabilities of nuclear explosives cannot be fully established without a strong dependence on the scientific judgment of the weapon scientists."[81]

According to our interviewees, that judgment goes beyond the explicit knowledge embodied in words, diagrams, equations, or computer programs. It rests upon knowledge that has not been, and perhaps could not be, codified. That knowledge is built up gradually, over the years, in constant engagement with theory, with the codes, with the practicalities of production, and with the results of testing. Knowing what approximations to make when writing a code requires experienced judgment, and some crucial phenomena simply cannot be expressed fully in the codes. One designer told us he had tried to make all this knowledge explicit by writing a classified "textbook" of nuclear weapons design and had been unable to do so: "It's too dynamic."[82] "Art," rather than "science," is a word that several nuclear weapons designers reached for to describe their trade: it is "very much an empirical art"[83]; it is "artsy."[84]

As a result, there is "a long learning curve"[85] for new designers. It takes a new designer, even one with a background in relevant areas of physics, "five years to become useful,"[86] and it may take ten years to "really train" one.[87] The number of fully experienced nuclear weapons designers is quite limited. In the late 1980s there were "about fifty good designers" in the United States; at its maximum, around 1976, the total was only eighty.[88] Another interviewee estimated the late-1980s total as "about forty" designers at Livermore and thirty at Los Alamos; they were the "only ones who understand nuclear explosions."[89] The numbers the interviewees would give for 1994 would be much lower.[90]

Designers' judgment is a communal as well as an individual phenomenon, and "community" is a reasonable term to use so long as it is not taken to imply harmony.[91] First, judgment is passed on, face-to-face and person-to-person, from "senior designers . . . to younger designers"[92] in what is essentially a relationship of apprenticeship as they work together in design and analysis. Second, judgment is collective and hierarchically distributed. Individuals may propose new approaches, and many a design is seen (like the Christy gadget) as bearing the imprint of a particular "lead designer." But no design goes into production without intensive and extensive peer review. As a result, to put it in idealized terms: "Our scientific judgment is broader than just the experience of each individual weapon scientist; the collective judgment of the entire weapon research infrastructure works synergistically to solve the problems we encounter."[93] More mundanely: "Younger designers take the output from their computer simulations and their interpretations of experimental results to test-seasoned senior designers for review and confirmation."[94] The process is competitive and highly charged. One designer told Hugh Gusterson, who has recently completed a remarkable anthropological study of the Livermore laboratory, that "for every twenty things people propose, maybe one is going to make it onto that shot schedule [i.e., full nuclear explosive testing]. . . . I've seen men all in tears [at the reaction to their proposals]."[95]

Thus, uncodified, personally embodied, and communally sanctioned knowledge plays, according to our interviewees, a continuing central role in the designing of nuclear weapons. Tacit knowledge is also important to the process of turning even the most detailed design into a physical artifact. Theory and computation deal with geometric abstractions such as cylinders and spheres; however, "nothing is truly a sphere," since there are always "little wobbles on the surface" and there is a "[small] difference in radius as you come out in different directions."[96] The quality of the machining—and thus the skill of the machinists—is crucial, and numerically controlled machine tools do not entirely remove the dependence on skill.[97]

Quality of machining can at least be checked independently without damaging the final product. But there are other aspects of nuclear weapons fabrication where such testing is impossible or impractical. An example is the solid-state bonding used in the W84 warhead for the ground-launched cruise missile: ". . . there is no adequate nondestructive testing technique that can evaluate the quality of the bonds."[98] "One of the key features of this process is the assured removal of all oxide from the surface before a layer of another metal is applied. . . . Simple things such as the way in which the part is clamped in its holding fixture can affect the rate of oxide removal. . . . Although we have tried several techniques to make this evaluation with instrumentation, we have found none equal the human eye . . . for detecting the change to a shiny, then slightly hazy, appearance that indicates a clear surface."[99]

Even with the careful handling that the components of nuclear weapons receive, it is inevitable that some of the thousands of separate parts that go into such a weapon will receive slight nicks and scratches as they are manufactured and assembled. Often these will be unimportant, but sometimes they would affect the performance of a weapon, and discarding or fully testing each slightly scratched part would be prohibitively expensive. So a procedure has had to be developed for reports on individual components with a nick or a scratch to be sent from production plants to the weapons labs, and for designers there to judge whether the defects matter. In the late 1980s, designers at Livermore were processing about 150–200 such evaluation requests per system per month.[100]

Yet another issue is that many aspects of manufacturing high explosives to the specifications required for an implosion weapon "are as much an art as a science."[101] Though another source suggests (in a private communication) that this may be putting matters too strongly, there is a potentially significant issue here, because nondestructive testing of explosives is hard to envisage (unless, of course, one sample of explosives can be relied upon to be the same as others).

However, where tacit knowledge is involved, judgments of "sameness" become problematic. Just as the dependence of scientific experimentation upon tacit skills can give rise to controversy over what is to count as a competent replication of an experiment,[102] so the products of a non-algorithmic production process cannot be relied upon consistently to be identical. In the production of nuclear weapons, "Documentation has never been sufficiently exact to ensure replication. . . . We have never known enough about every detail to specify everything that may be important. . . . Individuals in the production plants learn how to bridge the gaps in specifications and to make things work. Even the most complete specifications must leave some things to the individual's common knowledge; it would be an infinite task to attempt to specify all products, processes, and everything involved in their manufacture and use."[103]

Sameness has three aspects. First, "production weapons" can differ from laboratory-produced prototypes, because those involved in the manufacture and assembly of the former may lack the knowledge of those who made the latter. "The fellows who designed the circuits or the mechanical components almost had to be there when [the early bombs] were put together, because they were the only ones who understood how they worked."[104] Second, weapons produced to the "same design" at different times can differ: "Material batches are never quite the same, some materials become unavailable, and equivalent materials are never exactly equivalent; 'improved' parts often have new, unexpected failure modes; different people (not those who did the initial work) are involved in the remanufacturing; vendors go out of business or stop producing some products; new health and safety regulations prohibit the use of certain materials or processes."[105] Third, an individual weapon may change over time through radioactive decay, chemical decomposition, corrosion, and the "creeping" of materials.[106] Weapons are inspected regularly, and "if parts have deteriorated, they are replaced with parts that do not differ significantly from the original,"[107] but this again raises the question of how to judge the significance of differences, given that the production of parts cannot be wholly algorithmic.

Tacit Knowledge and the Spread of Nuclear Weapons

Perhaps, though, all this testimony on the current role of tacit knowledge needs to be taken with a pinch of salt. Some of it[108] has been part of a continuing struggle to ward off a comprehensive ban on nuclear testing; some of it might even be seen as the self-justification of an elite group whose occupation is threatened. More particularly, the current generation of American nuclear weapons designers has worked primarily on highly sophisticated weapons. The evolving military requirements and the competition between weapons labs have created both pressures and incentives to maximize yield/weight or yield/diameter ratios, and to economize on special materials such as the hydrogen isotope tritium (used in boosting). These pressures and incentives have pushed the design of boosted primaries "near the cliff," as some of those involved put it—that is, close to the region where performance becomes very sensitive to internal and external conditions, one potential result being that the explosion of a "primary" might fail to ignite the "secondary."

Near the cliff, the need for experienced judgment is conceded by all involved. But in the design of more basic, physically larger weapons, "much of the physics of nuclear weapons is quite forgiving,"[109] and the role of judgment is more disputable. Let us, therefore, turn to a third kind of evidence concerning the role of tacit knowledge: the record of the spread of design capability.

Why this record is relevant is straightforward. If explicit knowledge were sufficient for the design of basic nuclear weapons, acquiring them would be a straightforward matter for those who possessed both the necessary fissile material and the requisite knowledge—e.g., "public" nuclear physics plus a detailed diagram and instructions to cover the more practical side of design. If, on the other hand, tacit knowledge plays a key role, even the most detailed explicit knowledge would not, on its own, be enough. The recipients and the originators of such knowledge would have to be members of the same or similar technical cultures, so that the recipients could bring tacit background knowledge to bear in order to "repair" the insufficiency of the explicit instructions.[110]

In addition, whereas explicit knowledge can be copied, tacit knowledge (in the absence of prolonged, "hands-on," face-to-face interaction), has to be re-created. It is much easier to copy a book or a computer program than to write it in the first place, but there is no reason in principle[111] to expect the re-creation of tacit knowledge to be any easier than its original creation. Furthermore, precisely because tacit knowledge is not codified, both the skill and the product re-created may differ from the originals. Even if one sets out to copy, one may end up doing and building something that is, from some points of view, different from the original.

As we shall see, the spread of the ability to design nuclear weapons has generally taken place (at least in the well-documented cases) without extensive personal contact with previous successful programs. Furthermore, at least two programs have attempted to copy the results of previous programs, in at least one case on the basis of explicit knowledge alone. These two predictions—the difficulty of re-creation and the problematic nature of copying—can, therefore, be tested, at least within the limits of the available data.

Livermore

Livermore, the second American laboratory, was set up in September 1952. Although there were no formal security barriers between it and Los Alamos, relations between the two labs were troubled. Los Alamos staff members resented criticism of the laboratory by Livermore's founder, Edward Teller, and felt that they had been denied due credit for the first thermonuclear explosion.[112]

Only a small minority of those at the new lab seem to have had direct previous experience in nuclear weapons design. Teller himself had, in his wartime Los Alamos work, focused on research on the "Super," of which he was the main proponent, rather than on the practicalities of designing fission bombs. Teller aside, the core of Livermore's initial cadre was a group at the University of California at Berkeley of about forty people, including about twenty physics postdocs, set up in 1950 to study thermonuclear explosive phenomena experimentally.[113]

For Livermore staff members with the appropriate security clearances, there were no barriers to access to the stock of explicit knowledge (diagrams, data, and the like) generated at Los Alamos. "The Los Alamos administration treated the Livermore leadership formally correctly, and provided some much needed technological assistance to the new laboratory," Livermore's first director reports.[114] However, the tension between the two laboratories meant that face-to-face collaboration was not always easy.

The failure of the new laboratory's first efforts was due in part to a deliberate Livermore decision not to try to copy what Los Alamos had done. Livermore's first two tests (March 31 and April 11, 1953) were of fission bombs with cores of uranium hydride rather than metallic uranium or plutonium. The hope seems to have been that use of uranium hydride could help miniaturize atomic weapons.[115] Both tests were embarrassing fizzles. In the first, the weapon failed so badly that the tower supporting it was left standing. Although Livermore staffers tried to pull the tower down with a Jeep, they did not manage to do so before Los Alamos photographers had captured their rivals' humiliation.[116] Livermore's first hydrogen bomb test (April 6, 1954) was also a disappointment, producing less than a tenth of the expected yield. Not until March 1955 was a Livermore test successful, and not until 1956 was Livermore "beginning to be trusted as a nuclear weapons design organization."[117]

On the other hand, although overseas nuclear weapons programs were also typically to encounter fizzles at various points in their programs,[118] their first tests all seem to have been successful. (There have been rumors of a failed Indian test prior to the successful one in 1974, but an informed source has told us, in a private communication, that these rumors are false, although a serious problem was encountered.) Since this is certainly *a priori* evidence against a strongly "local knowledge" view of nuclear weapons design, let us now turn to these overseas efforts. Those that are believed to have ben successful are summarized in table 1.

The Soviet Union and the United Kingdom

The Soviet and British efforts are of particular interest from the viewpoint of tacit knowledge, since both began by trying to copy the Christy gadget. The Soviets did so on the basis of explicit knowledge alone. Although the Soviet Union had considerable strength in nuclear physics and had set up a small wartime project to investigate the possibility of an atomic weapon, no Soviet scientist took part in the Manhattan Project, nor did any member of the Manhattan Project join the Soviet bomb effort. Instead, the Soviet team worked from "a rather detailed diagram and description of the first American bomb," which had been given to the Soviet intelligence service by Klaus Fuchs, a

Table 1
Approximate chronologies of successful nuclear weapons development programs.

	Start of development program	Date of first atomic test explosion (*) or weapon(†)	Date of first thermo-nuclear test explosion (*) or weapon (†)	Significant personal contact with previously successful design team?	Began with attempt to copy previous design?
US	1942	1945*	1952*	No	No
USSR	1945	1949*	1953*	No	Yes
UK	1947	1952*	1957*	Yes	Yes
France	1955	1960*	1968*	No	?
China	c. 1955	1964*	1967*	No	No
Israel[a]	c. 1957(?)	c. 1968(?)†	? †	?	?
India[b]	c. 1964	1974*		?	?
South Africa[b]	1971	1979†		?	?
Pakistan[b]	c. 1974(?)	? †		?	Yes (?)

Sources: D. Albright and M. Hibbs, "Pakistan's bomb: Out of the closet," *Bulletin of the Atomic Scientists,* July–Aug. 1992: 38–43; Albright and Hibbs, "India's silent bomb," *Bulletin of the Atomic Scientists,* Sept. 1992: 27–31; J. Baylis, "The development of Britain's thermonuclear capability 1954–61: Myth or reality?" *Contemporary Record* 8 (1994): 159–174; M. Gowing, *Britain and Atomic Energy, 1939–45* (Macmillan, 1964); Gowing, assisted by L. Arnold, *Independence and Deterrence* (Macmillan, 1974); S. Hersh, *The Samson Option* (Faber & Faber, 1991); R. Hewlett and O. Anderson, Jr., *The New World, 1939/1946* (Pennsylvania State Univ. Press, 1962); R. Hewlett and F. Duncan, *Atomic Shield, 1947/1952* (Pennsylvania State Univ. Press, 1969); D. Holloway, "Entering the nuclear arms race: The Soviet decision to build the atomic bomb, 1939–1945," *Social Studies of Science* 11 (1981): 159–197; Holloway, *Stalin and the Bomb* (Yale Univ. Press, 1994); International Atomic Energy Agency, The Denuclearization of Africa: Report by the Director General (1993); Institut Charles-de-Gaulle, *L'Aventure de la Bombe* (Plon, 1984); Y. Khariton and Y. Smirnov, "The Khariton version," *Bulletin of the Atomic Scientists,* May 1993: 20–31; J. Lewis and Xue Litai, *China Builds the Bomb* (Stanford Univ. Press, 1988); D. Mongin, La Genèse de l'Armement Nucléaire Français, 1945–1988, Ph.D. thesis, Université de Paris I, 1991; L. Scheinman, *Atomic Energy Policy in France under the Fourth Republic* (Princeton Univ. Press, 1965); L. Spector, *Going Nuclear* (Ballinger, 1987); F. Szasz, *British Scientists and the Manhattan Project* (Macmillan, 1992).

a. It is not clear whether Israel has developed thermonuclear weapons.
b. These countries are not believed to have developed thermonuclear weapons.

German refugee physicist who was a member of the British mission to Los Alamos and who had been intimately involved with the design of the core and the initiator of the plutonium implosion weapon. In the second half of 1945, the leader of the Soviet fission bomb project, Yuli Khariton, and a small number of trusted colleagues were given the documents from Fuchs. Although they were already working on their own fission bomb design, they decided that it would be safer to make a "copy"[119] of the Christy gadget.

Despite the enormous priority their work was granted by Stalin, it took them four years from the receipt of the material from Fuchs, slightly longer than the original Manhattan Project: ". . . in order to build a real device from the American design, it was first necessary to perform a truly heroic feat that required nationwide mobilization: to create an atomic industry, corresponding technologies, a supply of unique, high-quality apparatus, and to train qualified people."[120] Although Fuchs's data and the Smyth Report gave them the confidence not to pursue as many approaches in parallel as the Americans had, the Soviet team ended up recapitulating much of the work of the Manhattan Project.

In particular, they found that building a "copy," even with the detailed diagram and description Fuchs had given them, was not easy. When Khariton named 70 people he wanted for the first Soviet nuclear weapons design facility, Arzamas-16, he was asked why he needed so many.[121] In reality he turned out to need many times that number. According to Khariton, "the information received from Fuchs did not lessen substantially the volume of experimental work. Soviet scientists and engineers had to do all the same calculations and experiments."[122] Although the requisite nuclear experiments were demanding and dangerous, the engineering aspects of the work seem to have caused the most problems: ". . . the scientists were too inexperienced and amateurish in the complex processes of mass production."[123] In 1948, an experienced mechanical engineer, General N. L. Dukhov, had to be brought in as Khariton's deputy to take charge of the engineering work at Arzamas.[124]

Many of the problems were not directly related to the fissile core; they were due to the practical difficulty of achieving successful implosion: "Even with the espionage material that had been made available, considerable effort was needed by Soviet chemical engineers to devise the technology to manufacture . . . large castings of homogeneous high explosive. Moreover, extensive testing was needed to ensure that the explosive charges detonated uniformly and predictably."[125] The electrical system required to achieve simultaneous detonation was another problem, and another senior engineer, V. I. Alferov, was brought to Arzamas take responsibility for it.[126] The device that was ultimately produced was not seen by those involved as entirely identical to the American original—although it was "very close," there were "minor differences."[127]

The British bomb team had both explicit knowledge of the American design and (unlike the Soviet team) a considerable degree of personal involvement in the processes leading to that design. British scientists (the native ones and, especially, those who were refugees from fascism) had indeed led the way in arguing that an atomic bomb was feasible. Particularly important were a 1940 memorandum by two of the refugee physicists, Otto Frisch and Rudolf Peierls, and the subsequent program of research in Britain under the "MAUD Committee" in 1940 and 1941. The British team played a subordinate role to the American team from 1942 on; however, a British mission was established at Los Alamos, and some of its members (including Peierls, Fuchs, Geoffrey Taylor, and the experimentalist James Tuck) played central roles in that laboratory's work.[128]

Anglo-American collaboration was ended by the U.S. Atomic Energy Act of 1946. When they began their atomic bomb project, in 1947, the British decided, as the Soviets had, to copy the Christy gadget. Under the agreement with the Americans, written records had been left behind at Los Alamos, but the former British team helped compile from memory "a working manual" which, they hoped, "would enable the American atomic bomb to be duplicated, without all the laborious Los Alamos work."[129] Particularly helpful was Klaus Fuchs, whose work on behalf of the Soviets meant that his memory of what had been done at Los Alamos was "outstanding" and who, unlike his colleagues, had removed written material from the American lab.[130]

Again, though, copying the Christy gadget turned out not to be straightforward. At the level of explicit knowledge, the former Los Alamos people were well placed: they "were able to list very clearly the bomb components and to set out the principle of the bomb."[131] At the practical level, however, their knowledge was more patchy. Although they had been over twenty in number, and widely dispersed through Los Alamos's divisions, members of the British mission had not had personal involvement in all the aspects of the laboratory's work. Furthermore, knowing what the final product should be like was not the same as knowing how it could be made. For example, although "some convenient plutonium handling tricks were . . . known," the British team's knowledge of plutonium metallurgy was "sketchy."[132] None of them knew how to make crucibles into which molten plutonium could be poured without its dissolving or reacting with the crucible material.[133] Similarly, much work had to be done on the chemistry of the initiator's polonium, and on how to manufacture and handle it.[134]

Indeed, the hope of avoiding "all the laborious Los Alamos work" was largely disappointed. The first (November 1944) plans for a postwar British atomic energy research establishment had envisaged a staff of less than 400, covering reactor development as well as weapons work.[135] By the start of 1952, however, the program's "non-industrial" staff numbered over 5000, with more than 1000 of these devoted to the weapons work alone.[136] Furthermore, the five years it took to make the intended copy was longer than it had taken to make the original. In part, that was because the atomic weapons program met with obstruction, especially over the release of skilled staff, from organizations within the British state whose priorities were different.[137] In part, it was because Britain had fewer resources to devote to the production of fissile material. In addition, the experiments whose detailed numerical results had been left behind at Los Alamos had to be replicated.

More generally, though, despite all the knowledge inherited from Los Alamos, the problems of designing, fabricating, and testing the components of weapons turned out to be "diverse and most intricate,"[138] and the work "dangerous and difficult."[139] Even in those areas (e.g., designing explosive lenses) in which the British felt confident of their knowledge, many practical problems arose: for example, despite much work on methods of casting, no way could be found of stopping the lenses from shrinking unevenly in their casts. Techniques for constructing the detonation circuitry "had very often to be invented, and then they had to be practiced and perfected" by the female production workers who had to implement them.[140] With a 1952 target date set for the first test explosion, the last year of the program became "a frantic race against time with serious problems solved only at the eleventh hour—questions of design, assembly systems, cavities in the castings for the uranium tamper, the firing circuit, the plating of various components, plutonium and polonium supply."[141]

Despite their initial intentions, and a strong, continuing desire not to lose "the safe feeling [of] making an object known to be successful,"[142] the British team found they could not successfully "duplicate" the Christy gadget. The Americans had assembled the Nagasaki bomb on the ground, but the British felt it unsafe for a bomber to take off with a complete weapon onboard and wanted the final assembly to take place in flight. However, they became worried that the weapon might inadvertently become supercritical while this was being done. In September 1950 the project's leader, William Penney, reluctantly "took the major decision to alter the design at the heart of the bomb."[143] As at Los Alamos, a team then set to work to grind out on mechanical calculators a numerical simulation of the likely results of an alternative design. In mid 1951 the design was changed once more to include a two-inch gap between the tamper and the core.[144] The momentum of the tamper moving inward through the gap intensified the compression of the core, but this third design involved a more complicated mechanical structure and was "more sensitive to implosion imperfections."[145] This sensitivity was particularly worrisome, since no way had been found to make explosive lenses of precisely correct shape. The team had to resort to "the use of PVC adhesive tape to fill up the clearance spaces [in the explosive lenses] and minimize settlement."[146] Only in the summer of 1952 did high-explosive firing trials provide reassurance that these imperfections would be small enough not to cause failure.

France and China

Less is known about the detailed history of the French atomic weapons program than about the British or even the Soviet effort. Like their Soviet and British counterparts, French physicists had considered the idea of an atomic bomb early in the Second World War.[147] Some of them had also taken part in the Manhattan Project, but they had been involved with the production of fissile materials rather than with the designing of weapons. In contrast with the United Kingdom and the Soviet Union, in France there was significant political opposition to a nuclear weapons program. There was also a feeling in France during the early postwar years that such a program was too ambitious an undertaking for any country but a superpower. The successful British test in October 1952 undermined the latter belief,[148] and in 1954 the French government made a commitment to develop nuclear weapons. In 1955, two research centers were established for that purpose. One, at Bruyères-le-Châtel, concentrated on the physics, metallurgy, and chemistry of nuclear materials; the other, at Vaujours, dealt with detonics (the study of high-explosive blast waves and similar matters).[149]

In February 1960 the first French atomic device was successfully exploded at Reggane in the Sahara. Like the Soviets and the British, the

French seem to have focused their efforts on a plutonium implosion weapon.[150] We have found no evidence that the French attempted to copy a previous weapon, and we presume that their design was developed by them. Their development effort was certainly considerable. In 1957 the project employed more than 750 (over and above those devoted to plutonium production), and that figure tripled in two years.[151] Solving practical problems was the main task: "the atomic bomb is to a large extent an engineering problem."[152]

The history of China's nuclear weapons program has been documented in broad outline in a remarkable study by Lewis and Xue.[153] Just as no member of the Soviet program had worked on the Manhattan Project, so it appears that no member of the Chinese project had been directly involved with either Soviet or Western nuclear weapons design. Although the 1957 Sino-Soviet Defense Technical Accord committed the Soviet Union to supply China with a prototype atomic bomb, the Soviets reneged on that promise, and they do not even seem to have provided design information at the level of detail that had been supplied to them by Klaus Fuchs. When the Soviet technical experts who had been sent to assist the Chinese were withdrawn, in 1960, the two nuclear weapons designers among them left behind shredded but legible and useful data on implosion. In general, though, their Chinese counterparts remember the Soviet weapons designers as "mute monks who would read but not speak."[154]

Although the Soviets were more helpful in other areas (notably in supplying a nuclear reactor and a cyclotron, in handing over design data for a uranium-separation plant, and in the general training of thousands of Chinese nuclear engineers), Chinese nuclear weapons design had to proceed without the benefit of contact with individuals who had "hands-on" experience in a successful program. Like the Soviet, British, and French programs, the Chinese program took longer than the original Manhattan Project—in this case, roughly nine years (1955–1964). It was a massive national effort involving several hundred thousand people, including tens of thousands of peasants who were given basic training in uranium prospecting and refinement.

Again, the obstacles met in this effort seem to have been predominantly practical engineering problems rather than, for example, deficits in explicit knowledge of nuclear physics. There is no evidence that the design of the weapon itself was an attempt to copy a previous device; indeed, the Chinese chose to begin their program differently from the Soviets, the British, and the French, constructing a uranium implosion weapon rather than a plutonium one. The design and fabrication difficulties encountered seem broadly similar to those faced by previous programs. Particularly problematic areas included the design and molding of the explosive lenses, the selection and production of the materials for the initiator, and bubbles in the uranium castings.[155]

More Recent Programs

All nuclear weapons programs since China's have been covert. Israel has never explicitly admitted to possessing a nuclear arsenal, and South Africa did so only in 1993. India maintains that its 1974 test in Rajasthan was of a "peaceful" nuclear explosive, not a bomb. Pakistan has admitted officially only to possessing the "components" of an atomic bomb.[156] In view of this desire for secrecy, it is not surprising that very little is known with any reliability about the sources of knowledge drawn on in these nuclear weapons programs. There have been widespread reports of assistance (notably by France to Israel, by Israel to South Africa, by China to Pakistan, and perhaps by the Soviet Union to India), but it is impossible to be sure of the nature of such assistance.

What little is known with any confidence seems broadly compatible with what has been learned from the histories of the better-documented programs. To the extent that we can determine their chronologies, all seem to have taken longer than the original Manhattan Project. The few specific development problems that have been reported with any authority were primarily practical ones; for example, the leader of the Indian program reports particular difficulties with the initiator.[157]

The most interesting program from the viewpoint of this chapter is Pakistan's, because it has been alleged to involve the direct supply of explicit design knowledge from a previous program. U.S. officials have stated that the Chinese government handed over to Pakistan the detailed design of an atomic bomb—reportedly a uranium-implosion missile warhead that had been exploded successfully in a Chinese nuclear test in 1966. Despite this, Pakistan apparently found copying the weapon far from trivial: "It took the Pakistanis several years to master an implosion system, even though they were working from a proven design."[158] One U.S. official reportedly commented that "[receiving a] cookbook design doesn't mean that you can make a cake on the first try."[159]

Discussion

Tacit Knowledge

All three forms of evidence we have examined suggest that tacit knowledge plays a significant role in atomic bomb design.

First, the task of the first atomic bomb designers at Los Alamos proved much harder than had been predicted on the basis of explicit knowledge of nuclear physics. Filling gaps in the latter (such as, most consequentially, the rate of spontaneous neutron emission in plutonium 240) was important, but many of the most demanding challenges faced were practical "engineering" challenges. These challenges were diverse enough to take their solution far beyond the capabilities of an individual or even a small group; a large, complex organization had to be constructed to tackle them.

Second, despite the huge amount of subsequent work to make fully explicit the knowledge needed for nuclear weapons design, and in particular to embody it in computer programs, current designers still argue strongly that this explicit knowledge alone is inadequate. They emphasize the ways in which even the best computer models are only approximations to reality. They note the consequent need in their work for non-algorithmic "judgment," forged by working alongside experienced designers and by long personal involvement in design and (crucially) testing. That judgment is communal and hierarchical: proposals by individuals are reviewed by senior colleagues. Furthermore, producing nuclear weapons, as well as designing them, requires tacit knowledge: it is not a matter simply of following explicit, algorithmic instructions. For example, the designer's judgment has to be called upon in deciding whether two nuclear weapons produced to "the same design" can actually be treated as identical.

Third, the record of the spread of atomic weapons is at least broadly compatible with the conclusion that tacit knowledge is involved in their design. In at least three cases (the Soviet Union, France, China) weapons appear to have been developed successfully without extensive personal contact with a previously successful design effort.[160] However, these efforts—and others—have had at least some of the characteristics of independent reinvention. All efforts since the Manhattan Project appear to have taken longer than that project. The possession of explicit information (such as diagrams and detailed descriptions) generated by previous efforts has not made the developers' task trivial, even where they were trying "simply" to construct a copy of a previous design. All development efforts about which details are known have had to struggle hard with a multiplicity of practical problems. As in the Manhattan Project, the solution of these problems has required not merely individual expertise but concerted effort by large staffs. The problems involved are so diverse that they require significant new work even when, as in the British case, it is possible to call on the knowledge and expertise of a number of individuals with direct experience of a previous successful program.

Of course, no individual aspect of this evidence is entirely compelling. Although it is clear that explicit knowledge of physics was inadequate for the original development of atomic weapons, it might still be that what was needed in addition was simply explicit knowledge from within the spheres of other disciplines—notably metallurgy and various branches of engineering. The historical record suggests that this was not the case; however, historical work on the topic has not been informed centrally by the issues addressed here, and thus there is a degree of tentativeness to this conclusion. And because the boundary between explicit and tacit knowledge shifts as some aspects of tacit skills become systematized and even embodied in machines, one cannot simply extrapolate from the experience of the Manhattan Project (or other early development efforts) to the present day.

Furthermore, as we have pointed out, the testimony of current designers may have been influenced by a desire to argue against a comprehensive test ban. Against this, we would note that the minority of members of nuclear weapons laboratories who favor such a ban do not deny the role of tacit knowledge.[161] Nor did the computer specialists from these laboratories whom we interviewed—who might be thought to have an interest in arguing for the adequacy of explicit, algorithmic knowledge—actually advance that argument; some, indeed, provided cogent grounds for regarding algorithmic knowledge as inadequate. However, the experience of all but the oldest of our interviewees was with the design of sophisticated, rather than simple, weapons. This experience—particularly the experience of boosted primary designs that are "near the cliff"—is not necessarily generalizable to the design of simpler weapons.

In addition, two issues are confounded in the record of the spread of nuclear weapons: the design of such weapons and the production of the necessary fissile material. With the exception of Livermore, which could call on the general U.S. stock of such material, all other nuclear weapons efforts so far have involved the production of fissile material as well as the designing of weapons. We have no way of knowing how long the design work alone might have taken had the fissile materials been available from the start of a program. Furthermore, the time taken to design a nuclear weapon will clearly be influenced by the urgency with which the task is pursued, the resources devoted to it, and the equipment and skills available. These considerations make the duration of the various development efforts a less than conclusive indicator of the "hardness" of the task, and they rule out any quantitative conclusion of the form "It takes *x* months or years to design a nuclear weapon."

Finally, the evidence suggests that tacit knowledge may have a significantly smaller role in designing a "secondary" (i.e., turning an atomic bomb into a hydrogen one) than it has in designing an atomic bomb. Our interviewees seemed more confident of the adequacy of explicit knowledge in understanding secondaries, and the record of the spread of the hydrogen bomb is different from that of the atomic bomb: three of the four countries known to have moved from an atomic to a hydrogen bomb since the United States did so took less time to make the transition than the United States did.[162] Thus, tacit knowledge may be more relevant to the first step in acquiring a nuclear arsenal than to subsequent steps.

Though all these qualifications are important, none of them seems to us to be decisive. The weight of the evidence, we believe, supports the conclusion that tacit knowledge plays an important role in nuclear weapons design. Nevertheless, before moving to the implications of this conclusion, we need to discuss four further issues raised by these qualifications or by other considerations.

Black Boxes

The record of the early nuclear weapons programs may be a poor guide to the future, because previously tacit knowledge has been made explicit, because that explicit knowledge is now available far more widely than it was in the 1940s or the 1950s, and, especially, because many technologies relevant to designing and producing nuclear weapons have been "black boxed."[163] What once had to be done by hand can now be done by machines, and those machines can simply be bought rather than having to be built. Much relevant information can be acquired simply by buying textbooks on nuclear physics and manuals of nuclear engineering. Computer programs helpful in nuclear weapons design can also be purchased.

The most obvious example of "black boxing" is that the development of digital computers, and their universal availability, mean that calculations that once had to be done by humans (at major costs in time and effort) can now be done automatically. Indeed, it is now neither difficult nor expensive to purchase computers as fast as those of the U.S. nuclear weapons laboratories of the early 1970s,[164] and a determined purchaser could acquire even more powerful machines while probably being able to disguise their intended application. Nor would the programs to run on these machines have to be developed entirely from scratch. Derivatives of computer programs developed at the nuclear weapons laboratories have been commercialized and are widely available.[165]

Furthermore, a variety of other relevant black boxes that early weapons programs had to design and construct are now available commercially, although they are more difficult to purchase than computers, and their purchase is likely to attract attention. These include specialized metallurgical equipment, diagnostic tools suitable for studying implosion and initiator behavior, and electrical and electronic equipment that could be used in detonation circuitry.[166]

That all this eases the task of developing nuclear weapons is undeniable. The question is how much it does so, and whether it eliminates or minimizes the need for specialized tacit knowledge.[167] Iraq's nuclear program—which was dissected in unprecedented detail by international inspectors after the 1991 Gulf War—serves as an experiment on precisely these points. It was a determined, high-priority, extremely well-resourced program conducted by a country with a relatively large scientifically and technically trained work force and ample computing power. The Iraqi team had conducted a thorough and successful literature search for relevant explicit knowledge, and had also obtained "weapons-relevant computer programs."[168] Some attempted purchases, particularly of precision electrical equipment for detonation circuitry, were intercepted. However, Iraq was able to buy much of what it needed from Western companies (especially German, but also American, British, Italian, French, Swedish, and Japanese). Among the project's successful acquisitions were vacuum furnaces suitable for casting uranium, plasma-coating machines that can coat molds for uranium, an "isostatic" press suitable for making high-explosive lenses, and high-speed oscilloscopes and "streak cameras" useful for the experimental investigation of implosion.[169] Iraq was also markedly successful in making purchases and obtaining explicit knowledge relevant to the production of fissile materials, and had enough spare electricity-generating capacity to support even the most energy-intensive route to uranium separation. Yet the Iraqi program, which seems to have begun in the mid 1970s, had still not been successful by 1991, and opinions vary on how close to success it was even then.[170] One reason for its slow progress is very specific. A 1981 Israeli bombing raid rendered inoperable the French-supplied nuclear reactor under construction at Osirak and shut off what may have been the Iraqi weapon program's intended source of plutonium.[171] Iraq was, therefore, having to concentrate on what is generally agreed to be the considerably more demanding task of uranium separation.

More generally, though, Iraq's "nuclear Achilles heel" was its "lack of skilled personnel."[172] This hampered both uranium separation (which never reached the necessary scale) and weapon design.[173] According to seized documents, the Iraqis' immediate goal was an implosion weapon with a solid uranium core, a beryllium/polonium initiator, a uranium 238 reflector, and an iron tamper. Extensive theoretical studies had been carried out, and at least five different designs had been produced. The designs were, in the judgment of one leading U.S. weapons designer, David Dorn, "all primitive," but "each one [was] an improvement over its predecessor."[174] However, a final, settled, fully "practical design had not been achieved."[175] Despite all their purchases, the Iraqis had to develop much of the requisite technology for themselves, relying on local competences in metallurgy, chemistry, and electronics. (Sources differ on their relative strengths in these fields.) The same was true for knowledge of detonics. Iraq's detonics program unquestionably benefited from explicit knowledge acquired from abroad, but extensive indigenous theoretical work and practical experimentation were still required. By 1991 this work had not yet reached the stage of testing a full three-dimensional implosion system. (Again, detailed assessments differ on how much further work was needed.) Crucially, the Iraqi designers seem to have been constrained to use much less high explosive than was used in early American designs, which were delivered to their targets by heavy bombers. The Iraqi design was probably meant to be carried by a Scud missile. Iraqi designers seem to have lacked confidence that, within that constraint, they could achieve a powerful, highly symmetrical implosion. As a result, the design of fissile core they were contemplating was far closer to criticality than Western experts believed wise—so much so that it could perhaps have been detonated by a fire or a minor accidental shock. "I wouldn't want to be around if it fell off the edge of this desk," said one inspector.[176]

The Hardness of Tasks

The Iraqi program seems to suggest that successful use of "black box" technology still requires tacit knowledge, which cannot be purchased unless skilled personnel can be recruited or sustained person-to-person contact can be achieved. Iraq's program, like all the other well-documented nuclear weapons programs except Britain's, had only very limited success in this. Learning from previous programs has thus had to proceed without direct transfer of specific tacit skills.

However, previous programs are not just a source of particular items of knowledge, tacit or explicit. They can also convey lessons about the hardness of the tasks involved.[177] Observing others riding bicycles does not enable one to learn their skills, but it shows one that cycling is possible. Knowing that older brothers or sisters have learned to ride can encourage younger siblings not to conclude from their early failures that the task is impossibly hard. Successful previous nuclear weapons programs have had analogous consequences. Thus, the confidence—indeed overconfidence—of wartime Anglo-American physicists (including Continental refugees) in the ease of development of a nuclear weapon does not seem to have been widely shared by their French, German, or Soviet colleagues, and the governments of the last two countries were unconvinced until 1945 that the task was feasible enough to be worth the kind of resources the Americans devoted to it.[178] Trinity, Hiroshima, and Nagasaki were dramatic demonstrations that the task was not impossibly hard, and this (as well as the perceived threat from the West) explains the Soviet Union's sudden shift in 1945 from a modest research effort to an all-out, top-priority program.[179]

As we have seen, Britain's test explosion in 1952, although no threat to France, contributed to the latter's weapons program by suggesting that developing an atomic bomb was easier than had previously been assumed. Likewise, China's explosion in 1964 showed other developing countries that the atomic bomb was not necessarily the preserve of the highly industrialized world. Furthermore, profound questions over the feasibility of early hydrogen bomb designs helped delay the American move from an atomic to a hydrogen bomb.[180] By contrast, all subsequent hydrogen bomb programs could proceed with confidence in the basic achievability of their goal, and, in words used in another context by a group of weapons designers, "the mere fact of knowing [that something] is possible, even without knowing exactly how, [can] focus . . . attention and efforts."[181]

Because of this, we need to qualify the inference from the role of tacit knowledge in nuclear weapons design to the possibility of uninvention. It is hard to imagine belief in the feasibility of atomic or thermonuclear weapons now disappearing, and that fact alone increases the probability of their reinvention. In addition, more was learned from the Manhattan Project (even by those without personal involvement in the project) than simply the feasibility of an atomic bomb. It was openly disclosed by Smyth that the project had produced two fissile materials, plutonium and uranium 235,[182] and it was in no meaningful sense a secret that both gun and implosion designs had been developed. The knowledge that both a uranium gun and a plutonium implosion weapon had worked meant that subsequent programs could save significant resources by focusing on only one fissile material and one design. For example, the confidence that it was safe to concentrate initially on plutonium production, and that it was not necessary to embark simultaneously on an equally rapid program of uranium separation, was of considerable help to the early Soviet project.[183]

Other Sources of Tacit Knowledge

Previous nuclear weapons programs are not the only possible source of tacit knowledge for the design of a nuclear weapon. The most important other source is the nuclear power industry. The literature on proliferation treats this industry primarily as a potential source of fissile material, but it is also clearly a potential source of knowledge. In South Africa, for example, overseas cooperation in the development of nuclear power, while not directly aiding the nuclear weapons program, was nevertheless helpful in increasing "the technical competence of South Africa's nuclear engineers, scientists, and technicians."[184]

Nuclear power plants can provide crucial experience in matters such as the chemistry, metallurgy, handling, and machining of fissile materials, and also in neutronics. Neutronics—the study of the behavior of neutrons in fissile materials—is clearly crucial for the nuclear weapons designer, who will want to ensure that a bomb will explode rather than fizzle, and also that a critical mass is not formed accidentally during the machining and assembly of fissile material. Designers of nuclear reactors use neutronics to find configurations that can be kept critical without becoming supercritical. Like designers of nuclear weapons, they use a combination of physical theory, experimental results, and computer modeling. The two tasks are similar enough[185] that one would expect the explicit and tacit knowledge of neutronics gained in reactor design to help considerably in weapons design.[186]

Because nuclear weapons integrate nuclear and non-nuclear technologies, tacit knowledge acquired in some of the latter is also relevant. Electrical and electronic engineering, needed for the design and construction of detonation circuitry, is obviously important. Perhaps most significant, however, is the field of detonics, and in particular the technology of achieving not simply explosions but blast waves of particular shapes. This is central to the "art of implosion design"[187] in nuclear weaponry. It is, however, also a technology with wider military uses (notably in the design of shaped charges for armor-piercing anti-tank weapons) and with some civilian applications (in diamond production, mining, and metallurgy).[188] Experience of detonics contributed to the development of nuclear weapons. The Los Alamos scientist James Tuck, who first suggested the use of explosive lenses, had previously worked in the United Kingdom on armor-piercing charges.[189] The leader of the Soviet atomic bomb project, Yuli Khariton, and his colleague Yakov Zel'dovitch had also done wartime work on detonation phenomena in chemical explosives.[190] Since the 1940s, detonics has developed into a sophisticated technical specialty. The largest concentrations of detonics expertise seem still to be in the nuclear weapons laboratories, but the technology is also practiced at a range of other military and civilian establishments, mainly in the industrialized countries.[191] The availability of experienced personnel from such establishments would ease the design and testing of an atomic bomb implosion system significantly.

Kitchen Bombs

To date, all demonstrably successful efforts to develop nuclear weapons have been major enterprises involving several years' work, design teams numbering (at least in the cases where this information is available) from several hundred to a thousand or more, and the dedication of major industrial facilities to the production of fissile materials. These efforts have had to acquire, often painstakingly, much of the knowledge and skills developed in the Manhattan Project or other previous efforts. Perhaps, though, all these programs (with the possible exception of South Africa's[192]) have simply been unnecessarily ambitious and laborious. Certainly, every well-documented effort since 1945 seems to have seen its first atomic bomb as a stepping stone to a more sophisticated arsenal (for example, one including hydrogen bombs). As two members of the French program put it, "the goal was not simply to make a device explode, but to measure the parameters controlling nuclear explosive reactions."[193] Even the Iraqi program was "grandiose" and "overdesigned" from the viewpoint of simply producing a crude weapon.[194]

Perhaps the need for tacit knowledge and reinvention could be circumvented by a modest program aiming simply to produce crude weapons as quickly and easily as possible. This issue will be made much more pressing if substantial quantities of fissile materials become available for illicit purchase. Up to now, all nuclear weapons programs, with the partial exception of Israel's,[195] have had to produce their own fissile materials. Typically, this activity has dwarfed weapons design in expense, in visibility, and in personnel and resource requirements. For example, the work force that built the nuclear reactors at Hanford numbered, at its peak, 45,000, and the uranium-separation plant at Oak Ridge consumed more electricity in 1945 than the whole of Canada produced during the Second World War.[196] There was, therefore, no incentive to skimp on weapons design, and enormous effort was devoted to increasing the chance that the first nuclear explosions would be successful. In 18 months of research, for example, more than 20,000 high-explosive castings were supplied for test implosions, and many more were rejected as inferior. Their cost was unimportant, given that Los Alamos management knew the cost of producing the necessary plutonium to have been of the order of a billion 1940s dollars.[197]

If fissile materials were to become available for illicit purchase, however, an aspirant nuclear weapons state, or even a terrorist group, might well decide to try a "quick and dirty" route to a nuclear weapon. Would they succeed? Twenty years ago, former Los Alamos designer Theodore Taylor sought to highlight the dangers of the diversion of fissile material, even in the forms in which it is commonly found in the civilian nuclear power program. He argued passionately that, if the fissile material can be obtained, a crude but workable nuclear weapon could be made using only readily available instruments, artifacts, and knowledge. His arguments were brought to wide public attention by the doyen of American reporting, John McPhee.

Taylor argued, for example, that a reflector could be built by soldering two wax-lined stainless steel kitchen mixing bowls together around a fissile core. Modern plastic explosive could be "kneaded and formed, by hand" around the mixing bowls, or perhaps on an "upturned salad bowl." The work could be done first by eye, then by poking a measured wire "into the high explosive until it hits the reflector." An initiator might not be necessary at all: what is normally thought of as the disadvantage of "reactor-grade" plutonium (its high level of plutonium 240, and the consequent high rate of spontaneous neutron emission) could be turned to advantage by doing away with the need for this traditionally troublesome component.[198] Nor, according to Taylor, need implosion circuitry and detonators go beyond what is commercially available: "If you don't care whether you get a tenth of a kiloton [of explosive yield] one time and five kilotons another time, you can be much less fussy about the way the high explosive is detonated."[199]

One cannot be sure that a "quick and dirty" route to a nuclear weapon would work. No one is known to have tried to build a bomb in this way.[200] Taylor is convinced that it could work; others deny it. Edward Teller peremptorily dismisses as a myth the idea that "a nuclear explosive could be secretly developed and completed in someone's garage"[201]; other sources offer more particular counterarguments.[202]

More recently, Theodore Taylor has put his name to a less alarming diagnosis.[203] The authors of that analysis still conclude that a terrorist group that had acquired fissile material could construct a nuclear weapon, but they place greater emphasis on the barriers (especially of knowledge and skill) such a group would encounter. They argue that the necessary detailed design would require "the direct participation of individuals thoroughly informed in several quite distinct areas: the physical, chemical, and metallurgical properties of the various materials to be used, as well as the characteristics affecting their fabrication; neutronic properties; radiation effects, both nuclear and biological; technology concerning high explosives and/or chemical propellants; some hydrodynamics; electrical circuitry; and others."[204] Nor would explicit knowledge alone be enough: "The necessary chemical operations, as well as the methods of casting and machining the nuclear materials, can be (and have been) described in a straightforward manner, but their conduct is most unlikely to proceed smoothly unless in the hands of someone with experience in the particular techniques involved, and even then substantial problems could arise."[205] We hope that this later account conveys the difficulties better than the earlier one; however, with (fortunately) no direct empirical evidence yet available there is no way to be certain. The feasibility of a low-skill route to a crude nuclear weapon cannot, therefore, be ruled out.

Uninventing the Bomb

There are thus at least three reasons not to overstate the extent to which lack of tacit knowledge would force full-scale reinvention of nuclear weapons even after a long hiatus in their development:

Knowing that the task is feasible would encourage and focus efforts.

Relevant tacit knowledge might be available from sources other than previous nuclear weapons programs.

The elaborate development path of currently existing programs might conceivably be avoided.

If nuclear weapons are to be uninvented, therefore, we have to add at least two elements to a "tacit knowledge" view of uninvention. The first point is familiar: control over fissile materials is the key component of the current nonproliferation regime, and one that clearly needs urgent reinforcement. The second is what actor-network theorists would call the "translation" of interests: the displacement of goals, invention of new goals, the creation of new social groups, and the like. To date actor-network theorists have looked at this primarily as a part of the process of invention,[206] but it must surely be a part of uninvention too. The physicist Wolfgang Panofsky is, unfortunately, right when he says that "ultimately, we can keep nuclear weapons from multiplying only if we can persuade nations that their national security is better served without these weapons."[207] However, verbal persuasion alone is not likely to be enough. Actor-network research on the translation of interests might well form a useful body of resources for addressing this issue.[208]

Issues of tacit knowledge, control over materials, and the translation of interests form a necessary three-sided approach to nonproliferation and uninvention. To date, public policy has tended to focus on the second of these alone, perhaps because of its physical concreteness. The first and the third also must be taken seriously.

In particular, despite all the reservations we have expressed, we feel that considerations of tacit knowledge (largely neglected hitherto because of the dominance of conventional images of science and technology) could be important to disarmament and nonproliferation. Successful nuclear weapons design, we have been arguing, depends not only on explicit knowledge and algorithmic instructions but also on tacit knowledge gained through processes such as attempts to fabricate real systems and trial-and-error experimentation with their components. These processes take time and effort. The requirement for tacit knowledge thus serves as the equivalent of friction in a physical system, slowing things down and perhaps adding a degree of stability to what might otherwise be unstable situations. For example, after a sufficiently long hiatus we would expect the effort needed to re-create nuclear arsenals to become quite considerable, even for those who possessed detailed documentary records from the original development. Especially if fissile materials have to be produced afresh, it begins to be imaginable that, in a world with open skies, open borders, and dedicated and sophisticated intelligence agencies, such reinvention efforts would be detected before they came to fruition.[209]

More generally, attention to tacit knowledge (and to its possible loss) can help counter the pessimism that can be engendered by the conventional view that nuclear weapons cannot be uninvented. We do not pretend even to have begun to sketch how an abandonment of nuclear weapons might be made durable and permanent, nor have we discussed its desirability. Nevertheless, we hope to have contributed to undermining one of the key barriers to starting to think about its possibility.[210]

An Accidental Uninvention?

A world in which the uninvention of nuclear weapons is pursued systematically may well seem utopian. The maintenance of nuclear arsenals by the existing nuclear powers, in continuing uneasy conjunction with attempts to restrain their proliferation, seems more likely. That world has at least the virtue of apparent familiarity, barring a sudden multiplication of attempts to develop nuclear weapons capabilities triggered by a breakdown of control over fissile materials.

However, as the word's etymology reminds us, a technology does not consist simply of artifacts; it includes knowledge and understanding of those artifacts. The reader familiar with the sociological studies of controversial scientific experiments[211] will have noted a crucial difference between them and the situation, hitherto, of nuclear weaponry. In the former case there is typically dispute as to what the correct substantive result of an experiment "should be"; however, because an experiment cannot be reduced to algorithmic procedures there is no other ultimate test of its competence. Substantive divides (over matters such as the existence of controversial physical phenomena) thus become utterly entangled with disagreements over the competence of the experimenters. However, in the case of nuclear weaponry there has seldom been doubt over what the success of a nuclear explosive test consists in.[212] Such debate is imaginable,[213] but controversies fully akin to those that are "classical" in the sociology of science have actually taken place only occasionally, and then only when a putative nuclear explosion has been "observed" only at a distance[214] or where particular controversial nuclear phenomena are concerned.[215] Nuclear testing, therefore, has placed an impersonal constraint on the designing of nuclear weapons that, as we have seen, the individuals involved have valued highly. There has been a great deal of room for arguing over why a particular test failed, and at least the wiser designers knew that a successful test did not *ipso facto* demonstrate the correctness of their knowledge. Over the fact of success or failure there has been less practical room for argument.

The testing of nuclear explosives may, however, soon be at an end. American, British, and Russian nuclear tests have ceased. A comprehensive nuclear test ban is a goal of the current U.S. and Russian administrations. During the 1995 talks on the extension of the Nuclear Nonproliferation Treaty, the existing declared nuclear powers committed themselves to negotiating such a ban. Failure to do so would make it much harder to dissuade other nations from seeking to acquire nuclear weapons.[216]

After a test ban, designing nuclear weapons will inevitably involve much greater reliance on computer modeling.[217] Older interviewees in the U.S. laboratories recalled for us the three-year voluntary test moratorium that began in 1958. During that period, dependence on computer programs and subjective confidence in their output increased, especially as, toward the end of the moratorium, some senior staff members left. One interviewee noted that "you start[ed] to *believe* your calculations, and young folks *really* believe them if the old timers have left."[218] According to another interviewee, "people start[ed] to believe the codes are absolutely true, to lose touch with reality."[219] This confidence then evaporated after the moratorium's end. The appearance of doubt about the validity of the modeling of the effects of radioactive decay of the tritium used in boosting was crucial. An underground nuclear test commissioned to investigate these so-called aging effects "showed that these effects had been so severely underestimated that a cloud of then unknown proportions immediately fell over many of our weapons."[220]

Today, the increase in subjective confidence in computer modeling that would follow a test ban would almost certainly be much greater, in view of the much more refined computer codes and the more powerful computers that now exist and especially in view of the capacity to display simulations visually by means of computer graphics. However, while those involved believe that the particular phenomena that caused problems after the earlier moratorium are now well understood, they also acknowledge that "weapons of that era were considered 'forgiving' relative to their more modern counterparts."[221]

The consequences of the continued pursuit of nuclear weapon design after a test ban is a topic that deeply concerns some of the current designers. In 1993, to mark the fiftieth anniversary of its founding, the Los Alamos National Laboratory gathered 22 leading current and retired members to discuss its future. Worries about the atrophying of designers' judgment were prominent in their discussion. Said one: "We'll find far too many people who are willing to certify new or modified nuclear weapons based on very little data, or maybe no data." "The scary part," said another, "is that there will be no shortage of people who are willing to certify untested weapons, especially if they are certifying their own designs, or if they want to please someone in Washington. . . . If the laboratories cannot conduct tests, the United States should consider the possibility of eliminating its capability to design and certify nuclear weapons."[222] It is surprising to hear that possibility aired in the establishment that first gave the world the capability whose elimination was being discussed. The record of the discussion, however, reveals no voice raised in dissent. Nor, indeed, would it necessarily be as radical a move as it sounds. The military situation has changed, budgetary constraints have tightened, and parts of the nuclear weapons production infrastructure in both the United States and the former Soviet Union are now either closed or in physically dangerous condition.[223] Add to these a possible ban on testing and it is far from clear that the governments of the major nuclear weapons states will commission new types of nuclear weapons in the foreseeable future. In the United Kingdom, for example, it seems probable that the Trident warhead, now about to enter service, will be the last nuclear weapons development program, at least for a generation.

The nuclear weapons laboratories may, therefore, face a future in which they are no longer developers of new weapons but custodians of past ones—quite possibly weapons they are unable to test. Custodianship may sound like an unproblematic task, but here again questions arise about tacit knowledge and the "sameness" of artifacts. Even if no new designs are ever introduced, the remaining arsenal will change through radioactive decay and other processes of aging and through maintenance and the replacement of aged components. As designers age, leave, and die, the number who have first-hand experience of development to the point of full nuclear testing will steadily diminish; yet they will have to decide whether the inevitable changes in the arsenal matter. In such a situation, will explicit knowledge be enough? Will tacit knowledge and judgment survive adequately? For how long?

Another senior figure at Los Alamos asks: "Imagine, twenty years from now, a stockpile-surveillance team noticing that one of the weapons being stored has changed in appearance. They will want to know, 'is this still safe, and would it work if needed?' They will call the Laboratory and ask the experts regarding this weapon. Will they be able to rely on the answer they get?"[224] We do not claim to be able to answer this question, which becomes even more pointed if the time span considered stretches from 20 years to more than a generation. That it can be asked, however, is an indicator that, even without disarmament, the nuclear future may in at least one respect be quite different from the past. Hitherto, nuclear weapons have been deeply controversial morally and politically, but the cognitive authority of their designers has seldom been questioned. If, in the years to come, some untoward event, such as a serious nuclear weapons accident,[225] were to generate large-scale public concern, then we would suggest that the challenge to that cognitive authority may well be profound, and its consequences major.[226]

Acknowledgments

This research was supported by the U.K. Economic and Social Research Council, mainly under the Science Policy Support Group Initiative on Science Policy Research on Defence Science and Technology (grant Y307253006). We would particularly like to acknowledge the help given to our research by David Holloway, William Walker, and, above all, our interviewees from Los Alamos, Livermore, and Aldermaston. Helpful comments on the first draft of this chapter were received from Harry Collins, Jeff Hughes, John Krige, Dominique Pestre, Annette Schaper, William Walker, Peter Zimmerman, and three anonymous referees for the *American Journal of Sociology.*

Appendix: List of Interviewees

Los Alamos National Laboratory, New Mexico: Current and Retired Staff Members

Harold Agnew, December 13, 1991

Ira Akins, Robert Frank, Roger Lazarus, and Bill Spack, April 12, 1989

Delmar Bergen, December 18, 1990

Bob Carr, December 18, 1990

Tom Dowler, April 12, 1989

Tom Dowler and Thurman Talley, December 18, 1990

Robert Glasser, December 16, 1991

John Hopkins, December 10, 1991

Harry Hoyt, March 14, 1990

Jim Jackson, April 12, 1989

Steve Maarenen, December 18, 1990

J. Carson Mark, March 15, 1990 and December 10, 1991

Norman Morse, April 12, 1989

Robert Osbourne, November 20, 1991

Raemer Schreiber, December 10, 1991

Thurman Talley, April 12, 1989

Don Westervelt, December 18, 1990

Roy Woodruff, December 11, 1991

Lawrence Livermore National Laboratory, California: Current and Retired Staff Members

Roger E. Anderson and George A. Michael, April 13, 1989

Roger E. Anderson, Norman Hardy, Cecil E. Leith, Jr., William A. Lokke, V. William Masson, George A. Michael, and Jack Russ, April 13, 1989

Roger Batzel, December 4, 1991

James Carothers, December 10, 1990

Hugh DeWitt, December 12, 1990

Sidney Fernbach, Tad Kishi, Francis H. McMahon, George A. Michael, and Harry Nelson, October 16, 1989

Norman Hardy and George A. Michael, April 14, 1989

John Harvey, April 3, 1990

Carl Haussmann, December 10, 1990

Art Hudgins, December 10, 1990

Ray Kidder, December 6, 1991

Charles McDonald, December 10, 1990

Francis H. McMahon, October 16, 1989

George A. Michael, October 15 and 16, 1989

George Miller, December 10, 1990

Peter Moulthrop, December 10, 1990

Milo Nordyke, December 10, 1990

Duane Sewell, December 10, 1990

Edward Teller, March 24, 1990

Lowell Wood, October 16, 1989

Lawrence Woodruff, December 10, 1990

In addition, we interviewed two retired staff members of the U.K. Atomic Weapons Establishment at Aldermaston, and a small number of people (not weapons laboratory staff) spoke to us on a nonattribution basis; comments from the latter are referenced as anonymous private communications.

Notes

Chapter 1

1. Details of Gold's meetings with Fuchs and Yakovlev are taken from his statement to FBI agents Richard E. Brennan and T. Scott Miller, Jr., July 10, 1950, as reprinted in Robert C. Williams, *Klaus Fuchs, Atom Spy* (Harvard University Press, 1987), pp. 203–220, with additional information from other parts of the book; Richard Rhodes, *The Making of the Atomic Bomb* (Simon & Schuster, 1986), p. 450; Peggy Pond Church, *The House at Otowi Bridge: The Story of Edith Warner and Los Alamos* (University of New Mexico Press, 1960), p. 4. Fuchs's independent confessions confirmed the broad accuracy of Gold's account, but only in 1993 was there an authoritative statement from Soviet scientists of the information passed to them: Yuli Khariton and Yuri Smirnov, "The Khariton version," *Bulletin of the Atomic Scientists,* May 1993: 20–31.

2. See chapters 7 and 8 below. See also *Charter Technologies Limited* v *the Secretary of State for Defence,* 1991 C No. 691, High Court of Justice, Queen's Bench Division; interview with Digby Dyke, Great Malvern, Worcs., November 22, 1990; *The Times* (London), January 29, 1988; *New Scientist* 16, October 1986; *Electronics Weekly,* October 15, 1986; *The Engineer,* February 4, 1988.

3. R. W. Apple, Jr., "Scud missile hits a U.S. barracks," *New York Times,* February 26, 1991; Patriot Missile Defense: Software Problem Led to System Failure at Dhahran, Saudi Arabia, Report GAO/IMTEC-92-26, General Accounting Office, 1992; Robert Skeel, "Roundoff error and the Patriot missile," *SIAM* [Society for Industrial and Applied Mathematics] *News* 25, part 4 (July 1992): 11. See chapter 9 below.

4. See chapter 3 below. For a helpful set of papers on technological determinism see *Does Technology Drive History? The Dilemma of Technological Determinism,* ed. M. R. Smith and L. Marx (MIT Press, 1994).

5. See, e.g., Cynthia Cockburn and Susan Ormrod, *Gender and Technology in the Making* (Sage, 1993).

6. See chapter 2 below.

7. See, in addition to the works by Cockburn and Noble cited in chapter 2, Noble's *Forces of Production: A Social History of Industrial Automation* (Knopf, 1984).

8. See Trevor J. Pinch and Wiebe E. Bijker, "The social construction of facts and artefacts: or how the sociology of science and the sociology of technology might benefit each other," *Social Studies of Science* 14 (1984): 399–441; also in *The Social Construction of Technological Systems: New Directions in the Sociology and History of Technology*, ed. W. Bijker et al. (MIT Press, 1987).

9. Even in the few years since chapters 5 and 6 were first published, the alternative technological tradition of massive parallelism has made substantial inroads into the territory occupied by mainstream supercomputing.

10. See W. Brian Arthur, "Competing technologies, increasing returns, and lock-in by historical events," *Economic Journal* 99 (1989): 116–131; Paul A. David, "Heroes, herds and hysteresis in technological history: Thomas Edison and 'the battle of the systems' reconsidered," *Industrial and Corporate Change* 1 (1992): 129–180.

11. Pinch and Bijker, "The social construction of facts and artefacts."

12. See also Harro van Lente, Promising Technology: The Dynamics of Expectations in Technological Developments, Ph.D. thesis, Universiteit Twente, Netherlands, 1993.

13. The classic discussion is Robert K. Merton's essay "The self-fulfilling prophecy," in his *Social Theory and Social Structure* (Free Press of Glencoe, 1949).

14. See Barry Barnes, "Social life as bootstrapped induction," *Sociology* 17 (1983): 524–545; Barnes, *The Nature of Power* (Polity, 1988).

15. Otis Port et al., "Wonder chips: How they'll make computer power ultrafast and ultracheap," *Business Week*, July 4, 1994, pp. 46–52.

16. Here I use the term "machine" in a wide sense to cover the "hardware" aspects of technology. "Knowledge," as I use the term here, means simply shared belief, not necessarily correct belief. See, e.g., Barry Barnes, *Scientific Knowledge and Sociological Theory* (Routledge and Kegan Paul, 1974).

17. *Logos* in Greek means speech, word, or reason; thus, strictly speaking, we should use the word "technology" to refer to knowledge of the practical arts rather than to "hardware." In English, however, "technology" has come to encompass machines themselves, as well as knowledge about machines, and in this book I make no attempt at etymological correctness.

18. Chapter 9 is an exception in this respect.

19. See, e.g., *Science Observed: Perspectives on the Social Study of Science*, ed. K. Knorr-Cetina and M. Mulkay (Sage, 1983).

20. The "strong program" was first laid out in David Bloor, "Wittgenstein and Mannheim on the sociology of mathematics," *Studies in the History and Philosophy*

of Science, 4 (1973): 173–191. See also Bloor, *Knowledge and Social Imagery* (Routledge and Kegan Paul, 1976); Barnes, *Scientific Knowledge and Sociological Theory*.

21. See, e.g., the criticisms cited (and responded to) in the postscript to the second edition of Bloor's *Knowledge and Social Imagery* (University of Chicago Press, 1991). For an early review of the empirical work, see Steven Shapin, "History of science and its sociological reconstructions," *History of Science* 20 (1982): 157–211. Among the best of the more recent studies are Jonathan Harwood, *Styles of Scientific Thought: The German Genetics Community, 1900–1933* (University of Chicago Press, 1993), and Steven Shapin, *A Social History of Truth: Civility and Science in Seventeenth-Century England* (University of Chicago Press, 1994).

22. For a more systematic presentation see D. MacKenzie, "How do we know the properties of artefacts? Applying the sociology of knowledge to technology," in *Technological Change*, ed. R. Fox (Harwood, 1996).

23. Mannheim, *Ideology and Utopia: An Introduction to the Sociology of Knowledge* (Harcourt, Brace & World, 1936), e.g., pp. 79, 293, and 298.

24. See chapter 10 for references to relevant work of Polanyi, Collins, and others.

25. In retrospect, only chapter 4 here seems to me to approach adequacy in its taking into account of economics. Even this chapter was handicapped by the inability of interviewees, for reasons of commercial confidentiality, to discuss the economic aspects of the laser gyroscope at the level of detail that would, e.g., be needed to develop the kind of "ethnoaccountancy" called for in chapter 3.

26. Nowadays the gulf is actually smaller as far as the analysis of *individuals* is concerned, given the new attention within sociology to "rational choice" theories of individual action; for an example, see Barnes, *The Nature of Power*. For useful reflections on theories of the firm by a historian of technology, see W. Bernard Carlson, "The coordination of business organization and technological innovation within the firm: A case study of the Thomson-Houston Electric Company in the 1980s," in *Coordination and Information: Historical Perspectives on the Organization of Enterprise*, ed. N. Lamoreaux and D. Raff (University of Chicago Press, 1995).

27. See chapters 2 and 3 of the present volume.

28. R. A. Buchanan, "Theory and narrative in the history of technology," *Technology and Culture* 32 (1991): 365–376. The volume Professor Buchanan particularly has in mind is Bijker, Hughes, and Pinch, *The Social Construction of Technological Systems*.

29. John Law, "Theory and narrative in the history of technology: Response," *Technology and Culture* 32 (1991): 377–384.

30. Chapter 10, of course, willingly acknowledges that existing historical work has, implicitly, a great deal to say about tacit knowledge. Particularly valuable in that light is Lillian Hoddeson et al., *Critical Assembly: A Technical History of Los Alamos during the Oppenheimer Years, 1943–1945* (Cambridge University Press, 1993).

31. I would not for a moment claim historical adequacy for chapters 7 and 8 (a more historical treatment is in preparation), nor would I wish to suggest that the history of the field is entirely unexamined. The details of computer arithmetic have not, to my knowledge, been the subject of any published historical work, but I was delighted to discover that on May 23, 1995, the San Francisco Bay Area Computer History Group held a panel discussion on the history of floating-point arithmetic. Among important historical works bearing on program verification are the following: Gabriele Lolli, *La Macchina e le Dimostrazioni: Matematica, Logica e Informatica* (Il Mulino, 1987); Eloína Peláez, A Gift from Pandora's Box: The Software Crisis, Ph.D. thesis, University of Edinburgh, 1988; Michael S. Mahoney, "Computers and mathematics: The search for a discipline of computer science," in *The Space of Mathematics: Philosophical, Epistemological and Historical Explorations*, ed. J. Echeverria et al. (de Gruyter, 1992); C. B. Jones, The Search for Tractable Ways of Reasoning about Programs, Technical Report UMCS-92-4-4, University of Manchester, 1992.

32. See John Law, "On the social explanation of technical change: The case of the Portuguese maritime expansion," *Technology and Culture* 28 (1987): 227–252; Law, "Technology and heterogeneous engineering: The case of Portuguese expansion," in *Social Construction of Technological Systems*, ed. Bijker et al.

33. See chapter 2 below and *The Social Shaping of Technology*, ed. D. MacKenzie and J. Wajcman (Open University Press, 1985).

34. Langdon Winner, "Do artifacts have politics?" *Daedalus* 109 (winter 1980): 121–136; also in MacKenzie and Wajcman, *The Social Shaping of Technology*.

35. See, e.g., S. C. Strum and Bruno Latour, "Redefining the social link: From baboons to humans," paper represented to International Primatological Society meeting, Nairobi, July 1984.

36. Michel Callon, "Some elements of a sociology of translation: Domestication of the scallops and the fishermen of St Brieuc Bay," in *Power, Action and Belief: A New Sociology of Knowledge?* (Sociological Review Monograph 32), ed. J. Law (Routledge, 1986), esp. pp. 200, 221.

37. H. M. Collins and Steven Yearley, "Epistemological chicken," in *Science as Practice and Culture*, ed. A. Pickering (University of Chicago Press, 1992).

38. David Bloor, *Knowledge and Social Imagery*, second edition (University of Chicago Press, 1991), pp. 173–174. For an essentially equivalent earlier statement see pp. 29–32 of the first edition (Routledge, 1976).

39. See Michael Mulkay, "Knowledge and utility: Implications for the sociology of knowledge," *Social Studies of Science* 9 (1973): 63–80; R. Kling, "Audiences, nar-

ratives, and human values in social studies of technology," *Science, Technology and Human Values* 17 (1992): 349–365; K. Grint and S. Woolgar, "Computers, guns, and roses: What's social about being shot?" ibid., pp. 366–380; R. Kling, "When gunfire shatters bone: Reducing sociotechnical systems to social relationships," ibid., pp. 381–385.

40. D. MacKenzie, "How good a Patriot was it?" *The Independent*, December 16, 1991, p. 13. See, e.g., T. A. Postol, "Lessons of the Gulf War experience with Patriot," *International Security* 16, no. 3 (1991–92): 119–171, and the subsequent debate: R. M. Stein and T. A. Postol, "Correspondence: Patriot experience in the Gulf War," *International Security* 17, no. 1 (1992): 199–240.

41. Barry Barnes and David Bloor, "Relativism, rationalism and the sociology of knowledge," in *Rationality and Relativism*, ed. M. Hollis and S. Lukes (MIT Press, 1982), p. 33.

42. Bruno Latour, *Science in Action* (Harvard University Press, 1987), p. 100.

43. See H. M. Collins, "Public experiments and displays of virtuosity: The core-set revisited," *Social Studies of Science* 18 (1988): 725–748, esp. 726; D. MacKenzie, *Inventing Accuracy: A Historical Sociology of Nuclear Missile Guidance* (MIT Press, 1990), esp. pp. 370–372 and 418–423.

44. J. S. Oteri, M. G. Weinberg, and M. S. Pinales, "Cross-examination of chemists in drugs cases," in *Science in Context: Readings in the Sociology of Science*, ed. B. Barnes and D. Edge (MIT Press, 1982). For relevant reflections upon trust, see Shapin, *A Social History of Truth*.

45. Langdon Winner, "Upon opening the black box and finding it empty: Social constructivism and the philosophy of technology," *Science, Technology, & Human Values* 18 (1993): 362–378, esp. 372.

46. See, e.g., the attack on symmetry in Hilary Rose and Steven Rose, "Radical science and its enemies," in *The Socialist Register, 1979*, ed. R. Miliband and J. Saville (Merlin Press, 1979). For a response see D. MacKenzie, "Notes on the science and social relations debate," *Capital and Class* 14 (summer 1981): 47–60.

47. See also Wiebe E. Bijker, "Do not despair: There is life after constructivism," *Science, Technology, and Human Values* 18 (1993): 113–138.

48. The controversies, particularly in France, that have followed the crashes of the highly computerized Airbus A320 are particularly interesting in this respect. See, e.g., Michel Asseline, *Le Pilote: Est-Il Coupable?* (Édition 1, 1992).

49. For a discussion of the possibilities of and the constraints upon challenges to test results, see D. MacKenzie "From Kwajalein to Armageddon? Testing and the social construction of missile accuracy," in *The Uses of Experiment: Studies in the Natural Sciences*, ed. D. Gooding et al. (Cambridge University Press, 1989).

50. Winner, "Social Constructivism," pp. 369–370. The same point was also made in Stewart Russell's paper "The social construction of artefacts: A

response to Pinch and Bijker," *Social Studies of Science* 16 (1986): 331–346. The point is, of course, a familiar one in sociological and political studies of power; see, e.g., Steven Lukes, *Power: A Radical View* (Macmillan, 1974).

51. Cynthia Cockburn, "Feminism/constructivism in technology studies: Notes on genealogy and recent developments," contribution to workshop on European Theoretical Perspectives on New Technology: Feminism, Constructivism and Utility, Brunel University, September 1993.

52. For an introduction to the literature on technology and gender see Judy Wajcman, *Feminism Confronts Technology* (Polity, 1991).

53. The main treatment of gender in this volume is in chapter 2.

54. John McPhee, "The Pine Barrens," in *The John McPhee Reader*, ed. W. Howarth (Random House, 1977), p. 118. See, e.g., Eric Schatzberg's beautiful study "Ideology and technical choice: The decline of the wooden airplane in the United States," *Technology and Culture* 35 (1994): 34–69, which attributes the triumph of the metal over the wooden airplane not to any intrinsic superiority but largely to a cultural preference for metal, with its modern, progressive connotations. Several years ago Cynthia Cockburn told me how often in her studies of technology she had found that men felt affinity for metal, as distinct from other materials such as wood; see, e.g., C. Cockburn, *Machinery of Dominance: Women, Men and Technical Know-How* (Pluto, 1985), p. 192. Schatzberg reports: "I know the gender dimensions are there, but I don't know how to get at them" (personal communication); I share that feeling of frustration.

55. See, e.g., James Fleck, "Configurations: Crystallising contingency," *International Journal on Human Factors in Manufacturing* 3 (1993): 15–36.

56. D. MacKenzie, "Militarism and socialist theory," *Capital and Class* 19 (spring 1983): 33–73. For particular examples of these shaping processes, see *Military Enterprise and Technological Change: Perspectives on the American Experience*, ed. M. Smith (MIT Press, 1985); MacKenzie, *Inventing Accuracy*, esp. pp. 165–239.

57. Again paradoxically, Marxist theory shows its strength in analyses of the faltering processes of "transition from socialism to capitalism." See, e.g., Michael Burawoy and Pavel Krotov, "The economic basis of Russia's political crisis," *New Left Review* 198 (March-April 1993): 49–69.

58. Karl Marx, *Capital: A Critique of Political Economy*, volume 1 (Penguin, 1976), p. 932.

59. Simon, of course, is also one of the founders of the modern sociology of organizations. True, there are tensions between some aspects of Simon's thought (notably his commitment to a "strong" form of artificial intelligence) and a sociological perspective on human action; see, e.g., H. M. Collins, *Artificial Experts: Social Knowledge and Intelligent Machines* (MIT Press, 1990). Nevertheless, these differences should not blind us to the existence of potentially productive common ground.

60. The topic, as it concerns global climate change, is currently being studied by Simon Shackley and Brian Wynne of the University of Lancaster. See also Mark E. Fernau et al., "Review and impacts of climate change uncertainties," *Futures* 25 (1993): 850–863.

61. The classic sociological treatment of "charisma" is Max Weber's. See, e.g., his essay "The social psychology of the world religions," in *From Max Weber: Essays in Sociology*, ed. H. Gerth and C. Mills (Routledge, 1970).

62. John Markoff, "Circuit flaw causes Pentium chip to miscalculate, Intel admits," *New York Times* New Service, November 24, 1994. That we can talk about and agree upon what constitutes "error" in this context shows that there is a consensual benchmark in human arithmetic against which computer arithmetics are judged. Nevertheless, as chapter 8 shows, consensus about human arithmetic was insufficient to create agreement on what was the best *general* form of computer arithmetic. (In respect to Pentium, there has also been sharp disagreement as to how frequently users might encounter the error.)

63. Indeed, since chapters 7 and 8 were written, my colleague Maggie Tierney and I discovered an earlier controversy in the U.S. over claimed proofs of the correctness of the design of a system called SIFT (Software-Implemented Fault Tolerance). This was intended not as an actual product but as a platform for research on highly reliable computing for aircraft control. No litigation ensued, and the episode seems not have become known outside the U.S. The main aspects of the controversy can, however, be seen in Peer Review of a Formal Verification/Design Proof Methodology, NASA Conference Publication 2377, 1985.

64. I spelled out to one group the definition used in chapter 9 and asked them individually and independently to estimate the number of computer-related accidental deaths, worldwide, to the end of 1992. Their answers ranged from 500 to 250,000. Unquestionably, the width of this range reflects in part the inherent ambiguity of the definition; I suspect, however, that it also reflects differences in risk perception.

65. Typically, even the best discussions, such as Lauren Ruth Wiener's *Digital Woes: Why We Should Not Depend on Software* (Addison-Wesley, 1993), describe particular incidents and offer (often, as in Wiener's case, cogent) general arguments, rather than attempting quantitative analyses of accidents.

66. Charles Perrow, *Normal Accidents: Living with High-Risk Technologies* (Basic Books, 1984).

Chapter 2

1. Karl Marx, *The Poverty of Philosophy* (International, 1971), p. 109.

2. Marc Bloch's work suggests that feudal lords actually tried to suppress hand mills, preferring centralized water milling with its greater potential for the

exacting of feudal dues. See Bloch, "Advent and triumph of the water mill," in his *Land and Work in Mediaeval Europe* (Routledge, 1967).

3. Alvin H. Hansen, "The technological interpretation of history," *Quarterly Journal of Economics* 36 (November 1921): 72–83; Robert L. Heilbroner, "Do machines make history?" *Technology and Culture* 8 (July 1967): 335–345; Tom Burns, *Industrial Man* (Penguin, 1969), p. 35.

4. Of the many pieces of work that could be cited, three stand out: David Dickson, *Alternative Technology and the Politics of Technical Change* (Fontana, 1974); Nathan Rosenberg, "Marx as a student of technology," *Monthly Review* 28 (1976): 56–77; Raniero Panzieri, "The capitalist use of machinery: Marx versus the 'Objectivists,'" in *Outlines of a Critique of Technology*, ed. P. Slater (Ink Links, 1980). My debt to each of these is clear. Monika Reinfelder's survey article, "Introduction: Breaking the spell of technicism," in *Outlines* (ed. Slater), provides a useful overall sketch of "technicism" and "antitechnicism" in twentieth-century Marxism, although her judgments sometimes seem unduly harsh.

5. William H. Shaw, "'The handmill gives you the feudal lord': Marx's technological determinism," *History and Theory* 18 (1979), p. 155.

6. Langdon Winner, *Autonomous Technology: Technics-out-of-Control as a Theme in Political Thought* (MIT Press, 1977), p. 79. It is not clear, however, that the "forces of production" should be read as equivalent to technology. See also Winner's comment (p. 39) that "one should not go too far in attributing a theory of autonomous technology to Karl Marx." Winner distinguishes between Marx's work, in which human beings (especially capitalists) are seen as in ultimate control of technology, and that of, for example, Jacques Ellul, according to which human control has been, in a crucial sense, lost. Several of the reservations expressed here about a technological-determinist interpretation of Marx can also be found in Winner's book, especially on pp. 80–85.

7. Heilbroner (n. 3 above).

8. Nikolai Bukharin, *Historical Materialism. A System of Sociology* (University of Michigan Press, 1969; first published in Russian in 1921), p. 124. Before his fall from power in the 1930s, Bukharin was one of the leading Bolshevik theorists.

9. Reinfelder, "Introduction: Breaking the spell of technicism," in *Outlines*, ed. Slater.

10. Karl Marx, *A Contribution to the Critique of Political Economy* (Lawrence and Wishart, 1971; first published in German in 1859), pp. 20–21.

11. For example, Marx's 1846 letter to P. V. Annenkov, in Burns (n. 3 above), pp. 35–36. There is also an interesting passage, seldom cited, at the end of volume 3 of *Capital* (pp. 883–884 of the edition cited above): "To the extent that the labor-process is solely a process between man and Nature, its simple elements remain common to all social forms of development. But each specific historical form of this process further develops its material foundations and social

forms. Whenever a certain stage of maturity has been reached, the specific his-torical form is discarded and makes way for a higher one. The moment of arrival of such a crisis is disclosed by the depth and breadth attained by the contradic-tions and antagonisms between the distribution relations, and thus the specific historical form of their corresponding production relations, on the one hand, and the productive forces, the production powers and the development of their agencies, on the other hand. A conflict then ensues between the material devel-opment of production and its social form." Shaw ("'The handmill gives you the feudal lord,'" p. 159) cites this passage but replaces "The moment of arrival of such a crisis is disclosed by the depth and breadth attained by" with "[because of]."

12. Rosenberg, "Marx as a student of technology," p. 74.

13. Marx, *A Contribution to the Critique of Political Economy*, p. 20.

14. Winner, *Autonomous Technology*, p. 78 (emphasis added).

15. Joseph Stalin, "Dialectical and historical materialism," in *The Essential Stalin: Major Theoretical Writings, 1905–52*, ed. B. Franklin (Croom Helm, 1973), p. 318; George Lukács, "Technology and social relations," *New Left Review* 39 (September-October 1966; first published in German in 1925), p. 29. Lukács, most famous as a literary critic, was in the early 1920s one of the leading oppo-nents of technicist and mechanical approaches to Marxism.

16. Shaw (n. 5 above), p. 158.

17. Karl Marx, *Economic and Philosophic Manuscripts of 1844* (Lawrence and Wishart, 1973), p. 113; *Capital: A Critique of Political Economy* (Penguin, 1976), 1: 284. This latter work (the Penguin/New Left Review translation of volume I of *Capital*) is cited below simply as *Capital* 1.

18. Stalin (n. 15 above), p. 321; Louis Althusser and Etienne Balibar, *Reading Capital* (New Left Books, 1970), p. 235; Lukács (n. 15 above), p. 30.

19. Raymond Williams, *Marxism and Literature* (Oxford University Press, 1977), pp. 83–89. For other useful comments on Marx's view of determinism see pp. 82–84 of Winner, *Autonomous Technology*.

20. Karl Marx, "The eighteenth Brumaire of Louis Bonaparte," in Marx and Engels, *Selected Works in One Volume* (Lawrence and Wishart, 1968), p. 97.

21. G. A. Cohen, *Karl Marx's Theory of History: A Defence* (Clarendon, 1978), quote on pp. 278–279; Erik Olin Wright, *Class, Crisis and the State* (New Left Books, 1978), chapter 1; E. P. Thompson, *The Poverty of Theory and Other Essays* (Merlin, 1978). See also Perry Anderson, *Arguments within English Marxism* (New Left Books, 1980).

22. *Capital* 1, parts 3 and 4, especially chapter 15, "Machinery and large-scale industry." Nathan Rosenberg has done most to bring this part of Marx's work to the attention of scholars within the English-language history of technology: see

Rosenberg

"Marx as a student of technology"; also see "Karl Marx on the Economic Role of Science," in *Perspectives on Technology* (Cambridge University Press, 1976). However, as Rosenberg writes in *Inside the Black Box: Technology and Economics* (Cambridge University Press, 1982), until quite recently "hardly anyone has . . . passed" through the "doors to the study of the technological realm" that Marx opened (p. viii). Marxist theory itself tended to neglect this part of *Capital* until Harry Braverman revived interest in it with his *Labor and Monopoly Capital: The Degradation of Work in the Twentieth Century* (Monthly Review Press, 1974). A noteworthy exception to that neglect is Panzieri's "The capitalist use of machinery," first published in the Italian journal *Quaderni Rossi* in 1961.

23. *Capital* 1 : 283–306. This section of my chapter owes a great deal to Iain Campbell. His manuscript "Marxist perspectives on the transformation of the labour process under capitalism" (unpublished as far as I know), and his comments on an earlier draft of this chapter were extremely helpful to me in developing my understanding of the structure of Marx's theory.

24. *Capital* 1: 279–280.

25. Ibid., pp. 283, 284, 290.

26. Ibid., p. 293.

27. See, particularly, ibid., pp. 302–303.

28. The unpublished chapter is now to be found in *Capital* 1; the quoted passage is on p. 990 (emphasis in the original deleted) and the phrase "material substratum" is on p. 981. For discussion of this point and further citations, see Campbell, "Marxist perspectives," p. 19 and n. 50. See also Brighton Labour Process Group, "The Capitalist Labour Process," *Capital and Class* 1 (spring 1977): 3–26.

29. *Capital* 1: 873–940. See also the three fascinating historical chapters of *Capital* 3 : "Historical facts about merchant's capital," "Pre-capitalist relationships," and "Genesis of capitalist ground-rent." The section of the *Grundrisse* on "Original Accumulation of Capital," which includes Marx's famous discussion of pre-capitalist economic formations, shows the development of Marx's thought on the topic. See Karl Marx, *Grundrisse: Foundations of the Critique of Political Economy (Rough Draft)* (Penguin, 1973), pp. 459–515.

30. *Capital* 1: 425.

31. Ibid., p. 645. (The Penguin translation of *Capital* prefers the more literal "formal subsumption.")

32. Ibid., pp. 383–389; E. P. Thompson, "Time, work-discipline and industrial capitalism," *Past and Present* 38 (1967): 56–97; N. S. B. Gras, *Industrial Evolution* (Oxford University Press, 1930), p. 77; Sidney Pollard, *The Genesis of Modern Management: A Study of the Industrial Revolution in Great Britain* (Arnold, 1965), pp. 33–34; Stephen A. Marglin, "What do bosses do? The origins and function

of hierarchy in capitalist production," in *The Division of Labour: The Labour Process and Class Struggle in Modern Capitalism*, ed. A. Gorz (Harvester, 1978).

33. *Capital* 1: 443.

34. Ibid., p. 447.

35. Ibid., p. 450.

36. See the works of Gras, Pollard, and Marglin cited in n. 32.

37. That is the eighteenth-century creation of "central workshops" in the British textile industry without any change in technique from that prevalent in putting-out. See Marglin, "What do bosses do?"; also see Jennifer Tann, *The Development of the Factory* (Cornmarket, 1972). Even this is not a pure example, for there were elements of a "manufacturing" division of labor involved. There is a short but useful discussion on p. 7 of *Technology and Toil in Nineteenth Century Britain*, ed. M. Berg (Conference of Socialist Economists, 1979).

38. *Capital* 1: 455–491.

39. Ibid., pp. 461–462, 463; Babbage as cited in Braverman's *Labor and Monopoly Capital*, pp. 79–80; *Capital* 1: 460.

40. Ibid., pp. 462 and 482; William Thompson, *An Inquiry into the Principles of the Distribution of Wealth* (London, 1824), p. 274, as cited by Marx, *Capital* 1: 482–483, n. 44.

41. *Capital* 1: 490.

42. Ibid., pp. 492–493 (emphasis added), p. 493.

43. Ibid., pp. 495, 494, 496–497, 508.

44. Ibid., pp. 517, 489.

45. Ibid., p. 531.

46. Ibid., pp. 545, 549.

47. Ibid., pp. 548, 442.

48. Within Marxism theoretical attention to sabotage as an important topic has come mainly from Italian "autonomists." See the collection *Working Class Autonomy and the Crisis: Italian Marxist Texts of the Theory and Practice of a Class Movement* (Conference of Socialist Economists, 1979). Raniero Panzieri was one of the intellectual founders of the "autonomist" tendency.

49. Ure, *The Philosophy of Manufactures* (London, 1835), p. 370, as quoted in *Capital* 1: 564; ibid., p. 563.

50. David S. Landes, *The Unbound Prometheus: Technological Change and Industrial Development in Western Europe from 1750 to the Present* (Cambridge University Press, 1969), p. 43.

51. Ibid., pp. 115–116.

52. Nathan Rosenberg, "The direction of technological change: Inducement mechanisms and focusing devices," in his *Perspectives on Technology*, particularly pp. 119–120; quote on p. 117. For the Fourdrinier machine, his sources are D. C. Coleman, *The British Paper Industry, 1495–1860: A Study in Industrial Growth* (Oxford University Press, 1958), pp. 258–259; and H. J. Habakkuk, *American and British Technology in the Nineteenth Century: The Search for Labour-saving Inventions* (Cambridge University Press, 1962), p. 153. While Habakkuk, of course, argues that the desire to replace skilled labor was less potent in Britain than in the United States, he does admit that "there are several instances where the desire to diminish the bargaining power of skilled craft labor provided a strong incentive to install machines" (ibid., pp. 152–153).

53. Anthony F. C. Wallace, *Rockdale: The Growth of an American Village in the Early Industrial Revolution* (Knopf, 1978), p. 193; see also p. 382.

54. Tine Bruland, "Industrial conflict as a source of technical innovation: Three cases," *Economy and Society* 11 (May 1982), p. 91.

55. Langdon Winner, "Do artifacts have politics?" *Daedalus* 109 (winter 1980), p. 125. See Robert Ozanne, *A Century of Labor-Management Relations at McCormick and International Harvester* (University of Wisconsin Press, 1967).

56. E. J. Hobsbawm's classic 1952 "revisionist" essay ''The machine breakers,'' is to be found in his collection *Labouring Men: Studies in the History of Labour* (Weidenfeld and Nicolson, 1968), pp. 5–22. See also E. P. Thompson, *The Making of the English Working Class* (Penguin, 1968), esp. pp. 515–659.

57. Maxine Berg, *The Machinery Question and the Making of Political Economy, 1815–1848* (Cambridge University Press, 1980). This book provides valuable background to the development of Marx's thought on machinery, particularly in the way Berg locates in its context Engels's book *The Condition of the Working Class in England* (Panther, 1969), first published in 1845, whose discussion of machinery was an important precursor of Marx's.

58. Merritt Roe Smith, *Harpers Ferry Armory and the New Technology: The Challenge of Change* (Cornell University Press, 1977), p. 292.

59. Carroll W. Pursell Jr. points out the significance of Smith's work in this respect in "History of technology," in *A Guide to the Culture of Science, Technology, and Medicine*, ed. P. Durbin (Free Press, 1980); the quote is from p. 84 of that book. Cf. John E. Sawyer, "The social basis of the American system of manufacturing," *Journal of Economic History* 14 (1954): 361–379.

60. G. N. von Tunzelmann, *Steam Power and British Industrialization to 1860* (Clarendon, 1978), p. 8 (see also pp. 217–225); *Capital* 1: 536.

61. Paul Mantoux, *The Industrial Revolution in the Eighteenth Century: An Outline of the Beginnings of the Modern Factory System in England* (Cape, 1928), p. 36. Mantoux's reliance on Marx was perhaps greater than he acknowledged. His

discussion of the difference between a "tool" and a "machine" (pp. 193–194) followed that of Marx very closely.

62. Rosenberg, "The direction of technological change," p. 118.

63. On "discovery accounts" in science see S. W. Woolgar, "Writing an intellectual history of scientific development: The use of discovery accounts," *Social Studies of Science* 6 (September 1976): 395–422.

64. *Capital* 1: 588–610, quote on p. 603; Raphael Samuel, "Workshop of the world: Steam power and hand technology in mid-Victorian Britain," *History Workshop* 3 (spring 1977): 6–72.

65. Outwork (renamed "teleworking") remains a subject of substantial discussion.

66. This is argued cogently, from a management point of view, in Sir Frederick Catherwood's "Shop floor power," *Production Engineer,* June 1976, pp. 297–301.

67. *Capital* 1: 545.

68. Braverman, *Labor and Monopoly Capital.* A useful summary of criticisms of the deskilling thesis can be found in *The Degradation of Work? Skill, Deskilling and the Labour Process,* ed. S. Wood (Hutchinson, 1982).

69. See, e.g., Raphael Samuel's comment in "Workshop of the world" (p. 59) that "nineteenth century capitalism created many more skills than it destroyed, though they were different in kind from those of the all-round craftsmen." See also Habakkuk's comments on skill requirements in nineteenth-century British industry (*American and British Technology,* pp. 153–156). For a balanced judgment see Tony Elger, "Valorisation and 'deskilling': A critique of Braverman," *Capital and Class* 7 (spring 1979), esp. pp. 72–78. This article, which is pertinent to several of the issues discussed here, is also to be found in *The Degradation of Work,* ed. S. Wood.

70. For an interesting though schematic account see Joan Greenbaum, *In the Name of Efficiency: Management Theory and Shopfloor Practice in Data-Processing Work* (Temple University Press, 1979).

71. Andrew L. Friedman, *Industry and Labour: Class Struggle at Work and Monopoly Capitalism* (Macmillan, 1977), p. 78.

72. For a useful survey, see Paul Uselding, "Studies of technology in economic history," *Research in Economic History,* suppl. 1 (1977), pp. 159–219.

73. For example, *Capital* 1: 513–517, which focuses on the level of wages as a determinant of choice of technique.

74. Ibid., p. 932.

75. Two very different articles that make this point are Thompson, "Time, work-discipline and industrial capitalism," and John Holloway and Sol Picciotto, "Capital, crisis and the state," *Capital and Class* 2 (summer 1977): 76–101.

76. William H. Lazonick, "Production relations, labor productivity, and choice of technique: British and U.S. cotton spinning," *Journal of Economic History* 41 (September 1981): 491–516. Cf. Lars C. Sandberg, "American rings and English mules: The role of economic rationality," *Quarterly Journal of Economics* 83 (1969): 25–43. Closely related work by Lazonick includes "Industrial relations and technical change: The case of the self-acting mule," *Cambridge Journal of Economics* 3 (1979): 213–262; "Factor costs and the diffusion of ring spinning in Britain prior to World War 1," *Quarterly Journal of Economics* 96 (February 1981): 89–109; and "Competition, specialization, and industrial decline," *Journal of Economic History* 41 (March 1981): 31–38.

77. *Capital* 1: 563; Lazonick, "Production relations, labor productivity and choice of technique" and "Industrial relations and technical change" (quote from p. 232 of latter); Sandberg, "American rings and English mules," p. 33.

78. See, e.g., Huw Beynon, *Working for Ford* (EP Publishing, 1975).

79. David F. Noble, "Social choice in machine design: The case of automatically controlled machine tools," in *Case Studies on the Labor Process*, ed. A. Zimbalist (Monthly Review Press, 1979), p. 41.

80. *Capital* 1: 545.

81. See Lazonick, "Industrial relations and technical change," esp. pp. 240–246 and 256–257.

82. Ruth Schwartz Cowan, "From Virginia Dare to Virginia Slims: Women and technology in American life," *Technology and Culture* 20 (January 1979): 51–63; Anne Phillips and Barbara Taylor, "Sex and skill: Notes towards a feminist economics," *Feminist Review* 6 (1980): 79–88.

83. Cynthia Cockburn, "The material of male power," *Feminist Review* 9 (autumn 1981), pp. 46 and 52. See also Cynthia Cockburn, *Brothers: Male Dominance and Technological Change* (Pluto, 1983).

84. George H. Daniels, "The big questions in the history of American technology," in *The State of American History*, ed. H. Bass (Quadrangle, 1970), pp. 200, 199; Thomas P. Hughes, "Emerging themes in the history of technology," *Technology and Culture* 20 (October 1979), p. 710; Pursell, "History of technology," p. 98; David A. Hounshell, "On the discipline of the history of American technology," *Journal of American History* 67 (March 1981), p. 863. Lest I be misunderstood, let me emphasize that I am saying only that Marx's approach converges with the question raised by this literature, not that it is identical with it. Marx studied in detail only one form of technology, that of production, and not, e.g., the equally important areas of domestic technology and military technology. And there clearly are non-Marxist ways of approaching the question of the influence of society on technical development.

85. Darwin Stapleton, "The discipline of the history of American technology: An exchange," *Journal of American History* 68 (March 1982), p. 899. For Marx's own remarks on the history of technology, see *Capital* 1: 493–494, n. 4.

86. *Capital* 1: 554–555 and 1024 (parentheses in original, emphasis deleted).

87. For the relevance of contingency in the history of *science*, see Steven Shapin, "History of science and its sociological reconstructions," *History of Science* 20 (1982): 157–211.

88. Winner, "Do artifacts have politics?" pp. 123–124. See Robert A. Caro, *The Power Broker: Robert Moses and the Fall of New York* (Knopf, 1974).

89. See, e.g., Habakkuk, *American and British Technology in the Nineteenth Century*, where the geographical comparison serves this function.

90. Once again, I have found Iain Campbell's ideas on this point very helpful.

91. David F. Noble, "Social choice in machine design: The case of automatically controlled machine tools, and a challenge for labor," *Politics and Society* 8 (1978), p. 337.

92. Seymour Melman, "Alternative criteria for the design of means of production," *Theory and Society* 10 (1981): 325–336.

93. This is clearest in a book by one of the former leaders of the Lucas work force: Mike Cooley, *Architect or Bee? The Human/Technology Relationship* (Hand and Brain, n.d.). For an overall account see Hilary Wainwright and Dave Elliott, *The Lucas Plan: A New Trade Unionism in the Making?* (Allison and Busby, 1982).

94. See the eleventh of Marx's "Theses on Feuerbach" in Marx and Engels, *Selected Works in One Volume*, pp. 28–30.

Chapter 3

1. Compare, for example, two pioneering monographs on domestic work and domestic technology: Ruth Schwartz Cowan, *More Work for Mother: The Ironies of Household Technology from the Open Hearth to the Microwave* (Basic Books, 1983), and Jonathan Gershuny, *Social Innovation and the Division of Labour* (Oxford University Press, 1983). Cowan's analysis rests fundamentally on value commitments. Gershuny, even though he has since held a chair in sociology, employs a typically economic model of rational, maximizing choice.

2. Sidney Winter, as quoted in Jon Elster, *Explaining Technical Change: A Case Study in the Philosophy of Science* (Cambridge University Press, 1983), pp. 139–140.

3. The classical counterargument is to be found in Milton Friedman, "The methodology of positive economics," in Friedman's *Essays in Positive Economics* (University of Chicago Press, 1953), pp. 3–43, esp. p. 22. According to this argument, even if firms do not consciously seek to maximize, a process akin to natural selection goes on. Firms that happen to hit on a maximizing or near-maximizing strategy will grow, while those that do not will shrink or fail, and therefore maximizing strategies will prevail, even if those who pursue them do so for reasons quite other than the knowledge that they are maximizing. If

the environment in which firms operate were unchanging, this defense would be perfectly plausible. The assumption of an unchanging, "given" environment is, however, far too unrealistic and restrictive, especially when technological change is being considered. If the environment is changing dramatically, it is far from clear that there is a stable maximizing strategy toward which selection will move populations. Game-theoretic elaborations to the neoclassical framework help, because they can model the way one firm's action changes another firm's environment, but even they rely on a certain stability of framework.

4. See R. M. Cyert and J. G. March, *A Behavioral Theory of the Firm* (Prentice-Hall, 1963); Richard R. Nelson and Sidney G. Winter, *An Evolutionary Theory of Economic Change* (Harvard University Press, 1982), pp. 107–112. Note the evident parallels with debates in political science over the explanation of national policy, especially in the field of defense and foreign affairs, where a "realist" position, akin to neoclassical economics, has contended with a "bureaucratic politics" position that has argued that policy is outcome rather than decision. Two classic discussions are Graham Allison, *Essence of Decision: Explaining the Cuban Missile Crisis* (Little, Brown, 1971) and John D. Steinbruner, *The Cybernetic Theory of Decision: New Dimensions of Political Analysis* (Princeton University Press, 1974).

5. Nelson and Winter, *Evolutionary Theory*.

6. David Noble, *Forces of Production: A Social History of Industrial Automation* (Knopf, 1984), p. 321.

7. Though there is no indication that it is a connection Noble would wish to draw, it is worth noting that "domination" is a category far more central to Max Weber's sociology than to Karl Marx's political economy.

8. Since capitalist domination can take a range of different forms, of which direct control is only one, it would be more correct to talk of a set of heuristics. The discussion in the text, for the sake of simplicity, deals with direct control alone.

9. Michael Piore, "The impact of the labor market upon the design and selection of productive techniques within the manufacturing plant," *Quarterly Journal of Economics* 82 (1986), as quoted by Noble, *Forces of Production*, p. 217n. For more recent evidence bearing upon the same issue, see Michael L. Dertouzos, Richard K. Lester, Robert N. Solow, and the MIT Commission on Industrial Productivity, *Made in America: Regaining the Productive Edge* (MIT Press, 1989).

10. A classic study of pricing behavior in the alternative economics tradition is the account of pricing in a department store in chapter 7 of Cyert and March, *Behavioral Theory*. See also Nelson and Winter, *Evolutionary Theory*, p. 410.

11. Interview with Neil Lincoln (formerly leading supercomputer designer at Control Data Corporation and ETA Systems), Minneapolis, April 3, 1990. My data on the prevalence of the 3:1 rule are strictly limited, and so what I say in the text can only be tentative.

12. Interview with John Rollwagen (chairman, Cray Research, Inc.), Minneapolis, April 3, 1990. The long-standing figure of 65 percent had been reduced to 60 percent by then.

13. Toshiro Hiromoto, "Another hidden edge—Japanese management accounting," *Harvard Business Review* 66 (1988), no. 4: 22–26.

14. See, e.g., Christopher Freeman, *The Economics of Industrial Innovation*, second edition (MIT Press, 1982), p. 163; Rod Coombs, Paolo Saviotti, and Vivien Walsh, *Economics and Technological Change* (Macmillan, 1987), p. 57.

15. Rollwagen interview.

16. R. Ball, R. E. Thomas, and J. McGrath, "A survey of relationships between company accounting and R&D decisions in smaller firms," paper read to ESRC meeting on New Technologies and the Firm Initiative, Stirling, 1991. The authors found that firms reported giving greater weight to the assessment of project costs and benefits than to simpler determinants such as previous year's R&D budget. It could be, however, that firms were reporting an idealized version of their practice; also, smaller firms might be expected to make more *ad hoc* decisions than larger ones.

17. See Nelson and Winter, *Evolutionary Theory*, pp. 251–254. Again, though, I have not been able to trace recent empirical work. Nelson and Winter's book in fact gives only passing attention to pricing and R&D budgets, and concentrates on developing quantitative, long-term economic growth models. Though these are impressive (and appear empirically successful), the assumptions built into them are too simple, and what is being explained too general, for them be of direct relevance here. There is a brief and clear summary of this aspect of Nelson and Winter's work in Paul Stoneman, *The Economic Analysis of Technological Change* (Oxford University Press, 1983), pp. 184–185.

18. Richard Nelson and Sidney Winter, "In search of useful theory of innovation," *Research Policy* 6 (1977): 36–76, esp. 56–60; Nelson and Winter, *Evolutionary Theory*, pp. 255–262; Giovanni Dosi, "Technological paradigms and technological trajectories: A suggested interpretation of the determinants of technical change," *Research Policy* 11 (1982): 147–162. A more recent discussion is Giovanni Dosi, "The nature of the innovative process," in *Technical Change and Economic Theory*, ed. Dosi et al. (Pinter, 1988). The concept is now in the textbooks. See Coombs et al., *Economics and Technological Change*.

19. For Moore's Law (named for Gordon E. Moore, director of research at Fairchild Semiconductor in 1964), see Robert N. Noyce, "Microelectronics," *Scientific American* 237, no. 3 (1977), reprinted in *The Microelectronics Revolution*, ed. T. Forester (MIT Press, 1980).

20. Nelson and Winter, *Evolutionary Theory*, p. 257.

21. It is perhaps of some significance that Nelson and Winter, whose overall project is framed in a biological metaphor, here (without discussion) change from biology to mechanics.

22. In his *Technical Change and Industrial Transformation: The Theory and an Application to the Semiconductor Industry* (Macmillan, 1984), Dosi argues that "technological trajectories are by no means 'given by the engineers' alone: we tried to show that they are the final outcome of a complex interaction between some fundamental economic factors (search for new profit opportunities and for new markets, tendencies toward cost-saving and automation, etc.) together with powerful institutional factors (the interests and structure of existing firms, the effects of government bodies, the patterns of social conflict, etc.)." (p. 192) This quotation may, however, be intended to describe only the initial selection of a technological paradigm, rather than the subsequent trajectory that the paradigm "determines" (ibid., p. 299). Another formulation (Dosi, "Technological paradigms and technological trajectories," p. 154) contains something of the first meaning of "natural" and grants a shaping role for economic factors *after* initial selection of a paradigm, but only within boundaries set by the latter: "A technological trajectory, i.e. to repeat, the 'normal' problem solving activity determined by a paradigm, can be represented by the movement of multi-dimensional trade-offs among the technological variables which the paradigm defines as relevant. Progress can be defined as the improvement of these trade-offs. One could thus imagine the trajectory as a 'cylinder' in the multidimensional space defined by these technological and economic variables. (Thus, a technological trajectory is a cluster of possible technological directions whose outer boundaries are defined by the nature of the paradigm itself.)" The usage of "paradigm" is an extension (by analogy) to technology of T. S. Kuhn's notion of scientific paradigm. For a useful discussion of the ambiguities of the latter, see Kuhn's postscript to the second edition of *The Structure of Scientific Revolutions* (Chicago University Press, 1970).

23. Dosi, *Technical Change and Industrial Transformation*, 68; Dosi, "Technological paradigms and technological trajectories," p. 153.

24. Thomas P. Hughes, "Technological momentum in history: Hydrogenation in Germany, 1898–1933," *Past and Present* 44 (1969): 106–132; Bruno Latour, *Science in Action* (Harvard University Press, 1987).

25. The closest to a comment on it I have found (again it may be my ignorance) is this remark in Dosi, "Innovative process" (p. 226): "To the extent that innovative learning is 'local' and specific in the sense that it is paradigm-bound and occurs along particular trajectories, but is shared—with differing competences and degrees of success—by all the economic agents operating on that particular technology, one is likely to observe at the level of whole industries those phenomena of 'dynamic increasing returns' and 'lock-in' into particular technologies discussed [by Brian Arthur and Paul David]."

26. See, e.g., Jack Worlton, "Some patterns of technological change in high-performance computers," in *Supercomputing '88*.

27. Lincoln interview. The ETA^{10} was designed with eight processors, not three, because processing speed does not go up linearly in the number of processors.

28. The blurring of the boundaries between these fields in the 1990s and the decline in the defense market for traditional supercomputers seem to have weakened the rule of thumb about speed.

29. This happened with the ambitious "MP" supercomputer project of Steve Chen of Cray Research, which was cancelled by Cray Research chairman John Rollwagen because it seemed likely to lead to an unduly expensive machine (Rollwagen interview). Chen then left Cray Research to pursue supercomputing development with funding (but no certainty of the ultimate marketing of a product) from IBM.

30. Nelson and Winter, *Evolutionary Theory*, pp. 259–261.

31. Paul A. David, "Understanding the economics of QWERTY: The necessity of history," in *History and the Modern Economist*, ed. W. Parker (Blackwell, 1986); see p. 43. I am grateful to Peter Swann for the reference.

32. See, e.g., W. Brian Arthur, "Competing technologies and economic predic-tion," *Options* (April 1984): 10–13. Among the examples discussed by Arthur and David are the QWERTY keyboard, the pressurized water reactor, and the gas-powered internal-combustion motor car.

33. Barry Barnes, "Social life as bootstrapped induction," *Sociology* 17 (1983): 524–545; Barnes, *The Nature of Power* (Polity Press, 1988).

34. See Christopher Freeman, "Induced innovation, diffusion of innovations and business cycles," in *Technology and Social Process*, ed. B. Elliott (Edinburgh University Press, 1988).

35. Christopher Freeman and Carlota Perez, "Structural crises of adjustment, busi-ness cycles and investment behaviour," in Dosi et al., *Technical Change*; see p. 48. These are the general conditions, according to Freeman and Perez, met by the "key factor" of all five successive paradigms they identify, but they clearly believe them to hold for microchip technology as key to the current, emerging, paradigm.

36. After the first draft of this paper was completed, I discovered an unpub-lished paper by Arie Rip, of the University of Twente, making essentially the same point about Moore's Law: Expectations and Strategic Niche Management in Technological Development (June 1989). See also Harro van Lente (1993), Promising Technology: The Dynamics of Expectations in Technological Developments, Ph.D. dissertation, Universiteit Twente, Netherlands, esp. pp. 79, 87, and 171, for useful remarks on Moore's Law. Van Lente's thesis contains case studies of the role of expectations in the development of an electrical insu-lator, of membrane technology, and of high-density television.

37. Dosi, *Technical Change and Industrial Transformation*, p. 68. I mention Dosi's study only because it explicitly makes use of the notion of trajectory. Other authors make the same assumption; see, for example, Ernest Braun and Stuart Macdonald, *Revolution in Miniature: The History and Impact of Semiconductor Electronics* (Cambridge University Press, 1982), pp. 103–104 and 217.

38. See Cyril Tomkins and Roger Groves, "The everyday accountant and researching his reality," *Accounting, Organizations and Society* 8 (1983): 361–374, esp. 364. There are, of course, interesting studies of nonindustrial societies to be found within economic anthropology and economic history, though not all by any means delve fully into ethnoaccountancy. See, e.g., Rhoda H. Halperin, *Economies across Cultures: Towards a Comparative Science of the Economy* (Macmillan, 1988), and Raymond W. Goldsmith, *Premodern Financial Systems: A Historical Comparative Study* (Cambridge University Press, 1987).

39. Although it does not deal with technological change, the sociological work that is closest to my argument here is R. J. Anderson, J. A. Hughes, and W. W. Sharrock, *Working for Profit: The Social Organisation of Calculation in an Entrepreneurial Firm* (Averbury, 1989), and Richard Harper, An Ethnographic Examination of Accountancy, Ph.D. thesis, University of Manchester, 1989. See also Harper, "Not any old numbers: An examination of practical reasoning in an accountancy environment," *Journal of Interdisciplinary Economics* 2 (1988): 297–306. Another intriguing study is Jean Lave, "The values of quantification," in *Power, Action and Belief*, Sociological Review Monograph 32, ed. J. Law (Routledge, 1986).

40. Aside from the material cited in the previous note, work on accountancy has also been done by sociologists Keith Macdonald and Colwyn Jones: see, e.g., Keith M. Macdonald, "Professional formation: The case of Scottish accountants," *British Journal of Sociology* 35 (1984): 174–189; Colwyn Jones, What is Social about Accounting? Bristol Polytechnic Occasional Papers in Sociology, no. 7, 1989.

41. Nelson and Winter, *Evolutionary Theory*, p. 411.

42. Alfred D. Chandler, Jr., *The Visible Hand: The Managerial Revolution in American Business* (Harvard University Press, 1977).

43. One reason why this is a dubious way of thinking is that an accountancy system does not come free. A balance has to be struck between the benefits of greater knowledge of one's operations and the costs of such knowledge. It may be that here is a minor replica of the general problem of maximization discussed earlier.

44. Judith A. McGaw, "Accounting for innovation: Technological change and business practice in the Berkshire County paper industry," *Technology and Culture* 26 (1985): 703–725. See also McGaw, *Most Wonderful Machine: Mechanization and Social Change in Berkshire Paper Making, 1801–1885* (Princeton University Press, 1987).

45. See, e.g., Anthony F.C. Wallace, *St. Clair: A Nineteenth-Century Coal Town's Experience with a Disaster-Prone Industry* (Knopf, 1987). There may be some purchase here on one of the classic debates of economic history: the explanation of the faster rate of mechanization in the nineteenth century United States compared to Great Britain, for which see H. J. Habakkuk, *American and British Technology in the Nineteenth Century: The Search for Labour-Saving Inventions*

(Cambridge University Press, 1962). However, it is not clear that British accounting practice was any different from American in the relevant respect. See Sidney Pollard, *The Genesis of Modern Management: A Study of the Industrial Revolution in Great Britain* (Edward Arnold, 1965).

46. Robert S. Kaplan, "The Evolution of Management Accounting," *Accounting Review* 59 (1984): 390–418, esp. 415.

47. Hiromoto, "Another hidden edge," p. 22.

48. See chapter 2 of this volume.

49. Freeman, *Industrial Innovation*, p. 167.

50. See chapter 1 of this volume for a brief exposition of actor-network theory. Note that actors involved include nonhuman entities as well as human beings: an actor-network is not a network in the ordinary sociological usage of the term. The concept is closer to philosophical monism than to sociometrics. See the usage of *le réseau* in Denis Diderot, "Le Rêve de d'Alembert," in Diderot, *Œuvres Complètes*, tome 17 (Hermann, 1987) (e.g. on p. 119).

51. It is perhaps significant that such success as neoclassical economics has enjoyed in the empirical explanation of technological change seems to be predominantly in the explanation of patterns of diffusion. Some degree of stabilization is a sine qua non of the applicability of the concept of diffusion, because there needs to be some sense in which it is the same thing (hybrid corn, or whatever) that is diffusing. For skeptical comments on the concept of diffusion, see Latour, *Science in Action*.

52. There have been calls for some time for bringing together the sociology of scientific knowledge and the study of technological innovation. See especially Trevor J. Pinch and Wiebe E. Bijker, "The social construction of facts and artefacts: or how the sociology of science and the sociology of technology might benefit each other," *Social Studies of Science* 14 (1984): 339–441. A first collection of studies exemplifying the connection was *The Social Construction of Technological Systems: New Directions in the Sociology and History of Technology*, ed. W. Bijker et al. (MIT Press, 1987). These studies, however, did not address the economic analysis of technological change very directly. The main effort to do so in *The Social Construction of Technological Systems*, by Henk van den Belt and Arie Rip ("The Nelson-Winter-Dosi model and synthetic dye chemistry") seems to me insufficiently critical of that model.

53. For "interpretative flexibility" see H. M. Collins, "Stages in the empirical programme of relativism," *Social Studies of Science* 11 (1981): 3–10. Pinch and Bijker, in "Facts and artefacts," develop the relevance of the concept for studies of technology, though drawing the more general analogy to "flexibility in how people think of or interpret artefacts [and] in how artefacts are designed" (p. 421).

54. My bias was reinforced by an eloquent presentation of the point by Bruno Latour in an informal seminar at the University of Edinburgh on February 6, 1990.

55. Coombs et al., *Economics and Technological Change*, pp. 6–7.

56. This is the central theme of an old but still valuable paper by Donald Schon, "The Fear of Innovation," as reprinted in *Science in Context: Readings in the Sociology of Science*, ed. B. Barnes and D. Edge (MIT Press, 1982). The original discussion is to be found in Frank H. Knight, *Risk, Uncertainty and Profit* (Houghton Mifflin, 1921).

57. There is no absolute way the distinction can be made *ex ante*. See Schon, "The fear of innovation," pp. 293–294.

58. I have argued elsewhere that there is a productive analogy to be drawn between the testing of technology and scientific experiment as analyzed by the sociology of scientific knowledge. See Donald MacKenzie, "From Kwajalein to Armageddon? Testing and the social construction of missile accuracy," in *The Uses of Experiment: Studies in the Natural Sciences*, ed. D. Gooding et al. (Cambridge University Press, 1989).

Chapter 4

1. See the following papers in *Technology and Culture* 17 (July 1976): Thomas P. Hughes, "The development phase of technological change: Introduction"; Lynwood Bryant, "The development of the diesel engine"; Thomas M. Smith, "Project Whirlwind: An unorthodox development project"; Richard G. Hewlett, "Beginnings of development in nuclear technology"; Charles Susskind, "Commentary." See also John M. Staudenmaier, *Technology's Storytellers: Reweaving the Human Fabric* (MIT Press, 1985), pp. 45–50. There are, of course, definite limits to the usefulness of dividing the process of technological change into separate phases of "invention," "development," "innovation," and "diffusion." Much important "invention," for example, takes place during "diffusion." See, for example, James Fleck, "Innofusion or diffusation? The nature of technological development in robotics," paper presented to workshop on Automatisation Programmable: Conditions d'Usage du Travail, Paris, 1987.

2. Much of the funding of laser gyroscope development in the United States (and elsewhere), like that of the laser itself, was conducted under military auspices and was thus subject to varying degrees of security classification. The recent Laser History Project addressed this problem by having two separate researchers, one using open materials, the other conducting classified interviews and working with classified archives: see Joan Lisa Bromberg, *The Laser in America, 1950–1970* (MIT Press, 1991), p. xii. As a foreign national, without security clearance, I have had to work solely with unclassified materials and have not, for example, enjoyed access to the holdings of the Defense Technical Information Center. However, some defense sector documents on the laser gyroscope have never been classified, and some originally classified material has now been cleared for public release. See the bibliographies produced by the National Technical Information Service—for example, Laser Gyroscopes (September 70–January 90): Citations from the NTIS Bibliographic Database (1990)—

although these are far from comprehensive. I am grateful to interviewees, particularly Professor Clifford V. Heer, for providing me with otherwise inaccessible documents from the early years of the laser gyroscope. Heer's own "History of the laser gyro" (*SPIE* [Society of Photo-Optical Instrumentation Engineers] 487 (1984) [*Physics of Optical Ring Gyros*]: 2–12) was of considerable help to me in preparing this chapter. The documentary record, though important, is not on its own sufficient to convey an understanding of the history of the laser gyro. This is not a result of security classification alone; it turns out to be equally the case for parts of the history where there is no direct military involvement, such as the adoption of the laser gyro in the civil market. Indeed, commercial confidentiality was, if anything, a greater constraint on the gathering of documentary sources for this paper than military classification. Therefore, essential to what follows are interviews with surviving pioneers of the laser gyroscope (and its competitor technologies). These interviews were cross-checked for mutual consistency and for consistency with documentary sources; several interviewees were also kind enough to comment by letter on the draft of this article.

3. O. Lodge, *Ether and Reality: A Series of Discourses on the Many Functions of the Ether of Space* (Hodder and Stoughton, 1925), p. 179. I owe the reference to Brian Wynne, "Physics and psychics: Science, symbolic action and social control in late Victorian England," in *Natural Order: Historical Studies of Scientific Culture*, ed. B. Barnes and S. Shapin (Sage, 1979). See also David B. Wilson, "The thought of late Victorian physicists: Oliver Lodge's ethereal body," *Victorian Studies* 15 (September 1971): 29–48, and *Conceptions of Ether: Studies in the History of Ether Theories, 1740–1900*, ed. G. Cantor and M. Hodge (Cambridge University Press, 1981).

4. See L. S. Swenson, Jr., *The Ethereal Aether: A History of the Michelson-Morley-Miller Aether-Drift Experiments, 1880–1930* (University of Texas Press, 1972).

5. If two beams from the same source of light cross, in the region of their crossing they sometimes reinforce each other and sometimes cancel each other out. The phenomenon, known as "interference," can be seen in the distinctive pattern of light and dark areas, called "fringes," thus produced. In an interferometer, such interference is deliberately created under closely controlled conditions. An interferometer can be used for optical experiments and also, for example, for highly accurate measurements of length. Interference is discussed in any text of physical optics: see, e.g., chapter 13 of F. A. Jenkins and H. E. White, *Fundamentals of Optics*, third edition (McGraw-Hill, 1957), which contains a good description of the Michelson interferometer.

6. A. A. Michelson and E. W. Morley, "Influence of motion of the medium on the velocity of light," *American Journal of Science*, third series, 31 (May 1886): 377–386; "On the relative motion of the Earth and the luminiferous æther," *Philosophical Magazine*, fifth series, 24 (December 1887): 449–463. The latter paper appeared also in *American Journal of Science*, third series, 34 (November 1887): 333–345.

7. Gerald Holton, "Einstein, Michelson, and the 'crucial' experiment," *Isis* 60 (summer 1969): 133–297.

8. See Swenson, *Ethereal Aether.*

9. This was suggested by Michelson and Morley themselves ("On the Relative Motion," pp. 458–459). Another possible explanation, reported by Oliver Lodge, was "suggested by Professor [George F.] Fitzgerald, viz., that the cohesion force between molecules, and, therefore, the size of bodies, may be a function of their direction of motion through the ether; and accordingly that the length and breadth of Michelson's stone supporting block were differently affected in what happened to be, either accidentally or for some unknown reason, a compensatory manner." See Lodge, "Aberration problems—a discussion concerning the motion of the ether near the Earth, and concerning the connexion between ether and gross matter; with some new experiments," *Philosophical Transactions,* series A, 184 (1893), pp. 749–750. Elaborated by the Dutch theoretical physicist H. A. Lorentz, this suggestion became known as the Lorentz-Fitzgerald contraction hypothesis.

10. G. Sagnac, "L'éther lumineux démontré par l'effet du vent relatif d'éther dans un interféromètre en rotation uniforme," *Comptes Rendus* 157 (1913): 708–710; Sagnac, "Effet tourbillonnaire optique: La circulation de l'éther lumineux dans un interférographe tournant," *Journal de Physique,* fifth series, 4 (March 1914): 177–195. Both Lodge and Michelson had earlier suggested the use of rotation to detect the ether, but neither had actually performed their proposed experiments. See Oliver Lodge, "Experiments on the absence of mechanical connexion between ether and matter," *Philosophical Transactions,* series A, 189 (1897): 149–166; A. A. Michelson, "Relative motion of Earth and æther," *Philosophical Magazine,* sixth series, 8 (December 1904): 716–719. The first actual experiment along these lines—using a ring of glass prisms, rather than the Earth—was described by Franz Harress in a 1911 Jena University doctoral thesis, published as *Die Geschwindigkeit des Lichtes in bewegten Körpern* (Erfurt, 1912). Harress's work, however, remained relatively unknown; for descriptions see B. Pogány, "Über die Wiederholung des Harress-Sagnacschen Versuches," *Annalen der Physik,* fourth series, 80 (1926): 217–231; Pogany, "Über die Wiederholung des Harresschen Versuches," *Annalen der Physik,* fourth series, 85 (1928): 244–256; André Metz, "Les problèmes relatifs à la rotation dans la théorie de la relativité," *Journal de Physique et le Radium* 13 (April 1952): 224–238. Note that some Anglo-Saxon writers have not had access to Harress's original work. E. J. Post ("Sagnac Effect," *Reviews of Modern Physics* 39 (April 1967): 475–495) writes: "Harress' objective was quite different from Sagnac's. Harress wanted to measure the dispersion properties of glasses . . . and he felt that a ring interferometer would be a suitable instrument." In fact, Harress's concern with questions of the ether and relativity is clear (see *Geschwindigkeit des Lichtes,* pp. 1–7).

11. Sagnac, "L'éther lumineux démontré," pp. 709–710.

12. Michel Paty, "The scientific reception of relativity in France," in *The Comparative Reception of Relativity,* ed. T. Glick (Boston Studies in the Philosophy of Science, volume 103 (Reidel, 1987)), pp. 113, 116.

13. P. Langevin, "Sur la théorie de relativité et l'expérience de M. Sagnac," *Comptes Rendus* 173 (1921): 831–835.

14. "La grandeur et le sens absolu du déplacement des franges sont conformes à la théorie de l'éther immobile de Fresnel et en constituent une vérification": Daniel Berthelot et al., "Prix Pierson-Perrin," *Comptes Rendus* 169 (1919), p. 1230; Alexandre Dufour et Fernand Prunier, "Sur l'observation du phénomène de Sagnac par un observateur non entrainé," *Comptes Rendus* 204 (1937): 1925–1927.

15. Swenson, *Ethereal Aether*, pp. 182, 237. An example is a skeptical footnote on p. 298 of L. Silberstein's "The propagation of light in rotating systems," *Journal of the Optical Society of America* 5 (July 1921): 291–307; see Heer, "History," p. 2.

16. Here I draw on Joseph Killpatrick's account of the relativistic explanation for a nonspecialist audience: "The laser gyro," *IEEE Spectrum* 4 (October 1967), p. 46.

17. A. A. Michelson and Henry G. Gale, "The effect of the Earth's rotation on the velocity of light, part II," *Astrophysical Journal* 61 (April 1925), p. 142.

18. Ibid., p. 143; Michelson as quoted in Swenson, *Ethereal Aether*, p. 218.

19. John E. Chappell, Jr., "Georges Sagnac and the discovery of the ether," *Archives Internationales d'Histoire des Sciences* 18 (1965), pp. 178, 190.

20. Sagnac, "Effet tourbillonnaire optique," p. 191.

21. On gyroscope work in this period see Thomas P. Hughes, *Elmer Sperry: Inventor and Engineer* (Johns Hopkins University Press, 1971).

22. D. MacKenzie, *Inventing Accuracy: A Historical Sociology of Nuclear Missile Guidance* (MIT Press, 1990).

23. David D. Dean to C. V. Heer, February 26, 1962.

24. Ibid.

25. See *Quantum Electronics: A Symposium*, ed. C. Townes (Columbia University Press, 1960); Townes, "Ideas and stumbling blocks in quantum electronics," *IEEE Journal of Quantum Electronics*, 20 (1984): 547–550; J. L. Bromberg, *The Laser in America*; Bromberg, "Engineering knowledge in the laser field," *Technology and Culture* 27 (October 1986): 798–818; Robert W. Seidel, "From glow to flow: A history of military laser research and development," *Historical Studies in the Physical and Biological Sciences* 18 (1987): 111–147; Seidel, "How the military responded to the laser," *Physics Today* (October 1988): 36–43; Paul Forman, "Behind quantum electronics: National security as basis for physical research in the United States," *Historical Studies in the Physical and Biological Sciences* 18 (1987): 149–229.

26. Henry A. H. Boot and John T. Randall, "Historical notes on the cavity magnetron," *IEEE Transactions on Electron Devices* 23 (July 1976): 724–729.

27. Bromberg, *The Laser in America*.

28. R. W. Ditchburn, *Light* (Blackie, 1952), pp. 337–340. The role of Ditchburn's text was confirmed to me by the two survivors of the trio: Clifford V. Heer

(telephone interview, March 27, 1985) and Warren Macek (interview, Great Neck, N.Y., April 5, 1985).

29. V. Heer to D. J. Farmer, Measurement of Angular Rotation by the Interference of Electromagnetic Radiation in Rotating Frames, Space Technology Laboratories interoffice correspondence, September 30, 1959, pp. 1, 3; interview with Heer conducted by Wolfgang Rüdig, Columbus, Ohio, January 19–21, 1987.

30. C. V. Heer, Measurement of Angular Velocity by the Interference of Electromagnetic Radiation in the Rotating Frame, Space Technology Laboratories Disclosure of Invention form, October 8, 1959. No actual patent was taken out, according to Heer, because staff members at Space Technology Laboratories "just didn't believe it would work" (interview with Heer by Rüdig). A fuller version of Heer's ideas was circulated as Measurement of Angular Velocity by the Interference of Electromagnetic Radiation in the Rotating Frame, Space Technology Laboratories Technical Memorandum STL/TM-60-0000-09007.

31. Heer, Disclosure of Invention, p. 2. With the invention of optical fiber, it later became possible to use this latter technique for light in the "fiber optic gyroscope." This device, though important in some fields (notably tactical missile guidance), has generally been held to be insufficiently accurate for aircraft inertial navigation, which has been the main market for the laser gyro.

32. C. V. Heer, "Interference of electromagnetic and matter waves in a nonpermanent gravitational field." The full paper was never published, but the abstract appears in *Bulletin of the American Physical Society* 6 (1961): 58. Heer submitted a longer version entitled "Interference of matter-waves in a non-permanent gravitational field" to *Physical Review*, but the paper was rejected on the basis of a referee's report that "the effect would be exceedingly difficult to observe," and that it "would not be very interesting" because of its analogy to the Sagnac-Michelson work and the theoretical ambiguity of the latter (report appended to letter from S. A. Goudsmit to C. V. Heer, September 18, 1960).

33. Research proposal, Ohio State University Research Foundation, March 16, 1961.

34. C. V. Heer to John E. Keto, October 26, 1961; Heer, "Interference of electromagnetic waves and of matter waves in a nonpermanent gravitational field," sent to NASA, Air Force Office of Scientific Research and Office of Naval Research, January 29, 1962. According to Heer ("History," pp. 3–4), this extension of his earlier paper was prompted by "the first indication of interest [in his proposal] by Dr. Chiu and Dr. Jastrow of the Institute for Space Studies on 5 January 1962."

35. A. H. Rosenthal, "Regenerative circulatory multiple-beam interferometry for the study of light propagation effects," *Journal of the Optical Society of America* 51 (December 1961): 1462. This is the abstract of the paper read by Rosenthal to the Optical Society's annual meeting. An extended version appeared in *Journal of the Optical Society of America* 52 (October 1962): 1143–1148.

36. Adolph H. Rosenthal, Optical Interferometric Navigational Instrument, U.S. Patent 3,332,314, July 25, 1967, filed April 8, 1963. A further patent from this period is Jack B. Speller, Relativistic Inertial Reference Device, U.S. Patent 3,395,270, July 30, 1968, filed June 28, 1962, which cites the experiments by Sagnac and Michelson and describes a range of devices that "utilize . . . the principles of the general relativity theory of Albert Einstein in order to detect angular motion by means of energy circulating in a loop path subjected to angular motion having a component in the plane of the loop." Speller went on to construct at least one such device, and he described its operation to a classified conference in 1963. It was not a laser gyroscope, though, but a "coaxial cable resonant cavity a few feet in length and with tunnel diode amplifiers," according to Heer ("History," p. 6).

37. Hughes, *Elmer Sperry*. Sperry Rand was formed in June 1955 by a merger of the Sperry Gyroscope and Remington Rand corporations.

38. Macek interview.

39. Aside from Macek, the group at Sperry working on the laser gyro consisted of Daniel T. M. Davis, Jr., Robert W. Olthuis, J. R. Schneider, and George R. White.

40. For descriptions see W. M. Macek and D. T. M. Davis, Jr., "Rotation rate sensing with traveling-wave ring lasers," *Applied Physics Letters* 2 (February 1, 1963): 67–68, and Philip J. Klass, "Ring laser device performs rate gyro angular sensor functions," *Aviation Week and Space Technology*, February 11, 1963: 98–103.

41. Macek and Davis, "Rotation rate sensing," p. 68.

42. For an explanation of this in terms of the general theory of relativity, see the introduction to this chapter.

43. Macek and Davis, "Rotation rate sensing," p. 68.

44. Heer, "History," pp. 5–6; P. K. Cheo and C. V. Heer, "Beat frequency between two traveling waves in a Fabry-Perot square cavity," *Applied Optics* 3 (June 1964): 788–789.

45. Robert C. Langford, "Unconventional inertial sensors," *Astronautica Acta* 12 (1966): 294–314.

46. Macek and Davis, "Rotation rate sensing"; Macek interview; Klass, "Ring laser"; *Aviation Week and Space Technology*, cover of issue of February 11, 1963. Klass coined the term "avionics" and was the leading journalist in that field.

47. Klass, "Ring laser," p. 98.

48. Macek interview. The main success enjoyed by Sperry laser gyro systems was in controlling the firing of naval guns. See R. W. McAdory, "Two decades of laser gyro development," in Proceedings of the Fourteenth Joint Services Data Exchange Group for Inertial Systems (Clearwater Beach, Florida, 1980).

49. There is a 1961 Soviet patent for the use for guidance purposes of laser light reflected by mirrors around a closed path, but the proposed device relies on displacement of the light beam on the surface of the mirrors rather than on the effect used in the laser gyro: B. N. Kozlov, Device for Stabilizing in Space the Course of a Moving Body, Soviet Patent 751787/26-10, November 15, 1961, described in *Soviet Inventions Illustrated* (February 1963), p. 22. (I owe the reference to Heer, "History," p. 19.) Early Soviet ring lasers are described in the following publications: S. N. Bagaev, V. S. Kuznetsov, Yu. V. Troitskii, and B. I. Troshin, "Spectral characteristics of a gas laser with traveling wave," *JETP Letters* 1 (1965): 114–116; I. L. Bershtein and Y. I. Zaitsev, "Operating features of the ring laser," *Soviet Physics JETP* 22 (March 1966): 663–667; E. M. Belenov, E. P. Markin, V. N. Morozov, and A. N. Oraevskii, "Interaction of travelling waves in a ring laser," *JETP Letters* 3 (1966): 32–34. Laser gyro work was being done in the U.K. in the 1960s at the Royal Aircraft Establishment, at the Services Electronic Research Laboratory, at the Admiralty Compass Observatory, and at EMI Ltd. (interviews with C. R. Milne and Sidney Smith, Farnborough, March 17, 1986; Michael Willcocks, Slough, March 19, 1986; Michael Wooley, Bracknell, March 21, 1986). Publicly available documentary records are scarce; two exceptions are A. F. H. Thomson and P. G. R. King, "Ring-laser accuracy," *Electronics Letters* 11 (November 1966): 417 and A. Hetherington, G. J. Burrell, and T. S. Moss, Properties of He-Ne Ring Lasers at 3.39 Microns, Royal Aircraft Establishment Technical Report 69099, Farnborough, Hampshire, 1969. French research began around 1964 in the laboratory of the Compagnie Générale d'Électricité and was then taken up by SFENA (Société Française d'Équipements pour la Navigation Aérienne), with military funding (interview with Bernard de Salaberry, Versailles, July 21, 1987).

50. For the work at Kearfott see Clement A. Skalski and John C. Stiles, Motion Sensing Apparatus, U.S. Patent 3,433,568, March 18, 1969, filed January 21, 1964. For that at Autonetics see F. Vescial, O. L. Watson, and W. L. Zingery, Ring Laser Techniques Investigation: Final Technical Report, report AFAL-TR-71-339, Air Force Avionics Laboratory, Wright-Patterson Air Force Base, Ohio, November 1971, and T. J. Hutchings, J. Winocur, R. H. Durrett, E. D. Jacobs, and W. L. Zingery, "Amplitude and frequency characteristics of a ring laser," *Physical Review* 152 (1966): A467–A473. For the work at Hamilton Standard see George Busey Yntema, David C. Grant, Jr., and Richard T. Warner, Differential Laser Gyro System, U.S. Patent 3,862,803, January 28, 1975, filed September 27, 1968. For that at the Instrumentation Laboratory see Joseph D. Coccoli and John R. Lawson, Gas Ring Laser Using Oscillating Radiation Scattering Sources within the Laser Cavity, U.S. Patent 3,533,014, October 6, 1970, filed June 4, 1968; Cynthia Whitney, "Contributions to the theory of ring lasers," *Physical Review* 191 (May 10, 1969): 535–541; Whitney, "Ring-laser mode coupling," *Physical Review* 191 (1969): 542–548.

51. Interview with Joseph Killpatrick, Minneapolis, March 7, 1985. Killpatrick recalls receiving an Air Force "request for proposal" for an investigation of the "Michelson-Gale effect." Being unable to figure out what this was, he called the relevant Air Force office; he was told that if they did not know what it was they were not the right people to investigate it!

52. The Republic Aviation Corporation was a further sponsor, and the Army Missile Command and NASA were also involved in the setting-up of the series. See Proceedings of the 1964 Symposium on Unconventional Inertial Sensors (Farmingdale, N.Y.), p. vi. For the general background of the military's interest in inertial systems see MacKenzie, *Inventing Accuracy*.

53. Interview with Tom Hutchings, Woodland Hills, Calif., February 20, 1985; interview with Charles Stark Draper, Cambridge, Mass., October 2 and 12, 1984.

54. In 1984 their laser gyro work won these three, together with Warren Macek, the Elmer Sperry award for advancing the art of transportation.

55. Early military contracts for Honeywell's laser gyro work included $500,000 from the Army Missile Command, $110,000 from the Naval Ordnance Test Station, and $51,000 from the Army's Frankford Arsenal. NASA also contributed $106,000. See Philip J. Klass, "Laser unit challenges conventional gyros," *Aviation Week and Space Technology*, September 12, 1966: 105–113. In general, NASA's support for laser gyroscope development was more modest than that of the armed services. See Jules I. Kanter, "Overview of NASA programs on unconventional inertial sensors," in Proceedings of the 1966 Symposium on Unconventional Inertial Sensors.

56. Interview with John Bailey, Minneapolis, March 7, 1985; letter from Bailey, April 12, 1990.

57. Donald MacKenzie and Graham Spinardi, "The shaping of nuclear weapon system technology: US fleet ballistic missile guidance and navigation," *Social Studies of Science* 18 (August 1988): 419–463; 18 (November): 581–624.

58. Klass, "Ring laser," p. 98.

59. Theodore J. Podgorski, Control Apparatus, U.S. Patent 3,390,606, July 2, 1968, filed March 1, 1965. The Sperry and Honeywell work can be traced in the companies' reports to their military sponsors. See, e.g., Electro-Optics Group, Sperry Gyroscope Company Division, Sperry Rand Corporation, Electromagnetic Angular Rotation Sensing: Final Report, report ALTDR 64-210, Air Force Systems Command, Research and Technology Division, Wright-Patterson Air Force Base, August 1964; Honeywell Inc., Aeronautical Division, Three-Axis Angular Rate Sensor, quarterly report 20502-QR1, Army Missile Command, Redstone Arsenal, Alabama, July 15, 1966.

60. W. M. Macek et al., "Ring laser rotation rate sensor," in Proceedings of the Symposium on Optical Masers (New York, 1963); q.v. Robert Adler, "A study of locking phenomena in oscillators," *Proceedings of the Institute of Radio Engineers and Waves and Electrons* 34 (June 1946): 351–357. Frederick Aronowitz (interview, Anaheim, Calif., February 27, 1985) cited Ali Javan, developer of the gas laser, as first having suggested this explanation in a private conversation.

61. Aronowitz interview; see Joseph E. Killpatrick, Laser Angular Rate Sensor, U.S. Patent 3,373,650, March 19, 1968, filed April 2, 1965. Dither was not

Killpatrick's first attempt to find a solution to lock-in—see Killpatrick, Apparatus for Measuring Angular Velocity having Phase and Amplitude Control Means, U.S. Patent 3,323,411, June 6 1967, filed June 29, 1964.

62. The results of the dither experiments were not presented publicly until 1971: Frederick Aronowitz, "The laser gyro," in *Laser Applications*, volume 1, ed. M. Ross (Academic Press, 1971). On noise see Joseph E. Killpatrick, Random Bias for Laser Angular Rate Sensor, U.S. Patent 3,467,472, September 16, 1969, filed December 5, 1966.

63. Killpatrick, "Laser gyro," pp. 48, 53.

64. F. Aronowitz, "Theory of traveling-wave maser," *Physical Review* 139 (1965): A635–A646.

65. Donald Christiansen, "Laser gyro comes in quartz," *Electronics* (September 19, 1966): 183–188; anonymous, Presentation of the Elmer A. Sperry Award for 1984 to Frederick Aronowitz, Joseph E. Killpatrick, Warren M. Macek, Theodore J Podgorski (no publisher or date of publication given), 14; Philip J. Klass, "Laser unit challenges conventional gyros," *Aviation Week and Space Technology* , September 12, 1966: 105–113.

66. Elmer Sperry Award, p. 15; Killpatrick interview. For an early overview of Navy support see John W. Lindberg, "Review of Navy activities on unconventional inertial sensors," in Proceedings of the 1966 Symposium on Unconventional Inertial Sensors.

67. Anonymous, "Laser gyro seen as applicable to missiles," *Aviation Week and Space Technology*, October 29, 1973: 60–62. By the end of the 1970s, military interest in laser gyro systems for tactical missiles seems to have cooled because of their relatively large size and high cost. In 1980, R. W. McAdory of the Air Force Armament Laboratory wrote ("Two Decades," p. 16) that laser gyro production cost "appears to be relatively constant regardless of accuracy," whereas with mechanical gyros low-accuracy systems could be produced quite cheaply. See also W. Kent Stowell et al., "Air Force applications for optical rotation rate sensors," *Proceedings of the Society of Photo-Optical Instrumentation Engineeers* 157 (1978): 166–171.

68. Philip J. Klass, "Laser gyro reemerges as INS contender," *Aviation Week and Space Technology*, January 13, 1975: 48–51; K. L. Bachman and E. W. Carson, "Advanced development program for the ring laser gyro navigator," *Navigation* 24 (summer 1977): 142–151. Laboratory tests of Autonetics and Sperry prototype laser gyroscopes were still yielding drift rates well in excess of $0.01°/$hour. See Central Inertial Guidance Test Facility, Ring Laser Gyro Test and Evaluation, report AFSWC-TR-75-34, Air Force Special Weapons Center, Kirtland Air Force Base, New Mexico, March 1975. For a discussion of how inertial components and systems are tested see MacKenzie, *Inventing Accuracy*, esp. pp. 372–378. As I outline there, such test results can be, and sometimes have been, challenged, but procedures for component and system testing were rela-

tively well established by the 1970s, and the results of laser gyro tests seem generally to have been taken to be "facts."

69. Central Inertial Guidance Test Facility, Developmental Test of the Honeywell Laser Inertial Navigation System (LINS): Final Report, report ADTC-TR-75-74, Armament Development and Test Center, Eglin Air Force Base, Florida, November 1975, quotes on pp. i and 20; see also Paul G. Savage and Mario B. Ignagni, "Honeywell Laser Inertial Navigation System (LINS) test results," paper presented at Ninth Joint Services Data Exchange for Inertial Systems, November 1975, p. 1.

70. Klass, "Laser gyro reemerges"; Stowell et al., "Air Force applications," p. 166.

71. Savage and Ignagni, "Test results," p. 11.

72. Philip J. Klass, "Honeywell breaks into inertial market," *Aviation Week and Space Technology*, November 19, 1979: 78–85.

73. Stowell et al., "Air Force applications"; M. C. Reynolds, "Keynote address," in Proceedings of the Fourteenth Joint Services Data Exchange Group for Inertial Systems (Clearwater Beach, Fla., 1980), p. 10; interview with Ronald G. Raymond, Minneapolis, March 6, 1985.

74. Phllip J. Fenner and Charles R. McClary, "The 757/767 Inertial Reference System (IRS)," in Proceedings of the Fourteenth Joint Services Data Exchange Group for Inertial Systems (Clearwater Beach, Fla., 1980); Raymond interview; letter from Raymond, April 8, 1990.

75. See David H. Featherstone, AEEC—The Committee that Works! Letter 82-000/ADM-218, Airlines Electronic Engineering Committee October 20, 1982. Insight into how the harmonization of interests is achieved can be found in a book by the committee's former chairman, William T. Carnes: *Effective Meetings for Busy People: Let's Decide It and Go Home* (McGraw-Hill, 1980).

76. Raymond interview; letter from Raymond, April 8, 1990.

77. The 90 percent claim comes from Litton Industries, which in 1990 filed a suit against Honeywell alleging antitrust violations and infringement of a Litton patent covering mirror-coating processes: anonymous, "Litton sues Honeywell over gyroscope," Minneapolis *Star Tribune*, April 5, 1990.

78. Anonymous, "Inertial navigation awards," *Aviation Week and Space Technology*, September 15, 1985: 61; anonymous, "Inertial navigation system to use laser gyro," *Defense Electronics*, November 1985: 17.

79. Philip J. Klass, "Litton tests laser-gyro inertial system," *Aviation Week and Space Technology*, December 1, 1980: 144–147.

80. A condition of the Honeywell-Litton settlement was that the terms of the settlement not be disclosed, so I have no further information on this case. There has also been litigation surrounding alleged violations of Speller's patent.

81. The Small ICBM was eventually cancelled. The requirements for precision azimuth alignment are given in Stowell et al., "Air Force applications," p. 167, and in W. D. Shiuru and G. L. Shaw, "Laser gyroscopes—The revolution in guidance and control," *Air University Review* 36 (1985): 62–66. The quantum limit is calculated in T. A. Dorschner et al., "Laser Gyro at quantum limit," *IEEE Journal of Quantum Electronics* 16 (1980): 1376–1379. For the MX episode, see also James B. Schultz, "En route to end game: Strategic missile guidance," *Defense Electronics*, September 1984: 56–63, and anonymous, "GE will compete for guidance system," *Aviation Week and Space Technology*, December 21, 1987: 31. Other military funding for high-accuracy laser gyros included $6.4 million from the U.S. Navy awarded to Honeywell and $5.8 million from the Air Force Avionics Laboratory awarded to Rockwell, both in 1983, with the goal of achieving a strapdown laser system with an error rate of 0.001 nautical mile per hour (comparable to that of the most accurate aircraft spinning-mass systems): anonymous, "Filter center," *Aviation Week and Space Technology*, September 26, 1983: 151.

82. See, e.g., Jay C. Lowndes, "British Aerospace pushing ring laser gyro effort," *Aviation Week and Space Technology*, November 19, 1984: 91–98.

83. Honeywell advertisement, *Aviation Week and Space Technology*, July 16, 1984: 58–59; also see Lowndes, "British Aerospace."

84. Interview with Polen Lloret, Paris, July 7, 1987; Lloret, "Centrales de reference inertielles: Perspectives et realités," paper read to École Nationale de l'Aviation Civile, May 20, 1987; Anthony King, private communication.

85. Raymond interview.

86. Interview with John Stiles, Wayne, N.J., September 25, 1986.

87. David Hughes, "Delco resonator gyro key to new inertial systems," *Aviation Week and Space Technology*, September 30, 1991: 48–49; interview with David Lynch, Goleta, Calif., September 11, 1986. The first discussion of the device that became the hemispherical resonator gyro was a report by A. G. Emslie and I. Simon of Arthur D. Little, Inc.: Design of Sonic Gyro: Report to AC Electronics Division, General Motors Corporation (July 1967). Emslie, an acoustics specialist, had been asked by Delco staff members to investigate the use of acoustic phenomena to detect rotation.

88. Lynch interview.

89. Forman, "Behind quantum electronics," pp. 201–202.

90. Bruno Latour, *Science in Action* (Harvard University Press, 1987), esp. pp. 174–175.

91. For the general debate between neoclassical and Schumpeterian approaches see, e.g., Rod Coombs, Paolo Saviotti, and Vivien Walsh, *Economics and Technological Change* (Macmillan, 1987).

92. For these revolutions more generally, see Edward W. Constant, II, *The Origins of the Turbojet Revolution* (Johns Hopkins University Press, 1980).

93. Stowell et al., "Air Force applications," p. 166.

94. Barry Barnes, *The Nature of Power* (Polity, 1988).

Chapter 5

1. See, e.g., Nathan Rosenberg, *Inside the Black Box: Technology and Economics* (Cambridge University Press, 19482).

2. For an explanation of the "floating-point" representation of numbers, see the section "Negotiating Arithmetic" in chapter 8.

3. Any definition of when "supercomputer" as a category emerged is somewhat arbitrary. The machine designed by John von Neumann and colleagues at the Princeton Institute for Advanced Study, Naval Ordnance Research Calculator (NORC) and Whirlwind, for example, would all have claims to be supercomputers. One difference between, NORC, say, and the first supercomputers I discuss, LARC and Stretch, is that the former was explicitly a unique, "once-off" machine, while LARC and Stretch were, at least potentially, commercial products to be made in multiple copies. Use of the term "supercomputer" seems to have become widespread only in the 1970s, though "super computer" (apparently as two words, not one) was used to describe the British project that became Atlas (Paul Drath, The Relationship between Science and Technology: University Research and the Computer Industry, 1945–1962, Ph.D. thesis, University of Manchester, 1973). I would be grateful to hear of other early uses of the term.

4. Werner Buchholz, *Planning a Computer System: Project Stretch* (McGraw-Hill, 1962), pp. 273–274.

5. Charles J. Bashe, Lyle R. Johnson, John H. Palmer, and Emerson W. Pugh, *IBM's Early Computers* (MIT Press, 1986), pp. 420–421.

6. Drath, The Relationship between Science and Technology; John Hendry, "Prolonged negotiations: The British Fast Computer Project and the early history of the British computer industries," *Business History* 26 (1984), no. 3: 286–300.

7. M. Bataille, "The Gamma 60," *Honeywell Computer Journal* 5 (1971), no. 3: 99–105; J. Jublin and J.-M. Quatrepoint, *French Ordinateurs—de l'Affaire Bull à l'Assassinat du Plan Calcul* (Moreau, 1976).

8. Peter Wolcott and Seymour E. Goodman, "High-speed computers of the Soviet Union," *IEEE Computer*, September 1988: 32–41.

9. See chapter 6 of the present volume. In summarizing in this way I am, of course, oversimplifying. But no firm in the United States other than Control Date and Cray Research maintained continuous development and production of a supercomputer series in a similar sustained way. Some of the more "one-off"

machines developed by other firms are nevertheless of considerable interest: see the remarks below on the Texas Instruments Advanced Scientific Computer.

10. During the 1960s and the early 1970s, IBM developed and marketed, in competition with Control Data, machines at the "top end" of the general-purpose 360 series: the 1967 IBM 360/91 and 1971 IBM 360/195. The mid-1970s Cray 1 and Cyber 205, with their vector processing facilities (see below), generated no direct IBM response, though in 1985 IBM introduced, as part of the IBM 370 series, the IBM 3090-VF (vector facility). For the history of IBM's involvement in "high-end computers," see chapter 7 of Emerson W. Pugh, Lyle R. Johnson, and John H. Palmer, *IBM's 360 and Early 370 Systems* (MIT Press, 1991).

11. Hugh Donaghue, "A business perspective on export controls," in *Selling the Rope to Hang Capitalism? The Debate on West-East Trade and Technology Transfer*, ed. C. Perry and R. Pfaltzgraff, Jr. (Pergamon-Brassey, 1987), p. 188.

12. A high-performance machine, the MU5, was designed at Manchester University beginning in 1966. Though never produced commercially, it influenced the design of the ICL 2900 series.

13. Wolcott and Goodman, "High-speed computers of the Soviet Union."

14. R. W. Hockney and C. R. Jesshope, *Parallel Computers 2: Architectures, Programming and Algorithms* (Hilger, 1988), p. 22.

15. Ibid., p. 3 fn.

16. Ibid., p. 4.

17. R. N. Ibbett and N. P. Topham, *Architecture of High Performance Computers* (Macmillan, 1989), volume 1, p. 1, quoting H. S. Stone, *Introduction to Computer Architecture* (Science Research Associates, 1975).

18. The IBM Selective Sequence Electronic Calculator (SSEC) had been "capable of working on three neighboring instructions simultaneously. Such concurency was later abandoned in the von Neumann-type machines, such as the IBM 701" (Buchholz, *Planning a Computer System*, p. 192).

19. Buchholz, *Planning a Computer System*, p. 192.

20. Hockney and Jesshope, *Parallel Computers 2*, p. 21.

21. In foreword to J. E. Thornton, *Design of a Computer: The Control Data 6600* (Scott, Foresman, 1970).

22. See especially Thomas P. Hughes, *Networks of Power: Electrification in Western Society, 1880–1930* (Johns Hopkins University Press, 1983).

23. The "clock" is part of the control logic of most computers. It produces signals at a regular rate which are used to synchronize the operations of different parts of the computer. Clock speed heavily affects overall processing speed, and is itself affected both by the size of gate delays and by the length of the interconnections between various parts of the computer.

24. Ibbett and Topham, *Architecture of High Performance Computers,* volume 1, p. 113.

25. The qualification "of the corresponding period" needs emphasizing. A "radical" architecture of the 1960s no longer looks so radical when set against a mainstream supercomputer of the late 1980s.

26. D. L. Slotnick, "The conception and development of parallel processors— A personal memoir," *Annals of the History of Computing* 4 (1982), no. 1: 20–30.

27. On the phenomenon's generality see Eugene S. Ferguson, "The mind's eye: Nonverbal thought in technology," *Science* 197 (1977), no. 4306: 827–836. For an analogous development in the biography of another important computer scientist (Edsger Dijkstra), see Eda Kranakis, "The first Dutch computers: A sociotechnical analysis," paper prepared for International Workshop on the Integration of Social and Historical Studies of Technology, University of Twente, the Netherlands, 1987.

28. "It was at Princeton that I first thought of building a parallel computer. The idea was stimulated by the physical appearance of the magnetic drum that was being built to augment the 1024-word primary memory of the IAS [Institute for Advanced Study] machine. The disposition of heads and amplifiers over the drum's 80 tracks (40-bit words in two 1024-word banks) suggested to me the notion of, first, inverting the bit-word relationship so that each track stored the successive bits of a single word (in fact, of several words) and, second, associating a 10-tube serial adder with each track so that in a single drum revolution an operation could be executed on the contents of the entire drum. The idea was to do, in parallel, an iterative step in a mesh calculation." (Slotnick, "Conception and Development of Parallel Processors," p. 20.)

29. C. Leondes and M. Rubinoff, "DINA—A digital analyser for Laplace, Poisson, diffusion and wave equations," *AIEE Transactions on Communications and Electronics* 71 (1952): 303–309; S. H. Unger, "A computer oriented towards spatial problems," *Proceedings of the Institute of Radio Engineers* (USA) 46 (1958): 1744–1750; Konrad Zuse, "Die Feldrechenmaschine," *MTW-Mitteilungen* 5 (1958), no. 4: 213–220.

30. Slotnick, "Conception and development of parallel processors," p. 20.

31. Ibid., p. 22.

32. D. L. Slotnick, C. W. Borck, and R. C. McReynolds, "The SOLOMON computer," in Proceedings of the 1962 Fall Joint Computer Conference.

33. Michael J. Flynn, "Some computer organizations and their effectiveness," *IEEE Transactions on Computers* C-21 (1972), no. 9: 948–960.

34. George H. Barnes, Richard M. Brown, Maso Kato, David J. Kuck, Daniel J. Slotnick, and Richard A. Stokes, "The ILLIAC IV computer," *IEEE Transactions on Computers* C-17 (1964), no. 8: 746–757; D. L. Slotnick, "The fastest computer," *Scientific American* 224 (1971), no. 2: 76–87.

35. Howard Falk, "Reaching for a gigaflop," *IEEE Spectrum* 13 (1976), no. 10: 65–69.

36. Ibid.; see also Slotnick, "Conception and development of parallel processors." Defense Department funding for ILLIAC IV, and its intended military applications, made it a natural target for protesters.

37. See the discussion of Moore's Law in chapter 3 of the present volume. See also Robert N. Noyce,"Microelectronics," in *The Microelectronics Revolution*, ed. T. Forester (Blackwell, 1980), pp. 33–34.

38. Marshall C. Pease, "The indirect binary *n*-cube microprocessor array," *IEEE Transactions on Computers* C-26 (1977), no. 5: 458.

39. See *Past, Present, Parallel: A Survey of Available Parallel Computing Systems*, ed. A. Trew and G. Wilson (Springer, 1991).

40. Interviews drawn on in this article were conducted by the author at Los Alamos and Livermore in April and October 1989. A full list of the individuals interviewed will be found in the acknowledgments.

41. Chuck Hansen, *U.S. Nuclear Weapons: The Secret History* (Aerofax, 1988), p. 11.

42. Ibid.

43. See, e.g., Herbert F. York, *The Advisors: Oppenheimer, Teller, and the Superbomb* (Freeman, 1976).

44. Los Alamos and Livermore interviews.

45. N. Metropolis and E. C. Nelson, "Early computing at Los Alamos," *Annals of the History of Computing* 4 (1982), no. 4: 350.

46. John Wilson Lewis and Xue Litai, *China Builds the Bomb* (Stanford University Press, 1988).

47. See, e.g., Matthew Evangelista, *Innovation and the Arms Race: How the United States and the Soviet Union Develop New Military Technologies* (Cornell University Press, 1988).

48. Talley interview.

49. Dowler interview. Because the weights, yields, and other characteristics of nuclear warheads are classified, Dowler could not go beyond this general statement. It is, however, consistent with data published in unclassified literature. For example, closely comparable 1960s and 1970s designs are the original W-62 and later W-78 warheads for Minuteman III. They are approximately the same weight, around 800 lbs. But the yield of the latter warhead, at around 335–350 kilotons, is twice that of its predecessor's 170 kilotons. (Data from Hansen, *U.S. Nuclear Weapons*, pp. 200–201.)

50. Nuclear Weapons Databook Project, "Nuclear notebook: Known nuclear tests by year, 1945 to December 31, 1987," *Bulletin of the Atomic Scientists* 44 (1988), no. 2: 56. See also chapter 10 of the present volume.

51. R. Courant, K. O. Friedrichs, and H. Lewy, "Über die partiellen Differenzengleichungen der mathematischen Physik," *Mathematische Annalen* 100 (1928): 32–74.

52. R. D. Richtmyer and K. W. Morton, *Difference Methods for Initial-Value Problems,* second edition (Wiley, 1967), p. vii (quotation from Richtmyer's 1957 preface). Richtmyer is referring to the hydrodynamics of implosion, discussed in chapter 10 of the present volume. See also Computing at LASL in the 1940s and 1950s, ed. R. Lazarus et al., Los Alamos Scientific Laboratory report LA-6943-H, 1978, p. 22.

53. John von Neumann, "The principles of large-scale computing machines," *Annals of the History of Computing* 3 (1981), no. 3: 267 (first published in 1946).

54. Livermore interviews.

55. Roger B. Lazarus, "Contributions to mathematics," in Computing at LASL in the 1940s and 1950s, p. 22.

56. Michael interview.

57. Los Alamos interviews.

58. Livermore interviews; N. R. Morse, C-Division Annual Review and Operating Plan, Los Alamos National Laboratory report LA-11216-MS, 1988, p. 59.

59. Michael interview.

60. Cuthbert C. Hurd, "A note on Early Monte Carlo computations and scientific meetings," *Annals of the History of Computing* 7 (1985), no. 2: 141–155; N. Metropolis, "The Los Alamos experience, 1943–1954," presented to ACM Conference on History of Scientific and Numeric Computation, 1987 (Los Alamos National Laboratory report LA-UR-87-1353), p. 10. See also N. Metropolis and S. Ulam, "The Monte Carlo method," *Journal of the American Statistical Association* 44 (1949), no. 247: 335–401.

61. Los Alamos interviews.

62. Michael interview.

63. Lazarus, "Contributions to mathematics," p. 23; interviews.

64. Morse, C-Division Annual Review and Operating Plan, p. 60.

65. Livermore interviews.

66. Michael interview.

67. See, e.g., N. Metropolis and E. C. Nelson, "Early computing at Los Alamos," *Annals of the History of Computing* 4 (1982), no. 4: 348–357.

68. ENIAC (Electronic Numerical Integrator And Computer) was built by the University of Pennsylvania's Moore School of Electrical Engineering for the Ballistic Research Lab at Aberdeen, Maryland.

69. Ibid., p. 355.

70. Nancy Stern, *From ENIAC to UNIVAC: An Appraisal of the Eckert-Mauchly Computers* (Digital Press, 1981), pp. 62–63.

71. W. Jack Worlton, "Hardware," in Computing at LASL in the 1940s and 1950s, p. 4.

72. N. Metropolis, "The MANIAC," in *A History of Computing in the Twentieth Century*, ed. N. Metropolis et al. (Academic Press, 1980), p. 459.

73. Herman H. Goldstine, *The Computer from Pascal to von Neumann* (Princeton University Press, 1972), pp. 318–319.

74. Francis H. Harlow and N. Metropolis, "Computing and computers: Weapons simulation leads to the computer era," *Los Alamos Science*, winter-spring 1983, p. 135.

75. In the late 1950s Metropolis also built a MANIAC III at the University of Chicago. See Mark B. Wells, "MANIAC," in Computing at LASL in the 1940s and 1950s.

76. Wells, ibid.; Los Alamos interviews.

77. Metropolis and Nelson, "Early computing at Los Alamos," p. 357.

78. Ibid., p. 351. A "bit parallel" computer processes in parallel all the binary digits constituting a "word" (e.g., representing a number). For von Neumann's influence more generally, see William Aspray, *John von Neumann and the Origins of Modern Computing* (MIT Press, 1990)

79. Metropolis and Nelson, "Early computing at Los Alamos," p. 135.

80. Herman Lukoff, *From Dits to Bits: A Personal History of the Electronic Computer* (Robotics Press, 1979), p. 128; M. R. Williams, "Pioneer Day 1984: Lawrence Livermore National Laboratory," *Annals of the History of Computing* 7 (1985), no. 2: 179.

81. Since the funding for purchases and development contracts has normally come from the Atomic Energy Commission, and now the Department of Energy, it is inexact to talk of the laboratories' "buying" computers or "sponsoring" developments. But the laboratories' preferences have carried considerable weight, and in the text I will write as it they were free agents, except when (as in the case of the SOLOMON computer) it is important that they were not.

82. John G. Fletcher, "Computing at LLNL," *Energy and Technology Review* (Lawrence Livermore National Laboratory), September 1982, pp. 28–39.

83. Lukoff, *From Dits to Bits*, p. 145.

84. John von Neumann, speech on the occasion of first public showing of IBM Naval Ordnance Research Calculator, December 2, 1954, reprinted as "The NORC and problems in high-speed computing," *Annals of the History of*

Computing 3 (1981), no. 3: 279; see Goldstine, *The Computer from Pascal to von Neumann* , p. 331.

85. Lukoff, *From Dits to Bits*, p. 145.

86. Emerson W. Pugh, *Memories That Shaped an Industry: Decisions Leading to the IBM System/360* (MIT Press, 1984), p. 162.

87. Lukoff, *From Dits to Bits*, p. 146.

88. Livermore interviews.

89. Lukoff, *From Dits to Bits*, pp. 147, 169.

90. Michael interview; Bashe et al., *IBM's Early Computers*, p. 448.

91. J. G. Fletcher, Computing at Lawrence Livermore National Laboratory, report UCID-20079, Lawrence Livermore National Laboratory, 1984.

92. Saul Rosen, "Electronic computers: A historical survey," *Computing Surveys* 1 (1969), no. 1: 26.

93. S. W. Dunwell, "Design objectives for the IBM Stretch Computer," in *Proceedings of the East Joint Computer Conference* (Spartan, 1956), p. 20.

94. Livermore interviews.

95. Los Alamos interviews.

96. Ibid.

97. Bashe et al. , *IBM's Early Computers*, pp. 465–458.

98. Michael interview.

99. Michael interview.

100. Livermore interviews. Stretch was designed to have four interleaved 16,384-word memories (Erich Bloch, "The engineering design of the Stretch computer," in *Proceedings of the Eastern Joint Computer Conference* (Spartan, 1959), p. 48). As will be seen from the calculation above, this is sufficient for a 50 × 50 two-dimensional mesh calculation with 15 quantities per cell, and would indeed permit resolution somewhat finer than 50 × 50.

101. Michael interview. Operation Dominic is described on pp. 81–89 of Hansen, *U.S. Nuclear Weapons*.

102. Bashe et al., *IBM's Early Computers*, p. 456. A ninth Stretch was retained by IBM; the figure of nine includes seven Stretches designated IBM 7030s. I am unclear to what extent the Saclay Stretch was used for military as distinct from civil atomic energy research and development.

103. Wood interview.

104. Los Alamos interviews; Fletcher, Computing at Lawrence Livermore; Morse, C-Division Annual Review and Operating Plan.

105. Omri Serlin, "Cray 3 in trouble? Stock market says 'Yes,'" *Serlin Report on Parallel Processing*, no. 18 (1988): 3.

106. Source: *The Guardian* (London and Manchester), May 25, 1989.

107. Los Alamos interviews.

108. Morse interview.

109. Morse, C-Division Annual Review and Operating Plan, p. 54.

110. Los Alamos and Livermore interviews.

111. Morse, C-Division Annual Review and Operating Plan, p. 111; Spack interview.

112. Slotnick, "Conception and development of parallel processors," p. 24.

113. Livermore interviews.

114. Slotnick, "Conception and development of parallel processors," p. 24.

115. Neil R. Lincoln, "Technology and design tradeoffs in the creation of a modern supercomputer," *IEEE Transactions on Computers* C-31 (1982), no. 5: 350.

116. Ibid.

117. Michael interview.

118. Ibid.

119. Slotnick, "Conception and development of parallel processors," p. 25.

120. See Lincoln, "Technology and design tradeoffs."

121. Michael interview.

122. Quoted in Michael interview.

123. Joel S. Wit, "Advances in anti-submarine warfare," *Scientific American* 244 (1981), no. 2: 27–37.

124. William J. Broad, *Star Warriors* (Simon and Schuster, 1985).

125. There was no single reason for the choice of the name. S-1 was "kind of a sacred label round here [Livermore]" (Wood interview), since it had been the name of the committee of the Office of Scientific Research and Development responsible for the atomic bomb project during the Second World War. The hydrophone systems were known as SOSUS (Sound Surveillance System) arrays, and this led to the proposed computer's being referred to in negotiations with the Navy as SOSUS-1; however, the abbreviation to S-1 was convenient for security reasons, since the existence of the SOSUS arrays was at that point classified.

Finally, McWilliams and Widdoes were registered for doctorates in the Computer Science Department at Stanford University and were trying to recruit fellow graduate students to the project, which they therefore referred to as Stanford-1.

126. L. C. Widdoes Jr. and S. Correll, "The S-1 project: Developing high performance digital computers," *Energy Technology Review* (Lawrence Livermore Laboratory), September 1979, pp. 1–15; P. M. Farmwald, "The S-1 Mark IIA supercomputer," in *High Speed Computation*, ed. J. Kowalik (Springer, 1984); Hockney and Jesshope, *Parallel Computers 2*, p. 39.

127. Hockney and Jesshope, ibid., p. 39.

128. Livermore interviews.

129. Ibid.

130. Alan Turing, "Computing machinery and intelligence," *Mind* 59 (1950): 433–460.

131. Ibbett and Topham, *Architecture of High Performance Computers* 1, p. 65.

132. Tracy Kidder, *The Soul of a New Machine* (Atlantic Monthly Press, 1981).

133. Bashe et al., *IBM's Early Computers*, p. 454.

134. Ibid., pp. 446, 440.

135. Ibid., p. 432.

136. David E. Lundstrom, *A Few Good Men from Univac* (MIT Press, 1987), p. 90.

137. Los Alamos interviews.

138. Ibid.

139. "By definition of ordinary normalized [floating-point] operations, numbers are frequently extended on the right by attaching zeros. During addition the n-digit operand that is not preshifted is extended with n zeros, so as to provide the extra positions to which the preshifted operand can be added. Any operand or result that is shifted left to be normalized requires a corresponding number of zeros to be shifted in at the right. Both sets of zeros tend to produce numbers smaller in absolute value than they would have been if more digits had been carried. In the *noisy mode* these numbers are simply extended with 1s instead of zeros (1s in a binary machine, 9s in a decimal machine). Now all numbers tend to be too large in absolute value. The true value, if there had been no significance loss, should lie between these two extremes. Hence, two runs, one made without and one made with the noisy mode, should show differences in result that indicate which digits may have been affected by significance loss." (Buchholz, *Planning a Computer System*, p. 102)

140. Los Alamos interviews.

141. Ibid.

142. Michael and Hardy interview. For the meanings of "significand" and "exponent," see chapter 8 of the present volume.

143. "Gather" refers to the construction of a vector out of items in a specified set of locations in memory, and "scatter" to its distribution to such a set. The Cray 1 and models of the Cray X-MP prior to the four-processor X-MP/4 "had no special hardware or instructions for implementation of scatter and gather operations" (Hockney and Jesshope, *Parallel Computers 2*, p. 137).

144. Francis H. McMahon, memorandum to Seymour Cray, March 1, 1983; Lawrence Berdahl, Francis H. McMahon, and Harry L. Nelson, letter to Seymour Cray, June 20, 1983.

145. Hockney and Jesshope, *Parallel Computers 2*, p. 21.

146. McMahon interview.

147. Livermore interviews.

148. Ibid.

149. Ibid.

150. See Tim Johnson and Tony Durham, *Parallel Processing: The Challenge of New Computer Architectures* (Ovum, 1986).

151. Even international comparison, the standby of much historical sociology, is of no use here. Though most countries have not had nuclear weapons laboratories, the international hegemony of the U.S. computing industry has been such that there is no proper comparative case. In particular, Japan entered the supercomputer business when supercomputer architecture was already highly developed.

152. Samuel S. Snyder, "Computer advances pioneered by cryptological organizations," *Annals of the History of Computing* 2 (1980), no. 1: 66.

153. McMahon interview; Francis H. McMahon, The Livermore Fortran Kernels: A Computer Test of the Numerical Performance Range, Lawrence Livermore National Laboratory report UCRL-53745, 1986.

154. McMahon interview.

155. Roger Shepherd and Peter Thompson, Lies, Damned Lies and Benchmarks, technical note 27, Inmos, Bristol, 1988.

156. Erwin Tomash and Arnold A. Cohen, "The birth of an ERA: Engineering Research Associates, Inc., 1946–1955," *Annals of the History of Computing* 1 (1979), no. 2: 83–97.

157. The Cray 2, which as noted above was not adopted for nuclear weapons design by Los Alamos and Livermore, appears to be an exception to this robust strategy. It has a very fast clock, but its very large main memory is slow relative to

the processors, and its performance is therefore heavily dependent on the programming of problems in ways that make as heavy use as possible of the processors' fast local memories (Hockney and Jesshope, *Parallel Computers 2*, p. 155).

Chapter 6

1. See, e.g., Samuel Smiles, *Lives of the Engineers: George and Robert Stephenson* (Murray, 1904).

2. Bill Gates of Microsoft has received much attention recently, but his field is software rather than the designing of computers. Perhaps the closest parallels to Seymour Cray in the latter area are the co-founders of Apple, Steve Jobs and Steve Wozniak. Because they (especially Jobs) have been less private than Cray, much more has been written about them; see, e.g., Michael Moritz, *The Little Kingdom: The Private Story of Apple Computer* (Morrow, 1984). But because of the lack of an eponymous product, the names of Jobs and Wozniak are less well known than that of Cray. Another prominent case of eponymy is the computer designer Gene Amdahl, but no public legend has attached itself to him.

3. We were not surprised to receive one of these polite but firm refusals.

4. We are grateful to George Michael of the Lawrence Livermore National Laboratory, who kindly arranged for us to be provided with two such videos, and to John Rollwagen of Cray Research, who provided a third.

5. Bradford W. Ketchum, "From start-up to $60 million in nanoseconds," *Inc.*, November 1980, p. 51.

6. *Business Week*, April 30, 1990.

7. Max Weber, "The social psychology of the world religions," in *From Max Weber: Essays in Sociology*, ed. H. Gerth and C. Mills (Routledge, 1970), p. 295.

8. See chapter 1 above.

9. Weber, "The social psychology of the world religions," p. 299.

10. For other patterns see B. Elzen and D. MacKenzie, "From megaflops to total solutions: The changing dynamics of competitiveness in supercomputing," in *Technological Competitiveness: Contemporary and Historical Perspectives on the Electrical, Electronics, and Computer Industries*, ed. W. Aspray (IEEE Press, 1993); Boelie Elzen and Donald MacKenzie, "The social limits of speed: The development and use of supercomputers," *IEEE Annals of the History of Computing* 16 (1994): 46–61.

11. Here we are primarily following the brief biographical details of Cray in chapter 18 of Robert Slater's *Portraits in Silicon* (MIT Press, 1987).

12. For the history of Engineering Research Associates see Erwin Tomash and Arnold A. Cohen, "The birth of an ERA: Engineering Research Associates, Inc., 1946–1955," *Annals of the History of Computing* 1 (1979), no. 2: 83–97.

13. David E. Lundstrom, *A Few Good Men from Univac* (MIT Press, 1987), p. 136.

14. Slater (*Portraits in Silicon*, p. 197) suggests that Cray told Norris he was leaving.

15. On "committee design" see Lundstrom, *A Few Good Men*.

16. Seymour Cray, as quoted in Russell Mitchell, "The genius," *Business Week*, April 30, 1990, p. 83.

17. Charles J. Bashe et al., *IBM's Early Computers* (MIT Press, 1986), p. 446.

18. See chapter 5 above (especially figure 1) for details.

19. Mitchell, "The genius," p. 84.

20. James E. Thornton, "The CDC 6600 project," *Annals of the History of Computing* 2 (1980), no. 4, p. 343.

21. Lundstrom, *A Few Good Men*, 214; also see chapter 5 above.

22. Interview with Bill Spack, Roger Lazarus, Robert Frank, and Ira Akins, Los Alamos, April 12, 1989.

23. Lundstrom, *A Few Good Men*, p. 135.

24. See chapter 5 above.

25. Kenneth Flamm, *Creating the Computer: Government, Industry and High Technology* (Brookings Institution, 1988), pp. 152–155. See Jacques Jublin and Jean-Michel Quatrepoint, *French Ordinateurs—de l'Affaire Bull à l'Assassinat du Plan Calcul* (Alain Moreau, 1976).

26. Cray's reports to Control Data's top managers were delightfully succinct and uninformative. Lundstrom (*A Few Good Men*, p. 138) quotes the entirety of one report: "Activity is progressing satisfactorily as outlined under the June plan. There have been no significant changes or deviations from the outlined June plan." The leaders of comparable parts of the corporation issued reports of 20 or 30 pages.

27. Emerson W. Pugh et al., *IBM's 360 and Early 370 Systems* (MIT Press, 1991), p. 383. See, e.g., Mitchell, "The genius," p. 84.

28. Interview with Les Davis, Chippewa Falls, May 3, 1990.

29. Interviews with Neil Lincoln (April 3), Chuck Purcell (April 4), and James Thornton (April 3), 1990, Minneapolis. Purcell worked as a salesperson for the STAR and also took part in its design.

30. *Chippewa Falls Herald-Telegram*, March 15, 1972, quoted in *Speed and Power: Understanding Computers* (Time-Life Books, 1987), p. 32.

31. John M. Lavine, "New firm here blueprinting most powerful computer in the world," *Chippewa Falls Herald-Telegram*, May 17, 1972.

32. Ibid.

33. See, e.g., the discussion of the work of guidance engineer Charles Stark Draper in Donald MacKenzie, *Inventing Accuracy: A Historical Sociology of Nuclear Missile Guidance* (MIT Press, 1990).

34. Cray acknowledged the debt of the Cray-1 to the STAR-100 in a lecture to a technical meeting in Florida on November 15, 1988. A videotape was made available to us by the Lawrence Livermore National Laboratory.

35. Interview with Les Davis, Chippewa Falls, May 3, 1990.

36. For a more detailed description of the Cray-1 see Richard M. Russell, "The Cray-1 computer system," *Communications of the ACM* 21 (1978), no. 1: 63–72.

37. Interviews with Carl Ledbetter (St. Paul, May 17) and Neil Lincoln (Minneapolis, April 3), 1990. Ledbetter was the last president of ETA systems.

38. John Rollwagen, quoted in "Gambling on the new frontier," *Interface* 9 (1986), no. 8, p. 7.

39. Ketchum, "From start-up to $60 million in nanoseconds," p. 56.

40. Cray Research, Inc., Annual Report 1978, p. 2.

41. "The seismic processing problem: Cray Research's response," *Cray Channels* 5 (1983), no. 2: 6–9; interview with Dave Sadler, Minneapolis, May 2, 1990. Sadler worked on the software for handling the tapes.

42. Cray Research, Inc., Annual Report 1984, p. 2.

43. Interview with Margaret Loftus, Minneapolis, May 4, 1990. Loftus was Cray's vice-president of software at the time.

44. "Cray Research fact sheet" giving marketing data as of December 31, 1989.

45. Ibid.

46. Cray, quoted in S. Gross, "Quicker computer? Cray leads the way," *Minneapolis Star*, December 22, 1981.

47. Ibid.

48. Interviews with Les Davis (Minneapolis, May 3) and Stu Patterson (Boulder, May 11), 1990. Patterson was head of the Cray Boulder Labs.

49. Interview with John Rollwagen, Minneapolis, April 2, 1990.

50. "What comes after the Y?," *Interface* 12 (1989), no. 6: 8.

51. Marc H. Brodsky, "Progress in gallium arsenide semiconductors," *Scientific American*, February 1990, p. 56.

52. In the videotape "Cray Private" (May 15, 1989), John Rollwagen explains the reasons for splitting off the Cray Computer Corporation from Cray

Research. See also Steve Gross, "Will Cray suffer without its founder?" *Star Tribune*, May 21, 1989.

53. Rollwagen, as quoted in Carla Lazzarechi, "Cray left his company to prevent internal conflict," *Los Angeles Times*, June 4, 1989, and in Russell Mitchell, "Now Cray faces life without Cray," *Business Week*, May 29, 1989, p. 27.

54. Cray, quoted in Mitchell, "Now Cray faces life without Cray."

55. T. Burns and G.M. Stalker, *The Management of Innovation* (Tavistock, 1961).

56. Interview with John Rollwagen, Minneapolis, April 2, 1990.

Chapter 7

1. Advisory Council for Applied Research and Development, *Software: A Vital Key to UK Competitiveness* (HMSO, 1986).

2. See, e.g., *Computer Weekly*, February 4, 1988; *The Times* (London), January 29, 1988.

3. W. J. Cullyer and C. H. Pygott, *IEE Proceedings* 134E (1987): 133–141.

4. Charter Technologies Ltd., VIPER Microprocessor Development Tools, December 1987.

5. See, e.g., *New Scientist*, October 16, 1986; *Electronics Weekly*, October 15, 1988; *The Engineer*, February 4, 1988.

6. M. Gordon, HOL: A Machine-Oriented Formulation of Higher-Order Logic, Technical Report 68, University of Cambridge Computer Laboratory, 1985.

7. A. Cohn, A Proof of Correctness of the Viper Microprocessor: The First Level, Technical Report 104, University of Cambridge Computer Laboratory, 1987.

8. A. Cohn, Correctness Properties of the Viper Block Model: The Second Level, Technical Report 134, University of Cambridge Computer Laboratory, 1988.

9. C. H. Pygott, Formal Proof of Correspondence between the Specification of a Hardware Module and its Gate Level Implementation, RSRE Report 85012, 1985.

10. B. Brock and W. A. Hunt, Jr., Report on the Formal Specification and Partial Verification of the VIPER Microprocessor, Computational Logic Technical Report 46, 1990.

11. J. Kershaw, foreword to Brock and Hunt (ibid.).

12. See the notes to chapter 8 of the present volume.

13. A. Cohn, *Journal of Automated Reasoning* 5 (1989): 127–139.

14. Ministry of Defence, Interim Defence Standard 00-55, part 1, issue 1, April 5, 1991.

Chapter 8

1. H. M. Collins, *Artificial Experts: Social Knowledge and Intelligent Machines* (MIT Press, 1990), chapter 5.

2. E. Peláez, J. Fleck, and D. MacKenzie, "Social research on software," paper read to National Workshop of Programme in Information and Communications Technologies, Manchester, 1987, p. 5.

3. K. Mannheim, *Ideology and Utopia* (Routledge & Kegan Paul, 1936). See D. Bloor, "Wittgenstein and Mannheim on the sociology of mathematics," *Studies in the History and Philosophy of Science* 4 (1973): 173–191.

4. This convention is known as "twos complement arithmetic," since if in this format "the sign is treated as if it were simply another digit, negative numbers are represented by their complements with respect to 2" (Robert F. Shaw, "Arithmetic operation in a binary computer," *Review of Scientific Instrumentation* 21 (1950), pp. 687–688). (Shaw's paper, along with other important papers in the area, is reprinted in Earl E. Schwartzlander, Jr., *Computer Arithmetic* (Dowden, Hutchinson & Ross, 1980).)

5. In common with other computer arithmetic, variants on this basic format (e.g., a 64-bit "double precision" mode) are permitted. For the sake of simplicity, this issue is ignored in the text.

6. The exponent is typically stored not in its natural binary representation, which would have to include a bit to represent its sign, but in a "biased" form without a sign.

7. W. Kahan, "Mathematics written in sand," *Proceedings of the American Statistical Association*, 1983 Statistical Computing section, pp. 12—26.

8. Imre Lakatos, *Proofs and Refutations: The Logic of Mathematical Discovery* (Cambridge University Press, 1976). Drawing on the work of Mary Douglas, David Bloor has suggested that different social circumstances may generate different typical reactions to anomalies. Unfortunately, evaluating this suggestion for the case under consideration would demand data that I do not possess. See D. Bloor, "Polyhedra and the abominations of Leviticus," *British Journal for the History of Science* 11 (1978): 245–272; M. Douglas, *Natural Symbols: Explorations in Cosmology* (Barrie & Rockcliff, 1970).

9. Interview with Prof. W. Kahan, Berkeley, October 19, 1989; Lakatos, *Proofs and Refutations*. In "forward error analysis," an upper bound is calculated for the accumulated error at each step, and thus eventually for the error in the final result. Backward analysis, on the other hand, "starts with the computed solution and works backward to reconstruct the perturbed problem of which it is an

exact solution." See P. J. L. Wallis, *Improving Floating-Point Programming* (Wiley, 1990), p. 12. For some remarks on the history of error analysis, see J. H. Wilkinson, "Modern error analysis," *SIAM Review* 13 (1971): 548–568.

10. Kahan uses this term in "Doubled-Precision IEEE Standard 754 Floating-Point Arithmetic" (typescript, 1987).

11. Lakatos, *Proofs and Refutations*; D. Bloor, *Wittgenstein: A Social Theory of Knowledge* (Macmillan, 1983), pp. 141–142.

12. Kahan interview.

13. Palmer was later to play an important role in the development of parallel computing in the United States when he and others left Intel in 1983 to establish the firm NCube.

14. Kahan interview; J. Palmer, "The Intel standard for floating-point arithmetic," in *Proceedings of IEEE COMPSAC 1977*, p. 107.

15. Letter to author from Robert G. Stewart, October 18, 1990.

16. IEEE Standard for Binary Floating-Point Arithmetic: American National Standards Institute/Institute of Electrical and Electronics Engineers Standard 754-1985, August 12, 1985.

17. Kahan interview.

18. Intel released the i8087 in 1980 in anticipation of the standard. For the i8087's arithmetic, see J. F. Palmer and S. P. Morse, *The 8087 Primer* (Wiley, 1984), 7. A third seriously considered proposal was drafted by Robert Fraley and J. Stephen Walther, but the main dispute seems to have been between the DEC and Kahan-Coonen-Stone proposals. For a comparison of the three proposals see W. J. Cody, "Analysis of proposals for the floating-point standard," *Computer* (March 1981): 63–66.

19. IEEE Standard for Binary Floating-Point Arithmetic, p. 14. The argument for two zeros is that they facilitate the removal of otherwise troublesome singularities in computations common in (e.g.) the study of fluid flow. With two zeros, "identities are conserved that would not otherwise be conserved" and capabilities not present in conventional mathematics are provided. An example from the complex numbers concerns the square root of -1, which conventionally can have two values: $+i$ and $-i$. Signed zeros make it possible to remove this discontinuity, the square root of $-1 + 0i$ being $+i$ and the square root of $-1 - 0i$ being $-i$ (Kahan interview).

20. On gradual underflow see I. B. Goldberg, "27 Bits are not enough for 8-Digit Accuracy," *Communications of the ACM* 10 (1967): 105–108. The computer pioneer Konrad Zuse was an early advocate, according to Cody ("Analysis of proposals for the floating-point standard," p. 63).

21. Jerome T. Coonen, "Underflow and the denormalized numbers," *Computer* (March 1981), p. 75.

22. Cody, "Analysis of proposals for the floating-point standard," p. 67.

23. Mary Payne and Dileep Bhandarkar, "VAX floating point: A solid foundation for numerical computation," *Computer Architecture News* 8, no. 4 (1980), pp. 28–29.

24. Bob Fraley and Steve Walther, "Proposal to eliminate denormalized numbers," *ACM Signum Newsletter* (October 1979), p. 22.

25. Cody, "Analysis of proposals for the floating-point standard," p. 67.

26. I draw these points from my interview with Kahan.

27. Ibid.

28. Mary H. Payne, "Floating point standardization," *COMPCON Proceedings* (fall 1979), p. 169.

29. Thus compliance with the standard is spreading into the field of dedicated digital signal processors (DSPs). One executive in the field writes: "Not being IEEE-compatible turns off users who want to replace an application running on an array processor with a board containing a 32-bit floating-point DSP chip. . . . Most array processors are IEEE-compatible." (quoted in Jonah McLeod, "DSP, 32-bit floating-point: The birth pangs of a new generation," *Electronics* (April 1989), p. 73)

30. W. Brian Arthur, "Competing technologies and economic prediction," *Options* (April 1984), p. 10.

31. Interview with Chuck Purcell, Minneapolis, April 4, 1990.

32. E. W. Dijkstra, "The humble programmer," *Communications of the ACM* 10 (1972), p. 864.

33. Ibid.

34. Trusted Computer System Evaluation Criteria, U.S. Department of Defense document DOD 5200.28-STD, December 1985 (first issued in 1983).

35. Cabinet Office, Advisory Council for Applied Research and Development, *Software: A Vital Key to UK Competitiveness* (HMSO, 1986), appendix B; Ministry of Defence Directorate of Standardization, Interim Defence Standard 00-55: The Procurement of Safety Critical Software in Defence Equipment (Glasgow, April 5, 1991).

36. Peláez, Fleck, and MacKenzie, "Social research on software," p. 5.

37. J. V. Grabiner, "Is Mathematical truth time-dependent?" *American Mathematical Monthly* 81 (1974): 354–365.

38. The chain of reasoning has subsequently been developed considerably. See, e.g., C. H. Pygott, "Verification of VIPER's ALU [Arithmetic/Logic Unit]," Technical Report, Divisional Memo (Draft), Royal Signals and Radar

Establishment, 1991; Wai Wong, "Formal verification of VIPER's ALU," typescript, University of Cambridge Computer Laboratory, April 15, 1993; J. Joyce, S. Rajan, and Z. Zhu, "A virtuoso performance becomes routine practice: A reverification of the VIPER microprocessor using a combination of interactive theorem-proving and B[inary] D[ecision] D[iagram]-based symbolic trajectory evaluation," typescript abstract of paper presented at Conference on the Mathematics of Dependable Systems, University of London, 1993. I have not seen the first of these items; I owe my knowledge of it to Wong ("Formal Verification," p. 4).

39. The case was Charter Technologies Limited vs. the Secretary of State for Defence, High Court of Justice, Queen's Bench Division, case 691, 1991.

40. R. S. Boyer and J. S. Moore, "Proof checking the RSA public key encryption algorithm," *American Mathematical Monthly* 91 (1984), p. 181.

41. J. Kershaw, foreword to Bishop Brock and Warren A. Hunt, Jr., Report on the Formal Specification and Partial Verification of the VIPER Microprocessor (technical report 46, Computational Logic, Inc., Austin, Texas, 1990).

42. M. Thomas, "VIPER Lawsuit withdrawn," electronic mail communication, June 5, 1991.

43. See Hilbert's 1927 address "The Foundations of Mathematics," in *From Frege to Gödel: A Source Book in Mathematical Logic, 1879–1931*, ed. J. van Heijenoort (Harvard University Press, 1967). A potentially significant difference, in view of the arguments by De Millo et al. discussed below, is that for Hilbert the array of formulas constituting a proof "must be given as such to our perceptual intuition" (ibid., p. 465).

44. Formalism's most famous opponent, the intuitionist L. E. J. Brouwer, wrote: "The question where mathematical exactness does exist, is answered differently by the two sides; the intuitionist says: in the human intellect, the formalist says: on paper" (quoted from "Intuitionism and Formalism," an English translation of Brouwer's 1912 Inaugural Address at the University of Amsterdam, in Brouwer, *Collected Works*, vol. 1, *Philosophy and Foundations of Mathematics* (North-Holland, 1975), p. 125).

45. For an interesting discussion of the significance of the machine analogy in Hilbert's mathematical program, see Herbert Breger, "Machines and mathematical styles of thought," paper read to conference on Mathematization of Techniques and Technization of Mathematics, Zentrum für interdisziplinäre Forschung, Universität Bielefeld, 1991.

46. R. A. De Millo, R. J. Lipton and A. J. Perlis, "Social processes and proofs of theorems and programs," *Communications of the Association for Computing Machinery* 22 (1979): 271–280.

47. Ibid., pp. 273–275. The phrase "'social' processes of proof" in the above quotation from Thomas is probably a reference to the argument of this paper.

48. Leslie Lamport of verification specialists SRI International, as quoted in Stuart S. Shapiro, Computer Software as Technology: An Examination of Technological Development (Ph.D. thesis, Carnegie Mellon University, 1990), p. 132. Shapiro's thesis contains a useful account of the debate surrounding the paper by De Millo, Lipton, and Perlis in the 1970s.

49. Edsger W. Dijkstra, "On a political pamphlet from the middle ages," *ACM SIGSOFT, Software Engineering Notes* 3, part 2 (April 1978), p. 14. Dijkstra was commenting on an earlier version of the De Millo-Lipton-Perlis paper published in the Proceedings of the Fourth ACM Symposium on Principles of Programming Languages (1977).

50. Edsger W. Dijkstra, "Formal techniques and sizeable programs," paper prepared for Symposium on the Mathematical Foundations of Computing Science, Gdansk, 1976, as reprinted in Dijkstra, *Selected Writings on Computing: A Personal Perspective* (Springer, 1982) (see pp. 212–213). For an interesting analysis of Dijkstra's overall position see Eloína Peláez, A Gift from Pandora's Box: The Software Crisis, Ph.D. thesis, University of Edinburgh, 1988.

51. The most recent focus of controversy was another article denying, on different grounds from those of De Millo, Lipton, and Perlis, the analogy between verification and mathematical proofs: James H. Fetzer, "Program verification: The very idea," *Communications of the ACM* 31 (1988), 1048–1063.

52. Kenneth Appel and Wolfgang Haken, as quoted in Philip J. Davis and Reuben Hersh, "Why should I believe a computer?" in Davis and Hersh, *The Mathematical Experience* (Harvester, 1981).

53. Ibid., pp. 385–386.

54. National Research Council, System Security Study Committee, *Computers at Risk: Safe Computing in the Information Age* (National Academy Press, 1991).

55. Ministry of Defence, Interim Defence Standard 00-55, Part 2, Guidance, p. 28. Margaret Tierney reviews the history of Def Stan 00-55 in "The evolution of Def Stan 00-55 and 00-56: An intensification of the 'formal methods debate' in the UK," working paper 30, Programme on Information and Communication Technologies, Edinburgh, 1991.

56. Ministry of Defence, Interim Defence Standard 00-55, Part 2, Guidance, p. 28.

57. See *From Frege to Gödel*, ed. van Heijenoort. Some of Brouwer's reasons for doubting an apparently obvious principle can be seen in the following quotation: "Now consider the principum *tertii exclusi* [law of excluded middle]: It claims that every supposition is either true or false; in mathematics this means that for every supposed imbedding of a system into another, satisfying certain given conditions, we can either accomplish such an imbedding by a construction, or arrive by a construction at the arrestment of the process which would lead to the imbedding. It follows that the question of the validity of the principum *tertii exclusi* is equivalent to the question whether unsolvable mathematical problems can exist.

There is not a shred of a proof for the conviction, which has sometimes been put forward that there exist no unsolvable mathematical problems . . . in infinite systems the principum *tertii exclusi* is as yet not reliable. . . . So long as this proposition is unproved, it must be considered as uncertain whether problems like the following are solvable: Is there in the decimal expansion of [pi] a digit which occurs more often than any other one? . . . And it likewise remains uncertain whether the more general mathematical problem: Does the principum *tertii exclusi* hold in mathematics without exception? is solvable. . . . In mathematics it is uncertain whether the whole of logic is admissible and it is uncertain whether the problem of its admissibility is decidable." (Brouwer, "The Unreliability of the Logical Principles," in Brouwer, *Collected Works*, vol. 1, pp. 109–111; emphases deleted)

58. I am drawing here on Bloor, *Wittgenstein*, pp. 124–136.

59. D. C. Makinson, *Topics in Modern Logic* (Methuen, 1973), pp. 27–28.

60. Cliff B. Jones, *Systematic Software Development using VDM*, second edition (Prentice-Hall, 1990), p. 24.

61. "Unless we have a philosophical commitment to intuitionism, maintaining constructiveness when it is not required can only make a proof system more cumbersome to use. We have seen that certain programs cannot be derived from their specifications in a constructive logic, but can be derived in a classical logic upon which minimal restrictions have been imposed." (Zohar Manna and Richard Waldinger, "constructive logic considered obstructive," undated typescript, p. 8)

Chapter 9

1. B. S. Dhillon, *Robot Reliability and Safety* (Springer, 1991), p. 38

2. C. Perrow, *Normal Accidents: Living with High-Risk Technologies* (Basic Books, 1984); C. V. Oster Jr., J. S. Strong, and C. K. Zorn, *Why Airplanes Crash: Aviation Safety in a Changing World* (Oxford University Press, 1992).

3. D. A. Norman, "Commentary: Human error and the design of computer systems," *Communications of the Association for Computing Machinery* 33 (1990): 4–7.

4. *Report of the Inquiry into the London Ambulance Service* (South West Thames Regional Health Authority, 1993).

5. C. Milhill, "Hospital error killed dozens of patients," *Guardian*, September 30, 1993.

6. Quoted in *Software Engineering Notes*, April 1992, p. 30.

7. See, e.g., M. Thomas, "Should we trust computers?" British Computer Society/Unisys Annual Lecture, Royal Society of London, July 4, 1988. Peter Neumann's own compilation of cases from this database, and his analysis of their implications, can now be found in his book *Computer-Related Risks* (Addison-Wesley, 1995).

8. Described in detail in N. G. Leveson and C. S. Turner, An Investigation of the Therac-25 Accidents, Technical Report 92-108, Information and Computer Science Department, University of California, Irvine, 1992. Also published in *Computer* 26 (1993): 18–41.

9. Leveson and Turner, An Investigation of the Therac-25 Accidents.

10. Patriot Missile Defense: Software Problem Led to System Failure at Dhahran, Saudi Arabia. Report GAO/IMTEC-92-26, General Accounting Office, Washington, 1992.

11. For an explanation of the "floating-point" representation of numbers, see the section "Negotiating Arithmetic" in chapter 8.

12. R. Skeel, "Roundoff error and the Patriot missile." *SIAM* [Society for Industrial and Applied Mathematics] *News* 25 (1992) no. 4, p. 11.

13. Patriot Missile Defense (note 10 above), p. 9.

14. Ibid.

15. The main uncertainty in this case is whether a successful interception would have taken place had defensive missiles been launched: the U.S. Army claims only a 70 percent success rate in interceptions of Scud attacks on Saudi Arabia, and critics (see, e.g., T. A. Postol, "Lessons of the Gulf War experience with Patriot," *International Security* 16, no. 3 (1991–92): 119–171) have alleged that the true success rate was substantially lower than that. However, a near miss that did not destroy or disable the Scud warhead might nevertheless have deflected it.

16. Report of the Independent Clinical Assessment commissioned by the North Staffordshire Health Authority on the Effects of the Radiation Incident at the North Staffordshire Royal Infirmary between 1982 and 1991 (1993).

17. Milhill, "Hospital error killed dozens of patients."

18. G. Sharp, Formal Investigation into the Circumstances surrounding the Attack on the USS *Stark* (FFG31) on 17 May 1987 (Commander, Cruiser-Destroyer Group Two, 1987); M. Vlahos, "The *Stark* Report," *Proceedings of the U.S. Naval Institute*, May 1988, pp. 64–67

19. Vlahos, "*Stark* Report," p. 65.

20. W. J. Crowe, endorsement added to W. M. Fogarty, Formal Investigation into the Circumstances surrounding the Downing of Iran Air Flight 655 on 3 July 1988 (Office of the Chairman, Joint Chiefs of Staff, 1988), p. 8.

21. G. I. Rochlin, "Iran Air Flight 655 and the U.S.S. *Vincennes*: Complex, large-scale military systems and the failure of control," in *Social Responses to Large Technical Systems: Control or Anticipation?* ed. T. La Porte (Kluwer, 1991).

22. There were two other A320 crashes prior to the end of 1992: one at Habsheim in Alsace in 1988 (three deaths) and one near Bangalore, India, in 1990 (92 deaths). Their interpretation, particularly that of the former, has been

a matter of dispute. Peter Mellor ("CAD: Computer Aided Disaster!" Typescript, Centre for Software Reliability, City University, London, 1994) argues that both should be seen as computer-related, primarily in the context of human-computer interaction. In particular, it was suggested in the report of the official inquiry into the Bangalore crash that the aircraft's pilots may have had undue confidence in the capacity of an automated protection facility ("alpha-floor," triggered by the angle between the aircraft's pitch axis and the air flow) rapidly to increase engine thrust and bring their aircraft back to a safe condition. Generally, though, the case for the computer-relatedness of these accidents seems to me to be weaker than in the Strasbourg crash, and therefore only the latter is included in the data set.

23. P. Sparaco, "Human factors cited in French A320 Crash," *Aviation Week and Space Technology,* January 3, 1994, p. 30.

24. Ibid., p. 31.

25. See, e.g., V. M. Altamuro, "Working safely with the iron collar worker," *National Safety News,* July 1983, pp. 38–40.

26. The official investigation of the sole reported U.S. robot-related fatality speculates about a further possible factor: that workers may perceive robots as "something more than machines" and may "personalize" them. Thus the worker in this case had nicknamed his robot "Robby." They suggest that "this personalization may cause the worker's mind to be more focused on the 'teamwork' with the robot rather than upon relevant safety issues" (L. M. Sanderson, J. W. Collins, and J. D. McGlothlin, "Robot-related Fatality involving a U.S. manufacturing plant employee: Case report and recommendations," *Journal of Occupational Accidents* 8 (1986), p. 20).

27. R. Edwards, Accidents on Computer Controlled Manufacturing Plant and Automated Systems in Great Britain 1987–91, typescript, Health and Safety Executive, Birmingham, n.d.

28. Ibid., p. 7.

29. J. P. Vautrin and D. Dei-Svaldi. 1989. "Accidents du Travail sur Sites Automatisés: Évaluation d'une Prévention Technique," *Cahiers de Notes Documentaires* no. 136 (1989): 445–453; Edwards, Accidents on Computer Controlled Manufacturing Plant and Automated Systems in Great Britain 1987–91.

30. Edwards, p. 26.

31. Computers are now so central to modern long-range civil aviation that the comparison would in effect be of two different epochs in the history of flight.

32. R. Smithers, "Road deaths at new low," *The Guardian,* March 26, 1993, p. 3.

33. Major injury is defined in U.K. safety-at-work legislation as including, for example, "amputation of a joint of a finger, fracture of any bone in the skull,

spine, neck, arm or leg (but not in the hand or foot) and a penetrating or burn injury to an eye." Minor injury is injury which is not (in this sense) major but which causes "a person to be incapable of doing his normal work for more than 3 consecutive days" (Edwards, Accidents on Computer Controlled Manufacturing Plant and Automated Systems in Great Britain 1987–91, p. 3).

34. The overall ratio of non-fatal serious injuries in air travel to fatalities seems typically to be less than one (Oster, Strong, and Zorn, *Why Airplanes Crash*, p. 23).

35. See, e.g., J. Rushby, Formal Methods and the Certification of Critical Systems, Technical Report 93-07, SRI International Computer Science Laboratory, Menlo Park, California, 1993, pp. 127, 128.

36. E. Peláez, A Gift from Pandora's Box: The Software Crisis. Ph.D. thesis, University of Edinburgh, 1988.

37. Skeel, "Roundoff error and the Patriot missile."

38. Patriot Missile Defense (note 10 above).

39. B. Randell, "Technology doesn't have to be bad." *Software Engineering Notes* 14 (1989), no. 6, p. 21.

40. P. Neumann, "Letter from the editor: Are risks in computer systems different from those in other technologies?" *Software Engineering Notes* 13 (1988), no. 2: 3.

41. S. S. Brilliant, J. C. Knight, and N. G. Leveson, "Analysis of faults in an N-version software experiment," *IEEE Transactions on Software Engineering* 16 (1990), p. 238.

42. For example, the North Staffordshire radiation therapy incident involved an incorrect mental model of a computerized system. In the Therac-25 case an error hidden amongst the logical complexity of even only modestly large software manifested itself not in gradual deterioration of performance but in a sudden and fatal switch in mode of operation. In the Dhahran incident a tiny cause (an uncorrected rounding error of 0.0001 percent) led to system failure.

43. Perrow, *Normal Accidents*.

44. Policy Statement on Safety-Related Computer Systems (British Computer Society, 1993).

45. Leveson and Turner, An Investigation of the Therac-25 Accidents, pp. 37–38.

46. For which see Perrow, *Normal Accidents*.

47. Rochlin, "Iran Air Flight 655 and the U.S.S. *Vincennes*."

48. In defense against ballistic missiles, firing is generally automatic because of the extremely limited decision time.

49. Rochlin, "Iran Air Flight 655 and the U.S.S. *Vincennes* ," p. 119.

50. Ibid.

51. See, e.g., Rushby, Formal Methods and the Certification of Critical Systems.

52. B. D. Nordwall, "GPWS to improve regional safety," *Aviation Week and Space Technology*, April 26, 1993, pp. 53–54. The general point about overconfidence applies here too: wise use would involve measures to make sure that pilots do not rely exclusively on ground-proximity warning systems to avoid "controlled flight into terrain"!

53. D. L. Parnas, A. J. van Schouwen, and S. P. Kwan, "Evaluation of safety-critical software," *Communications of the Association for Computing Machinery* 33 (1990), no. 6, p. 636.

54. N. G. Leveson, "High-pressure steam engines and computer software," International Conference on Software Engineering, Melbourne, 1992.

Chapter 10

1. Michael Lynch, *Art and Artifact in Laboratory Science: A Study of Shop Work and Shop Talk in a Research Laboratory* (Routledge and Kegan Paul, 1985), p. 5.

2. See, e.g., Karin Knorr-Cetina, "The couch, the cathedral, and the laboratory: On the relationship between experiment and laboratory in science," in *Science as Practice and Culture*, ed. A. Pickering (University of Chicago Press, 1992); Andrew Pickering, "The mangle of practice: Agency and emergence in the sociology of science," *American Journal of Sociology* 99 (1993): 559–589. There are, of course, evident connections to developments in other areas, in particular the ethnomethodological critique of structural-functionalist sociology (for which see John Heritage, *Garfinkel and Ethnomethodology* (Polity, 1984)) and the "situated action" critique of the symbol-processing paradigm in cognitive science (see Donald A. Norman, "Cognition in the head and in the world: An introduction to the special issue on situated action," *Cognitive Science* 17 (1993): 1–6, and the subsequent papers in that issue of *Cognitive Science*).

3. Bruno Latour, "Visualization and cognition: Thinking with eyes and hands," *Knowledge and Society* 6 (1986): 1–40; *Science in Action: How to Follow Scientists and Engineers through Society* (Open University Press, 1987)

4. For useful surveys from the early 1980s and the 1990s, respectively, which indicate some of the tensions within the alternative view as well as common ground, see *Science Observed: Perspectives on the Social Study of Science*, ed. K. Knorr-Cetina and M. Mulkay (Sage, 1983) and *Science as Practice and Culture*, ed. A. Pickering (University of Chicago Press, 1992).

5. Hubert L. Dreyfus, *What Computers Can't Do: The Limits of Artificial Intelligence* (Harper & Row, 1979); H. M. Collins, *Artificial Experts: Social Knowledge and Intelligent Machines* (MIT Press, 1990).

6. For more general weaknesses in the view of technology as "applied science," see *Science in Context: Readings in the Sociology of Science,* ed. B Barnes and D. Edge (Open University Press, 1982), pp. 147–185.

7. See Michael Polanyi, *Personal Knowledge* (Routledge and Kegan Paul, 1958); Polanyi, *The Tacit Dimension* (Routledge and Kegan Paul, 1967); Kenneth J. Arrow, "The economic implications of learning by doing," *Review of Economic Studies* 29 (1962): 155–173; Tom Burns, "Models, images and myths," in *Factors in the Transfer of Technology,* ed. W. Gruber and D. Marquis (MIT Press, 1969); Jerome R. Ravetz, *Scientific Knowledge and its Social Problems* (Clarendon, 1971); H. M. Collins, "The TEA set: Tacit knowledge and scientific networks," *Science Studies* 4 (1974): 165–186; Collins, "The seven sexes: A study in the sociology of a phenomenon, or the replication of experiments in physics," *Sociology* 9 (1975): 205–224; Collins, *Changing Order: Replication and Induction in Scientific Practice* (Sage, 1985); Collins, *Artificial Experts: Social Knowledge and Intelligent Machines* (MIT Press, 1990); Eugene S. Ferguson, "The mind's eye: Nonverbal thought in technology," *Science* 197 (1977): 827–836; Ferguson, *Engineering and the Mind's Eye* (MIT Press, 1992); Karin Knorr-Cetina, *The Manufacture of Knowledge: An Essay on the Constructivist and Contextual Nature of Science* (Pergamon, 1981); Knorr-Cetina, "The couch, the cathedral, and the laboratory: On the relationship between experiment and laboratory in science," in *Science as Practice and Culture,* ed. A. Pickering (University of Chicago Press, 1992); Michel Callon, "Is science a public good?" *Science, Technology and Human Values* 19 (1994): 395–424.

8. Collins, "The seven sexes" and *Changing Order.*

9. Collins, "The TEA set"; Collins, "The place of the 'core-set' in modern science: Social contingency with methodological propriety in science," *History of Science* 19 (1981): 6–19.

10. The best-known argument against the cumulative nature of science is that of Thomas Kuhn, which highlights the incommensurability of successive scientific "paradigms." See Kuhn, *The Structure of Scientific Revolutions,* second edition (University of Chicago Press, 1970).

11. A degree of knowledge of how the latter knapped can sometimes be recovered by the technique of "remontage," in which the original stone is gradually and painstakingly reconstructed from the flint implement and the discarded fragments. We owe our information on knapping to discussions with archaeologists at a conference at the Fondation des Treilles in June 1992.

12. Harvard Nuclear Study Group (Albert Carnesale et al.), *Living with Nuclear Weapons* (Bantam, 1983), p. 5.

13. On reinvention see Collins, "The TEA set," p. 176. The Harvard Nuclear Study Group also talk of reinvention, writing that "even if all nuclear arsenals were destroyed, the knowledge of how to reinvent them would remain" (*Living with Nuclear Weapons,* p. 5). The difference between their position and that explored in this chapter lies in the assumption that the necessary knowledge would still exist intact.

14. Karin Knorr-Cetina, "Epistemic cultures: Forms of reason in science," *History of Political Economy* 23 (1991): 105–122.

15. See, e.g., "Schock am Flughaven," *Der Spiegel*, August 15, 1994: 18–20. For a skeptical opinion see Josef Joffe, "Nuclear black market: Much ado about not much," *International Herald Tribune*, August 26, 1994.

16. Interviewees are listed in the appendix to this chapter. Not all interviews were tape recorded, and the quotations below from the Bergen, Dowler and Talley, Hudgins, McDonald, Miller, Sewell, and Westervelt interviews are from notes rather than transcripts. However, all interviewees whom we wished to quote were sent drafts of intended quotations and given the opportunity to correct errors or to withdraw permission for quotation. Only three interviewees exercised that latter right. We would not claim representativeness for our sample, which was constructed by "snowballing" from laboratory members who were well known in the outside world. Indeed, the sample is clearly biased towards more senior figures, both by the way it was constructed and by the need for interviewees to possess the confidence and experience to embark upon unclassified discussions that might stray onto sensitive matters. Furthermore, the course of interviews was to a considerable degree dictated by what interviewees were prepared to talk about, and they dealt with many matters other than those discussed here. It was, therefore, impossible to ensure that all interviewees were asked the same questions. Nevertheless, there appeared to be a degree of consensus on the inadequacy in nuclear weapons design of explicit knowledge alone.

17. Trevor Pinch, H. M. Collins, and Larry Carbone, "Cutting up skills: Estimating difficulty as an element of surgical and other abilities," in *Between Technology and Society: Technical Work in the Emerging Economy*, ed. S. Barley and J. Orr (ILR Press, forthcoming).

18. Langdon Winner, "Upon opening the black box and finding it empty: Social constructivism and the philosophy of technology," *Science, Technology and Human Values* 18 (1993): 362–378.

19. A chemical element (such as uranium) often exists in the form of more than one "isotope." The nucleus of any atom of a given element will always contain the same number of positive particles (protons), but different isotopes will contain different numbers of electrically neutral particles (neutrons). Isotopes are conventionally distinguished by their mass number, the total of protons and neutrons in their nuclei. Differences between isotopes are crucial to atomic physics. Thus uranium 235 is highly fissile (its nuclei readily split when struck by a neutron), while the more common isotope uranium 238 is relatively inert.

20. Government Printing Office, 1945.

21. For example, Margaret Gowing, assisted by Lorna Arnold, *Independence and Deterrence: Britain and Atomic Energy, 1945–52* (Macmillan, 1974), volume 2, p. 457.

22. Notably Chuck Hansen, *U.S. Nuclear Weapons: The Secret History* (Aerofax, 1988).

23. Albert Friendly, "Atom control fight centers on secrecy," *Washington Post*, March 25, 1946, p. 3. See Alice K. Smith, *A Peril and a Hope: The Scientists' Movement in America, 1945–47* (MIT Press, 1970), p. 84.

24. Harvard Nuclear Study Group, *Living with Nuclear Weapons*, p. 219.

25. Seymour M. Hersh, *The Samson Option: Israel, America and the Bomb* (Faber & Faber, 1991), p. 155.

26. Eventually published in the November 1979 issue of *The Progressive* (pp. 14–23).

27. A. DeVolpi, G. E. Marsh, T. A. Postol, and G. S. Stanford, *Born Secret: The H-Bomb, the "Progressive" Case and National Security* (Pergamon, 1981).

28. We have not seen UCRL-4725. The acronym stands for University of California Radiation Laboratory; the University of California manages the Los Alamos and Livermore laboratories. For the document's significance, see DeVolpi et al., *Born Secret*, and Hansen, *U.S. Nuclear Weapons*.

29. Hansen, *U.S. Nuclear Weapons*.

30. We write "mainstream" because of recent reports, emanating from the former Soviet Union, that fusion could be initiated not by a fission bomb but by enhancing the detonation of chemical explosives with a substance called "red mercury." One article that takes this claim seriously suggests that the substance is produced by dissolving mercury antimony oxide in mercury, heating and irradiating the resultant amalgam, and then evaporating off the elemental mercury (Frank Barnaby, "Red mercury: Is there a pure-fusion bomb for sale?" *International Defense Review* 6 (1994): 79–81). Russian weapons designers, however, report that red mercury was simply the Soviet code name for lithium-6, which tends to get colored red by mercuric impurities during its separation (Mark Hibbs, "'Red mercury' is lithium-6, Russian weaponsmiths say," *Nucleonics Week*, July 22, 1993: 10), and that it is therefore merely a component in the standard solid thermonuclear fuel, lithium-6 deuteride.

31. Now available as *The Los Alamos Primer: The First Lectures on How to Build an Atomic Bomb* (University of California Press, 1992).

32. See Kuhn, *The Structure of Scientific Revolutions*.

33. Kuhn, *The Structure of Scientific Revolutions*, passim.

34. David Hawkins, Manhattan District History. Project Y: The Los Alamos Project. Vol. 1.: Inception until August 1945, Los Alamos Scientific Laboratory report LAMS-2532, 1946, p. 9.

35. Lillian Hoddeson, Paul W. Hendriksen, Roger A. Meade, and Catherine Westfall, *Critical Assembly: A Technical History of Los Alamos during the Oppenheimer Years* (Cambridge University Press, 1993), p. 41.

36. Hawkins, Manhattan District History, p. 15.

37. Richard Rhodes, *The Making of the Atomic Bomb* (Simon and Schuster, 1986), pp. 417–419; Hoddeson et al., *Critical Assembly*, pp. 45–46.

38. Edward Teller, "The laboratory of the atomic age," *Los Alamos Science* 21 (1993), p. 33.

39. Quoted in Hoddeson et al., *Critical Assembly*, p. 42.

40. Ibid., p. 58.

41. Ibid., p. 400.

42. John H. Manley, "A new laboratory is born," in *Reminiscences of Los Alamos, 1943–1945*, ed. L. Badash et al. (Reidel, 1980), p. 33.

43. Smyth, *Atomic Energy*, p. 127.

44. Manley, "A new laboratory," p. 33.

45. Hoddeson et al., *Critical Assembly*, chapters 7 and 13.

46. Manley, "A new laboratory," p. 33.

47. The cyclotron, in which charged particles such as protons are accelerated, was a key experimental tool of early nuclear physics, and one which could be used to produce small quantities of fissile materials.

48. Hoddeson et al., *Critical Assembly*, p. 35.

49. Rhodes, *The Making of the Atomic Bomb*, p. 549.

50. Manley, "A new laboratory," p. 33.

51. Hawkins, Manhattan District History, p. 29.

52. Ibid.

53. Richard P. Feynman, "Los Alamos from below," in *Reminiscences of Los Alamos, 1943–1945*, ed. L. Badash et al. (Reidel, 1980), p. 125; Nicholas Metropolis and E. C. Nelson, "Early computing at Los Alamos," *Annals of the History of Computing* 4 (1982), p. 359.

54. Unless otherwise stated, details in this and the following four paragraphs are drawn from Hoddeson et al., *Critical Assembly*.

55. Hawkins, Manhattan District History, p. 77.

56. Serber, *Los Alamos Primer*, p. 52.

57. Hoddeson et al., *Critical Assembly*, pp. 270–271.

58. Teller, "The laboratory of the atomic age," p. 33.

59. In its early years, Livermore seems to have placed greater emphasis on computer modeling than did Los Alamos and to have regarded itself as techni-

cally less conservative. Any remaining differences in these respects are now not great, and on the matters discussed in the text we could detect no systematic difference between the two laboratories.

60. DeWitt interview.

61. Miller interview.

62. See Barry Barnes, "On the implications of a body of knowledge," *Knowledge: Creation, Diffusion, Utilization* 4 (1982): 95–110.

63. Bergen interview.

64. Haussmann interview.

65. See chapter 5 of the present volume.

66. Miller interview.

67. Hudgins interview.

68. Bergen interview.

69. W. Van Cleave, "Nuclear technology and weapons," in *Nuclear Proliferation Phase II*, ed. R. Lawrence and J. Larus (University Press of Kansas, 1973).

70. Donald R. Westervelt, "The role of laboratory tests," in *Nuclear Weapon Tests: Prohibition or Limitation*, ed. J. Goldblat and D. Cox (Oxford University Press, 1988), p. 56.

71. Ibid.

72. Bergen interview.

73. Ibid.

74. George H. Miller, Paul S. Brown, and Carol T. Alonso, Report to Congress on Stockpile Reliability, Weapon Remanufacture, and the Role of Nuclear Testing, Lawrence Livermore National Laboratory report UCRL-53822, 1987, p. 4.

75. Bergen interview.

76. R. E. Kidder, Maintaining the U.S. Stockpile of Nuclear Weapons during a Low-Threshold or Comprehensive Test Ban, Lawrence Livermore National Laboratory report UCRL-53820, 1987, p. 6.

77. Hoyt interview.

78. Ibid.

79. Ibid.

80. See, e.g., Miller et al., Report to Congress, p. 4 and passim.

81. Ibid., p. 4.

82. Hudgins interview.

83. Westervelt interview.

84. McDonald interview.

85. Hudgins interview.

86. Westervelt interview.

87. Hudgins interview.

88. Dowler and Talley interview.

89. Hudgins interview.

90. Source: private communications from interviewees.

91. See Knorr-Cetina, *The Manufacture of Knowledge*, pp. 69–70. There has, e.g., often been fierce rivalry between Los Alamos and Livermore.

92. Miller et al., Report to Congress, p. 4.

93. Ibid., p. 12.

94. Ibid., p. 26.

95. Hugh Gusterson, Testing Times: A Nuclear Weapons Laboratory at the End of the Cold War. Ph.D. thesis, Stanford University, 1991, p. 258. See also Gusterson, "Keep building those bombs," *New Scientist*, October 12, 1991: 30–33; Gusterson, "Coming of age in a weapons lab: Culture, tradition and change in the house of the bomb," *The Sciences* 32, no. 3 (1992): 16–22; Gusterson, "Exploding anthropology's canon in the world of the bomb," *Journal of Contemporary Ethnography* 22 (1993): 59–79.

96. Mark interview II.

97. David F. Noble (*Forces of Production: A Social History of Industrial Automation* (Knopf, 1984)) has found this to be the case more generally.

98. Miller et al., Report to Congress, p. 55.

99. Ibid.

100. Ibid.

101. Ibid., p. 28.

102. Collins, *Changing Order.*

103. Miller et al., Report to Congress, p. 25.

104. Mark interview II.

105. Miller et al., Report to Congress, p. 3.

106. Jack W. Rosengren, Some Little-Publicized Difficulties with a Nuclear Freeze, report RDA-TR-112116-001, R&D Associates, Marina del Rey, Calif., 1983.

107. Tom Z. Collina and Ray E. Kidder, "Shopping spree softens test-ban sorrows," *Bulletin of the Atomic Scientists,* July-August 1994, p. 25.

108. In particular, Miller et al., Report to Congress.

109. Hausmann interview.

110. Collins, *Artificial Experts.*

111. As will be discussed in the conclusion to this chapter, there are some particular contingencies that may affect ease of re-creation.

112. Norris Bradbury, press statement to *The New Mexican* (Santa Fe), September 24, 1954, reprinted in *Los Alamos Science* 4, no. 7 (1983): 27–28.

113. Herbert York, *The Advisors: Oppenheimer, Teller and the Superbomb* (Freeman, 1976), p. 126.

114. Ibid., p. 134.

115. Thomas B. Cochran, William M. Arkin, Robert S. Norris, and Milton M. Hoenig, *Nuclear Weapons Databook, Volume II: U.S. Nuclear Warhead Production* (Ballinger, 1987), pp. 153–154; Hansen, *U.S. Nuclear Weapons*, pp. 32–33, 39.

116. Lowell Wood and John Nuckolls, "The development of nuclear explosives," in *Energy in Physics, War and Peace: A Festschrift Celebrating Edward Teller's 80th Birthday,* ed. H. Mark and L. Wood (Kluwer, 1988), p. 316.

117. Duane C. Sewell, "The Branch Laboratory at Livermore during the 1950's," in *Energy in Physics, War and Peace,* ed. Mark and Wood, p. 323.

118. See, e.g., Leonard S. Spector, *Going Nuclear* (Ballinger, 1987), p. 35.

119. Yuli Khariton and Yuri Smirnov, "The Khariton version," *Bulletin of the Atomic Scientists,* May 1993, p. 22.

120. Ibid.

121. David Holloway, personal communication, February 15, 1994.

122. David Holloway, *Stalin and the Bomb: The Soviet Union and Atomic Energy, 1939–1956* (Yale University Press, 1994), p. 199.

123. Steven J. Zaloga, *Target America: The Soviet Union and the Strategic Arms Race, 1945–1964* (Presidio, 1993), p. 53.

124. Holloway, *Stalin and the Bomb*, p. 199.

125. Ibid., p. 54.

126. Holloway, *Stalin and the Bomb*, p. 199.

127. D. Holloway, personal communication, September 20, 1994. Holloway was, unfortunately, unable to elicit what these differences were.

128. Ferenc M. Szasz, *British Scientists and the Manhattan Project: The Los Alamos Years* (Macmillan, 1992).

129. Gowing and Arnold, *Independence and Deterrence*, volume 2, p. 456.

130. Brian Cathcart, *Test of Greatness: Britain's Struggle for the Atom Bomb* (Murray, 1994), p. 105.

131. Gowing and Arnold, *Independence and Deterrence*, p. 456.

132. Ibid., p. 458.

133. Szasz, *British Scientists*, pp. 51–52.

134. Cathcart, *Test of Greatness*, p. 132.

135. Margaret Gowing, *Britain and Atomic Energy, 1939–45* (Macmillan, 1964), p. 330.

136. Gowing and Arnold, *Independence and Deterrence*, p. 90.

137. Cathcart, *Test of Greatness*.

138. Gowing and Arnold, *Independence and Deterrence*, p. 459.

139. Ibid., p. 72.

140. Cathcart, *Test of Greatness*, p. 71.

141. Gowing and Arnold, *Independence and Deterrence*, p. 474.

142. Ibid., p. 472.

143. Ibid.

144. Cathcart, *Test of Greatness*, p. 140.

145. Gowing and Arnold, *Independence and Deterrence*, p. 472.

146. Ibid., p. 462..

147. Jacques Chevallier and Pierre Usunier, "La mise en œuvre scientifique et technique," in *L'Aventure de la Bombe: De Gaulle et la Dissuasion Nucléaire (1958–1969)* (Plon, 1984), p. 127.

148. Bertrand Goldschmidt, "La genèse et l'héritage," in *L'Aventure de la Bombe*, p. 29.

149. Chevallier and Usunier, "La mise en œuvre scientifique et technique," pp. 128–129.

150. Albert Buchalet, "Les premières étapes (1955–1960)," in *L'Aventure de la Bombe*, pp. 40–41.

151. Ibid., pp. 51, 57.

152. Chevallier and Usunier, "La mise en œuvre scientifique et technique," p. 130 (our translation). No details are available, although Buchalet ("Les premières étapes," p. 48) writes that plutonium metallurgy caused particular concern.

153. John W. Lewis and Xue Litai, *China Builds the Bomb* (Stanford University Press, 1988).

154. Ibid., p. 160.

155. Ibid., pp. 87–88, 106, 150–169.

156. David Albright and Mark Hibbs, "Pakistan's bomb: Out of the closet," *Bulletin of the Atomic Scientists*, July-August 1992, p. 38.

157. David Albright and Mark Hibbs, "India's silent bomb," *Bulletin of the Atomic Scientists*, September 1992, p. 29.

158. David Albright and Mark Hibbs, "Iraq and the bomb: Were they even close?" *Bulletin of the Atomic Scientists*, March 1991, p. 19.

159. Quoted in Albright and Hibbs, "Pakistan's bomb," p. 42. In a private communication, one source has questioned the veracity of this account of the Pakistani program. Unfortunately, we lack the data to clarify matters further.

160. Available data leave unclear the extent to which this is true of Israel, India, South Africa, and Pakistan; the British case is one of extensive previous personal contact.

161. See, e.g., Kidder, Maintaining the U.S. Stockpile, pp. 7–8; J. Carson Mark, "The purpose of nuclear test explosions," in *Nuclear Weapon Tests: Prohibition or Limitation*, ed. J. Goldblat and D. Cox (Oxford University Press, 1988), pp. 40–41.

162. There are no detailed histories of the nature of the technical problems encountered in the various hydrogen bomb programs equivalent to those now available for the atomic bomb. The relative slowness of the U.S. effort may in part be accounted for by opposition to hydrogen bomb development; as noted below, there was also doubt about the feasibility of designs prior to the Teller-Ulam configuration. There has also been debate over whether the first claimed Soviet and British thermonuclear explosions deserve such a categorization. Thus "Joe 4," the Soviet test explosion on August 12, 1953, was not of a Teller-Ulam device, but of one with alternating layers of thermonuclear fuel and uranium 238 sandwiched between the high explosive and core of an implosion bomb (Holloway, *Stalin and the Bomb*, pp. 298, 307–308). In table 1 we class "Joe 4" as a thermonuclear explosion, but we accept that there is a case for regarding the

device as more akin to a boosted fission weapon. For Britain we follow John Baylis, "The development of Britain's thermonuclear capability 1954–61: Myth or reality?" *Contemporary Record* 8 (1994): 159–174.

163. See Bruno Latour, *Science in Action: How to Follow Scientists and Engineers through Society* (Open University Press, 1987).

164. See chapter 5 of the present volume.

165. The best-known example is "DYNA3D," a Livermore-designed program to assist three-dimensional analysis of the response of mechanical structures to stresses and impacts (David Allen, "How to build a better beer can: DYNA does it all, from DOE to R&D," *Supercomputing Review,* March 1991: 32–36).

166. Numerically controlled machine tools are sometimes included in such lists, but our impression is that the tolerances required for the fabrication of simple nuclear weapons are achievable with manually controlled precision machine tools commercially available (e.g. from Swiss suppliers) even in the late 1940s and the 1950s. The availability of skilled machinists to operate such tools has, however, on occasion been a constraint, e.g. on the British program.

167. In one sense, of course, tacit knowledge is required for the successful operation of all such equipment: Collins (*Artificial Experts*) describes the tacit knowledge needed even to perform arithmetic on a simple pocket calculator. If, however, the relevant tacit knowledge is widely distributed—if, say, it is of a kind that any physics or engineering graduate might be expected to possess—then the need for it would not be a real constraint on nuclear weapons development.

168. David Albright and Mark Hibbs, "Iraq's bomb: Blueprints and artifacts," *Bulletin of the Atomic Scientists,* January-February 1992, p. 33.

169. Gary Milhollin, "Building Saddam Hussein's bomb," *New York Times Magazine,* March 8, 1992: 30–36.

170. Technologies Underlying Weapons of Mass Destruction, Office of Technology Assessment report OTA-BP-ISC-115, U.S. Government Printing Office, 1993, pp. 150–151.

171. Spector, *Going Nuclear,* p. 161.

172. Albright and Hibbs, "Iraq and the bomb," p. 16.

173. Clearly there is a possibility that achievements may successfully have been hidden from the inspectors (although, had Iraqi leaders believed the program to have been on the verge of producing a usable weapon, it is difficult to see why they did not postpone the invasion of Kuwait until it had done so). In general, the results of the inspections seem plausible. The inspectors were skilled, determined, persistent and intrusive, and they succeeded in finding not just physical artifacts but also extensive (and credible) documentary material on the program. Some of the latter was in the form of progress reports (see the report from the Al-Athir research establishment reproduced in English translation in Peter Zimmerman, Iraq's Nuclear Achievements: Components, Sources, and

Stature, Congressional Research Service report 93-323F, 1993), and there is thus the opposite possibility that they may have been over-optimistic, as progress reports to sponsors often are.

174. Zimmerman, Iraq's Nuclear Achievements, pp. 11–12.

175. International Atomic Energy Agency Action Team for Iraq, Fact Sheet, January 1994, p. 6.

176. Albright and Hibbs, "Iraq's bomb: Blueprints and artifacts," pp. 31, 33, 35; Zimmerman, Iraq's Nuclear Achievements; Milhollin, "Building Saddam Hussein's bomb," p. 33.

177. See the discussion of the "hardness" of surgery in Pinch et al., "Cutting up skills."

178. See, e.g., David Holloway, "Entering the nuclear arms race: The Soviet decision to build the atomic bomb, 1939–1945," *Social Studies of Science* 11 (1981): 159–197; Goldschmidt, "La genèse et l'héritage," p. 24. The reasons for these divergent beliefs are unclear to us. Some of the physicists remaining in Germany may have allowed their view of the feasibility of an atomic bomb to be influenced by their desire to deny such a weapon to Hitler. The evidence for this hypothesis is, however, not entirely compelling, and in any case the views of French and Soviet physicists cannot be accounted for in this way.

179. Holloway, "Entering the nuclear arms race."

180. Hans Bethe, "Comments on the history of the H-Bomb," *Los Alamos Science*, fall 1982: 43–53.

181. J. Carson Mark, Theodore Taylor, Eugene Eyster, William Maraman, and Jacob Wechsler, "Can terrorists build nuclear weapons?" in *Preventing Nuclear Terrorism*, ed. P. Leventhal and Y. Alexander (Lexington Books, 1987), p. 64.

182. Smyth, *Atomic Energy*.

183. Zaloga, *Target America*.

184. J. W. De Villiers, Roger Jardine, and Michell Reiss, "Why South Africa gave up the bomb," *Foreign Affairs* 72 (1993), p. 105.

185. Annette Schaper, The Transferability of Sensitive Nuclear Weapon Knowledge from Civil Science to Military Work, paper prepared for Fifth International Symposium on Science and World Affairs, 1993.

186. Curiously, the historical record known to us contains no evidence that it *has* done so. This may be because, in the relatively well-documented programs, weapons design went on simultaneously with reactor design, but was largely the province of distinct groups.

187. John McPhee, *The Curve of Binding Energy* (Farrar, Straus and Giroux, 1974), p. 215.

188. Schaper, The Transferability of Sensitive Nuclear Weapon Knowledge.

189. Szasz, *British Scientists and the Manhattan Project*, p. 23.

190. David Holloway, personal communication, February 15, 1994.

191. See, e.g., the registration list of the Ninth International Symposium on Detonation (Portland, Oregon, 1989), which notes attendance by representatives from the Al Qaqaa State Establishment in Baghdad, a detonics research establishment at the heart of the Iraqi nuclear weapons program. We owe the reference to Schaper, The Transferability of Sensitive Nuclear Weapon Knowledge.

192. Unlike all previous programs for which the information is available, South Africa settled for the simpler of the Manhattan Project designs, the gun weapon, and seems to have concentrated innovative effort on developing a new uranium separation process. (The perceived relative difficulty of uranium separation compared with plutonium production, and the fact that gun designs are believed to require more fissile material and physically to be larger, probably account for other countries' eschewal of the gun.) Unfortunately, we have no information on how difficult the South African team found it to design their uranium gun.

193. Chevallier and Usunier, "La mise en œuvre scientifique et technique," p. 129 (our translation).

194. International Atomic Energy Agency Action Team for Iraq, Fact Sheet, January 1994, p. 2.

195. In the mid 1960s, Israel is believed to have obtained illicitly around 100 kilograms of enriched uranium from a privately owned plant in the United States, and there have also been allegations that Israel was supplied with plutonium by France (Spector, *Going Nuclear*, p. 131). However, Israel also had substantial plutonium-production capability from its French-supplied reactor at Dimona, which became operational in 1962 (Hersh, *The Samson Option*, p. 119).

196. Gowing and Arnold, *Independence and Deterrence,*, p. 76; David Albright and Mark Hibbs, "Iraq's nuclear hide-and-seek," *Bulletin of the Atomic Scientists*, September 1991, p. 19.

197. Hoddeson et al., *Critical Assembly*, pp. 175, 297.

198. In standard usage, "reactor-grade" plutonium contains, as well as plutonium 239, upwards of 18 percent plutonium 240 (and, indeed, significant quantities of other isotopes as well) while "weapons-grade" plutonium contains less than 7 percent: see, e.g., David Albright, Frans Berkhout, and William Walker, *World Inventory of Plutonium and Highly Enriched Uranium 1992* (Oxford University Press, 1993). There has, however, been at least one successful test (in 1962) of a nuclear device constructed out of "reactor-grade" plutonium (Technologies Underlying Weapons of Mass Destruction, p. 133; Richard Norton-Taylor, "Reactor fuel used in N-weapons test," *The Guardian*, June 29, 1994). The design and fabrication of such a weapon is, nevertheless, conventionally seen as harder

than working with weapons-grade plutonium (Gerhard Locke, Why Reactor-Grade Plutonium is no Nuclear Explosive Suitable for Military Devices, paper prepared for Workshop on the Disposition of Plutonium, Bonn, 1992; J. Carson Mark, "Explosive properties of reactor-grade plutonium," *Science and Global Security* 4 (1993): 11–38). There is believed to be a greater risk of a fizzle, and reactor-grade plutonium is a significantly more powerful heat source than weapons-grade, which can give to rise to problems when it is enclosed within a shell of high explosive (which is a thermal insulator).

199. McPhee, *Curve of Binding Energy*, pp. 214–218.

200. Taylor suggested that "Los Alamos or Livermore [should] build and detonate a crude, coarse, unclassified nuclear bomb—unclassified in that nothing done in the bomb's fabrication would draw on knowledge that is secret" (McPhee, *Curve of Binding Energy*, p. 123). To our knowledge, that suggestion was not taken up.

201. Teller, "The laboratory of the atomic age," p. 33.

202. One source pointed out to us in a personal communication that relatively small amounts of high explosive, in an appropriately shaped charge, can blow a hole through armor plate. Plutonium would be significantly less resistant, so a misjudged implosion design could easily simply blow the fissile mass apart.

203. J. Carson Mark, Theodore Taylor, Eugene Eyster, William Maraman, and Jacob Wechsler, "Can terrorists build nuclear weapons?" The only point of detail on which Taylor apparently changed his mind is a radical upward revision of the quantities needed if a bomb is to be built with "plutonium oxide powder seized from a fuel fabrication plant" (ibid., p. 61). The overall tone of the later piece is, nevertheless, different.

204. Mark et al., "Can terrorists build nuclear weapons?" p. 58.

205. Ibid., p. 60. This passage assumes the fissile material to be available as oxide, rather than as metal. If the latter were the case, the need for chemical knowledge and skill is greatly reduced, but the demands of casting and machining remain.

206. See, e.g., Latour, *Science in Action*, pp. 108–132. See chapter 1 of this volume for the main tenets of actor-network theory.

207. Wolfgang K. H. Panofsky, "Safeguarding the ingredients for making nuclear weapons," *Issues in Science and Technology*, spring 1994, pp. 67–68.

208. For some relevant considerations, see Steven Flank, "Exploding the black box: The historical sociology of nuclear proliferation," *Security Studies* 3 (1993–94): 259–294.

209. The recent South African and Iraqi inspections and disclosures allow us retrospectively to assess the accuracy of intelligence estimates of the progress of these programs. The broad accuracy of quantitative estimates of the likely status of the South African program (as reported, e.g., in Spector, *Going Nuclear*, pp.

220–239) have been confirmed (International Atomic Energy Agency, The Denuclearization of Africa: Report by the Director General, 1993). It appears that in the late 1980s U.S. intelligence agencies underestimated the seriousness of Iraq's nuclear ambitions, but there was nevertheless ample evidence of the program's existence (see Spector, *Going Nuclear*, pp. 161–163). Of course, the political arrangements that would permit appropriate action to be taken to prevent programs coming to fruition are quite another matter.

210. Among the few concrete discussions of the feasibility of the abandonment of nuclear weapons are Jonathan Schell, *The Abolition* (Picador, 1984) and Regina C. Karp, *Security without Nuclear Weapons? Different Perspectives on Non-Nuclear Security* (Oxford University Press, 1992).

211. See, e.g., Collins, *Changing Order.*

212. There has, on the other hand, been much debate as to what a *test* consists in—e.g., over the precise threshold of nuclear yield above which an event becomes a (detectable) nuclear test.

213. For example, events classed as "fizzles" can be quite substantial explosions, the equivalents of tens or even hundreds of tons of high explosive.

214. Recordings by a U.S. nuclear detection satellite suggested two extremely bright flashes of light in rapid succession over the south Indian Ocean on September 22, 1979—a pattern generally regarded as characteristic of first the initial detonation and then the fireball of a nuclear explosion. However, no other evidence (seismological, radiological, etc.) of an explosion could be found. A U.S. government-appointed group of outside experts, chaired by Jack Ruina of the Massachusetts Institute of Technology, concluded that the twin peaks in the satellite data were probably not the result of a nuclear explosion and may indeed not have represented genuine external flashes of light. However, some members of the U.S. Nuclear Intelligence Panel (notably those from Los Alamos) remained convinced that, to quote former Los Alamos Director Harold Agnew, "If it looks like a duck, it's got to be a duck." There was intense speculation that there had been a nuclear test by either South Africa or Israel, or possibly both nations in concert (see Hersh, *The Samson Option*, pp. 271–283). Hersh also quotes anonymous Israeli interviewees as claiming that Israel had indeed conducted a small series of covert tests. To our knowledge, no definitive resolution of the controversy has been reached.

215. The clearest instance is the controversy that took place in the U.S. in the 1980s over whether nuclear tests intended to demonstrate substantial x-ray lasing had indeed done so, or whether positive results were artifacts of the instrumentation employed. See William J. Broad, *Teller's War* (Simon and Schuster, 1992).

216. The 1993 Energy and Water Development Appropriations Act, passed in September 1992, prohibits U.S. nuclear testing after September 30, 1996, unless another state conducts a test after that date. It also banned U.S. nuclear testing until July 1993, and subsequent presidential decisions have extended this mora-

torium until September 1996 (John R. Harvey and Stefan Michalowski. "Recent weapons safety: The case of Trident," *Science and Global Security* 4 (1994): 261–337). Recent international negotiations over a test ban have hinged on the precise definition of "testing" and on the date of such a ban. France and China wish the ban to be postponed to 1996; China wishes exemption for "peaceful" nuclear explosions, and until recently France and Britain wanted exemption for "safety tests."

217. In the U.S., at least, there has already been considerable investment in sophisticated facilities to expand the range of relevant "hydronuclear" and non-nuclear experiments that are possible (see, e.g., Philip D. Goldstone, "An expanding role for AGEX: Above-ground experiments for nuclear weapons physics," *Los Alamos Science* 21 (1993): 52–53; Timothy R. Neal, "AGEX I: The explosives regime of weapons physics," *Los Alamos Science* 21 (1993): 54–59). However, those involved do not yet see such experiments as the full equivalent of nuclear explosive testing.

218. McDonald interview.

219. Sewell interview.

220. Donald R. Westervelt, "Can cold logic replace cold feet?" *Bulletin of the Atomic Scientists*, February 1979, p. 62.

221. Ibid.

222. Harold Agnew et al., "Taking on the future," *Los Alamos Science* 21 (1993), p. 9.

223. Nuclear warhead production in the U.S. was suspended early in 1992 following health and safety concerns about the Rocky Flats plant near Denver, Colorado, where the plutonium components for nuclear weapons are fabricated. According to Harvey and Michalowski ("Nuclear weapons safety," p. 286), "prospects are dim for a return to operations in the near term" at Rocky Flats.

224. Goldstone, "An expanding role for AGEX," p. 52.

225. Scott D. Sagan, *The Limits of Safety: Organizations, Accidents, and Nuclear Weapons* (Princeton University Press, 1993).

226. For example, Harvey and Michalowski ("Nuclear weapons safety," p. 279) argue that a serious accident to the Trident program "would almost certainly result in [its] extended suspension or termination," a measure that would radically change the strategic landscape, since Trident warheads will probably soon make up half of the U.S. strategic arsenal and all of the British one. On the wider vulnerability to challenge of scientific expertise, see, e.g., Barnes and Edge, *Science in Context*, pp. 233–335.

Index

Gallium arsenide, 107, 153–154
Gender, and technological change, 6, 18, 34, 42–43
Grabiner, Judith, 177
Groups, relevant social, 6
Gusterson, Hugh, 232
Gyroscope, 9, 67
 electrostatically supported, 79, 81
 hemispherical resonator, 93–94
 laser, 13, 18, 20, 67–97

Heer, Clifford V., 74–76, 78, 85
Heuristics, 52–54. *See also* Pricing
Hilbert, David, 178, 181
History of technology, 12–13, 43–46
 contingency in, 46–47
HOL (Higher Order Logic), 161, 163, 178, 182
Honeywell, 80–96
Hounshell, David, 44
Hughes, Thomas P., 44, 55, 104, 134
Hunt, Warren, 163–164, 177
Hydrogen bomb, 108, 221, 224, 229, 235, 246, 251. *See also* Nuclear weapons

IBM
 and computer arithmetic, 170, 175
 and Control Data Corporation, 140–141
 dominance of computer market, 135
 and LARC contract, 115
 and Los Alamos, 116–117, 122
 punched-card machines, 109
 and Steve Chen, 155
 Stretch computer, 100, 115–116, 121–122, 137, 140
 and supercomputing, 101, 118
 System/360, 100, 116, 136, 140–142
ILLIAC IV scheme, 105–106
India, nuclear weapons program of, 236–237, 243
Instrumentation Laboratory, MIT, 79–81, 92
Intel Corporation, 21, 118, 170–171, 174, 182–183
Interference, electromagnetic, 189, 193–194

Internet, 7, 190. *See also* RISKS Forum
Iraq
 and Gulf War, 3, 195, 201–202
 nuclear weapons program of, 219, 247–248, 251
Israel, nuclear weapons program of, 237, 243

Kahan, W., 168–175, 182–183
Khariton, Yuli, 238, 251
Killpatrick, Joseph, E., 80, 83–85, 87–88
Klass, Philip J., 79–81, 83, 88
Knowledge
 explicit, 215
 scientific, 9–11, 14–16, 50, 62–64
 sociology of, 9–11, 13, 15, 21, 176, 183
 tacit, 11, 22, 215–258
Kuhn, Thomas, 223

Lakatos, Imre, 168–169
Landes, David, 36
Langevin, Paul, 70
LARC (Livermore Automatic Research Computer), 100, 114–116
Laser gyroscope. *See* Gyroscope
Laser, invention of, 74
Latour, Bruno, 9, 13–16, 55, 58, 94, 134, 139, 146
Law, John, 13, 55, 134, 139, 146
Lawrence Livermore National Laboratory. *See* Livermore
Lazonick, William, 41–42
Lewis, Clarence I., 181
Lewis principles, 181–182
Lincoln, Neil, 142
Linotype, 43
Lipton, Richard, 178–180
Livermore Automatic Research Computer, 100, 114–116
Livermore Fortran Kernels, 123–124, 126–127
Livermore National Laboratory
 and access to fissile material, 245